Lecture Notes in Physics

Founding Editors: W. Beiglböck, J. Ehlers, K. Hepp, H. Weidenmüller

The Lecture Notes in Physics

The series Lecture Notes in Physics (LNP), founded in 1969, reports new developments in physics research and teaching – quickly and informally, but with a high quality and the explicit aim to summarize and communicate current knowledge in an accessible way. Books published in this series are conceived as bridging material between advanced graduate textbooks and the forefront of research and to serve three purposes:

- to be a compact and modern up-to-date source of reference on a well-defined topic

- to serve as an accessible introduction to the field to postgraduate students and nonspecialist researchers from related areas

- to be a source of advanced teaching material for specialized seminars, courses and schools

Both monographs and multi-author volumes will be considered for publication. Edited volumes should, however, consist of a very limited number of contributions only. Proceedings will not be considered for LNP.

Volumes published in LNP are disseminated both in print and in electronic formats, the electronic archive being available at springerlink.com. The series content is indexed, abstracted and referenced by many abstracting and information services, bibliographic networks, subscription agencies, library networks, and consortia.

Proposals should be sent to a member of the Editorial Board, or directly to the managing editor at Springer:

Christian Caron
Springer Heidelberg
Physics Editorial Department I
Tiergartenstrasse 17
69121 Heidelberg / Germany
christian.caron@springer.com

P. Hájíček

An Introduction
to the Relativistic
Theory of Gravitation

Translated by Frank Meyer and Jan Metzger

 Springer

Author

Petr Hájíček
Universität Bern
Inst. Theoretische Physik
Sidlerstraße 5
3012 Bern
Switzerland
hajicek@itp.unibe.ch

Translators

Frank Meyer
MPI für Physik
Werner-Heisenberg-Institut
Föhringer Ring 6
80805 München
Germany

Jan Metzger
Weissdornweg 3
14469 Potsdam
Germany

Hájíček, P., *An Introduction to the Relativistic Theory of Gravitation*, Lect. Notes Phys. 750 (Springer, Berlin Heidelberg 2008), DOI 10.1007/978-3-540-78659-7

ISBN: 978-3-642-09742-3   e-ISBN: 978-3-540-78659-7

DOI 10.1007/978-3-540-78659-7

Lecture Notes in Physics ISSN: 0075-8450

Cover design: eStudio Calamar S.L., F. Steinen-Broo, Pau/Girona, Spain

Printed on acid-free paper

9 8 7 6 5 4 3 2 1

springer.com

Contents

Introduction

The contemporary theoretical physics consists, by and large, of two independent parts. The first is the quantum theory describing the micro-world of elementary particles, the second is the theory of gravity that concerns properties of macroscopic systems such as stars, galaxies, and the universe. The relativistic theory of gravitation which is known as general relativity was created, at the beginning of the last century, by more or less a single man from pure idea combinations and bold guessing. The task was to "marry" the theory of gravity with the theory of special relativity. The first attempts were aimed at considering the gravitational potential as a field in Minkowski space–time. All those attempts failed; it took 10 years until Einstein finally solved the problem. The difficulty was that the old theory of gravity as well as the young theory of special relativity had to be modified. The next 50 years were difficult for this theory because its experimental basis remained weak and its complicated mathematical structure was not well understood. However, in the subsequent period this theory flourished. Thanks to improvements in the technology and to the big progress in the methods of astronomical observations, the amount of observable facts to which general relativity is applicable was considerably enlarged. This is why general relativity is, today, one of the best experimentally tested theories while many competing theories could be disproved. Also the conceptual and mathematical fundamentals are better understood now. Nowadays general relativity serves as one of the fundamental theories for modern astrophysics and space research. More and more people get interested in its fascinating results.

General relativity is a very fruitful theory; it predicted many new and surprising effects. The most famous examples are:

Gravitomagnetic phenomena The origin of gravity is not mass alone but every form of energy, momentum, angular momentum, and pressure, too. Several components of the relativistic gravitational field are similar to those of the electromagnetic field. In particular the motion of mass creates the components that are called *gravitomagnetic field*. This phenomenon can be observed for instance around rotating stars. A satellite orbiting in the opposite direction to that of the star rotation is less attracted than a satellite orbiting in the same direction.

Gravitational radiation In Newton's theory the gravitational potential $\Phi(t, \mathbf{r})$ at the time t is determined by the mass density $\rho(t, \mathbf{r})$ given at the same time through the Poisson equation

$$\Delta\Phi = 4\pi G\rho \; .$$

Hence a change of the density ρ manifests itself immediately in the form of the gravitational potential all over the equal-time surface. The field is, however, measurable; the information about the movement of the source therefore propagates infinitely fast in the gravitational field. In a relativistic theory only the speed of light can be considered for the propagation of signals. The gravitational signals, which carry energy and information with finite speed, are called gravitational waves.

Dynamical cosmology The solution of the Einstein equations that describe cosmology are not static. The first were found by Friedmann. They mostly start in a singular point where all matter is brought together: in the so-called *Big Bang*. This means: our universe is the remnant of a gigantic explosion. The Big Bang cosmology theory is a very successful one. Moreover, the gravitational field is characterized by two coupling constants: not only Newton's constant but also the so-called cosmological constant. Only today do we start to understand the role of the latter.

Black holes Let us consider a source of mass M and gravitational potential $\Phi(r) = -GMr^{-1}$. In this field a force of attraction that can perform work acts on a particle of mass μ. If we move the particle from $r = \infty$ to a finite distance r from the source, in principle the work $A(r) = \mu\Phi(\infty) - \mu\Phi(r) = -\mu\Phi(r) = GM\mu r^{-1}$ can be applied. At some value of radius, it is possible to obtain the full rest energy of the particle; this radius is given by the equation

$$GM\mu r^{-1} = \mu c^2 \; .$$

It is called the *gravitational radius* of the mass M,

$$R_G := \frac{G}{c^2}M \; . \tag{1}$$

The gravitational radius appears also in another estimation. Let us calculate the distance from our spherically symmetric source, where the escape velocity admits the value v. The escape velocity is the velocity corresponding to the circular orbit at a given radius. The orbit satisfies the equation

$$G\frac{M\mu}{r^2} = \frac{\mu v^2}{r}$$

(gravitational centripetal force = centrifugal force). Then the following holds for the distance r that we want to determine:

$$r = \frac{G}{v^2}M \; ,$$

and for $v = c$ we have $r = R_G$. Hence, no energy- or information-carrying signal coming from below the gravitational radius can reach us. The objects of the size given by the gravitational radius are called black holes. The assumption that black holes exist in the center of most galaxies and in other star systems leads to a wonderful order and a good understanding for a big class of apparently mysterious objects in the cosmos.

The two pillars of general relativity are the geometric interpretation of gravitation and the Einstein equations. In the first part of these Notes (Chaps. 1, 2, 3 and 4), the geometrization is carefully motivated and the necessary mathematical tools from differential geometry are introduced. The second part (Chaps. 4, 5, 6 and 7) contains the most important applications: gravitomagnetism, gravitational waves, cosmology, gravitational collapse, and black holes. There is plenty of mostly easy exercises throughout. The Notes focus on those ideas that are most frequently used in the contemporary astrophysics, although no astrophysics worth mentioning is explicitly presented. At the points, where such information could be felt to be missing, other books or papers are mentioned.

Over the many years an earlier version of these Notes was written, students helped me with their questions during the lectures as well as with advice on phrasing and grammar of the German language. In particular, I would like to mention Martin Schön, Matthias Zürcher, and Matthias Peter Burkhardt. The Notes also gained a lot from discussions with J. Bičák, J. Ehlers, G. W. Gibbons, S. W. Hawking, K. V. Kuchař, W. Kundt, and B. G. Schmidt. The present text is based on a translation of the German version by two young researchers, Dr. Frank Meyer and Dr. Jan Metzger—their thorough work is gratefully acknowledged.

Chapter 1
Geometrization of Mechanics

1.1 Selected Facts

Let us first look at what is known about the nature of gravitation from observations and experiments. We want to stay in the framework of Newton's notions of space, time, and dynamics, and will mention only those aspects that are of direct importance for the development of general relativity.

1.1.1 The Cavendish Experiment

In 1789 H. Cavendish measured the force of attraction between two massive spheres using a torsion balance. The result was in accordance with

$$F_G = -G\frac{m_1 m_2}{r^2} ,$$

where m_1 and m_2 denote the masses, r the radius, and where G is Newton's constant. In this way, Newton's constant was measured. The measurement has been improved several times since then; a more accurate value and the account of the corresponding experiments can be found in [1, 2, 3]. For us it will be sufficient to remember the value

$$G \approx 10^{-10} \, \mathrm{N\,kg^{-2}\,m^2} .$$

The gravitational law is formally analogue to the Coulomb law

$$F_C = \frac{1}{4\pi\varepsilon_0} \frac{q_1 q_2}{r^2} ,$$

where q_1 and q_2 are the charges and ε_0 is the dielectric constant for vacuum:

$$\frac{1}{4\pi\varepsilon_0} \approx 10^{10} \, \mathrm{N\,C^{-2}\,m^2} .$$

Hájíček, P.: *Geometrization of Mechanics*. Lect. Notes Phys. **750**, 1–37 (2008)
DOI 10.1007/978-3-540-78659-7_1   © Springer-Verlag Berlin Heidelberg 2008

Differences:

1. The sign. Two masses attract each other, two identical charges repel. This is why we seldom find accumulations of electric charges in nature, whereas high concentrations of mass are absolutely normal (gravitational instability, gravitational collapse).
2. The gravitational "charge" is the mass, which is positive for all bodies. This leads to the so-called universality of gravitation. It acts on all matter and is produced by all matter.
3. Comparison of the two forces for two protons:

$$\frac{F_G}{F_C} \approx -10^{-36} \; .$$

Thus, gravitation seems to be negligible in the world of elementary particles.

1.1.2 Eötvös Experiment

Another important difference to electrodynamics is that the mass in physics not only takes the role of the gravitational charge (gravitational mass), but also appears in Newton's second law as the so-called inertial mass:

$$F = ma \; .$$

At first glance, this property of the mass has nothing to do with gravitation and is measured in a completely different way. We can actually only talk of the proportionality of the inertial and the gravitational mass of one and the same body. The universal coefficient of proportionality depends on the choice of units and can be reduced to 1.

How well is this proportionality tested experimentally? The first time R. Eötvös (1889, 1922) undertook such measurements ($1 : 2 \times 10^8$). More recent experiments are discussed in [1, 2, 3]. These experiments are fantastically accurate ($1 : 10^{12}$).

From this proportionality it follows that trajectories of bodies that are falling in a gravitational field depend on the field only, and not on the body. This is because the acceleration of the body is inversely proportional to its mass, whereas the force is directly proportional. The motion in a gravitational field is hence a lot simpler than, for example, in an electrostatic field. In the first case the field, the initial point, and the initial speed fully determine the movement. In the second case we have to know in addition the relation q/m of the charge to the mass of the test particle.

1.1.3 Deflection of Light

Everything that was said so far is only valid for massive bodies. How do photons react to the gravitational field? They are deflected by the field, much as happens to massive bodies. The existing observations concern the field of the Sun; the photons move tangentially to the surface and the deflection angle δ is measured. Very precise measurements were done with radio signals coming from a pair of quasars (for details, see [1, 2, 3]). The result is

$$\delta \approx 1.77'' .$$

Electromagnetic waves are affected by the gravitational field also in another way.

1.1.4 Redshift

If a photon ascends in a homogeneous gravitational field it must lose energy. Otherwise it would be possible to build a perpetuum mobile. It is possible to show that the redshift of a photon that ascends up to a height l must admit the value $gl c^{-2}$. An adequate experiment was first done by Pound, Rebka and Snider in 1960 (for discussion, see [1, 2, 3]). In this experiment, photons ascended to a height of 22.5 m. Then the frequency was measured by means of the Mössbauer effect.

1.2 Equivalence Principle

The underlying idea of general relativity is Einstein's answer to the question of the origin of the proportionality between inertial and gravitational masses. He noticed that the same proportionality holds for the so-called apparent forces. Centrifugal force, Coriolis force, etc., are all proportional to the inertial mass of the sample. The assumption that the gravitational force is an apparent force[1] (i.e., that it is due to an accelerated motion of the reference system) is—roughly speaking—what is known as the so-called equivalence principle (the equivalence of gravitation and phenomena in accelerated systems).

Let us consider for simplicity a linearly accelerated system. Its axes x, y, and z move with the acceleration g with respect to the axes $\bar{x}, \bar{y}$, and $\bar{z}$:

$$x = \bar{x} , \quad y = \bar{y} , \quad z = \bar{z} - \frac{1}{2} g t^2 . \tag{1.1}$$

Imagine a physicist who is locked in a box that is accelerated in such a way. He cannot see, for instance, that a force outside the box is responsible for his acceleration. But he can investigate the phenomena inside the box. This is the so-called Einstein box (Gedankenexperiment).

The observer sees that all objects fall to the ground (in direction of the negative z-axis) with an acceleration g, which is independent of the object's mass and characteristics. If he wants to interpret this effect as coming from a force $\vec{F}$ he must set

---

[1] Such "explicit" formulation can also be found in [4].

$$\vec{F} = m\vec{g} \,,$$

where $\vec{g}$ is a vector with components $(0, 0, -g)$. Here, $\vec{g}$ may be called the "intensity of the gravitational field" and m the "gravitational charge of the test body".

How does light react to this gravitation? A light ray with the trajectory

$$\bar{x} = ct \,, \quad \bar{y} = 0 \,, \quad \bar{z} = 0$$

with respect to $(\bar{x}, \bar{y}, \bar{z})$ moves in the frame (x, y, z) according to the transformation (1.1) as follows

$$x = ct \,, \quad y = 0 \,, \quad z = -\frac{1}{2} g t^2 \,.$$

Hence, it reveals a deflection in the direction of the gravitational acceleration.

Do we observe a redshift also if the light moves from the bottom of the box $(z = 0)$ to the top $(z = l)$? Yes, and it is possible to show (exercise) that the corresponding Doppler effect leads to a redshift given by

$$\frac{\Delta \lambda}{\lambda} = g l c^{-2} \,.$$

We have found an explanation for the redshift.

Does this mean that the gravitation at the Earth's surface can be explained by an acceleration of this surface in the direction away from the Earth's center? Then also all distances at the Earth's surface have to grow with the same acceleration, don't they? This would be a paradox.

This paradox, however, does not follow by assuming only that the gravitational force is an apparent force. It results rather from the combination of this assumption with the Newtonian picture of space and time. It is even possible to show exactly and explicitly how the assumption that gravitation is an apparent force changes the picture of the space–time geometry, in such a way that a specific curvature of space–time appears.

In this part of the lecture we want to introduce the necessary geometric notions and properties. We will work directly with dynamics—we geometrize dynamics so to speak—so that geometry takes on a physical meaning. Previous knowledge about differential geometry is not assumed. In this way both the relevant geometric quantities and facts and the central idea of general relativity will be introduced based on known material from non-relativistic dynamics.

1.3 Newton–Galilei Space–Time

It is helpful to bring together the three-dimensional space and the one-dimensional time in a four-dimensional construction, the so-called space–time. This step is essential for the geometrization. We now want to translate the Newtonian picture of time and space in postulates about this space–time (which we will denote in the following by $\mathscr{M}$).

Postulate 1.1 *On $\mathcal{M} \times \mathcal{M}$ a function $\Delta T: \mathcal{M} \times \mathcal{M} \mapsto \mathbb{R}$ is defined, which is called time distance. Its properties are:*

1. $\Delta T(p,q) = -\Delta T(q,p)$, *for all p and q,*
2. $\Delta T(p,q) + \Delta T(q,r) = \Delta T(p,r)$, *for all p, q, and r.*

It follows that $\Delta T(p,p) = 0$ for all p. The time distance is to be measured with an Newtonian *ideal clock*.

We can then define a time function $T: \mathcal{M} \mapsto \mathbb{R}$ by choosing an event p_0 and set

$$T(q) := \Delta T(p_0, q).$$

Two different time functions can only differ by a constant, which is equal to $\Delta T(p_1, p_2)$ if the two time functions are related to the two events p_1 and p_2.

Also the future (or the past) of an event p can be defined as $\{q \in \mathcal{M} \,|\, \Delta T(p,q) > (<)0\}$. More interesting is the definition of simultaneity: two events p and q are *simultaneous* if $\Delta T(p,q) = 0$. This is an equivalence relation and this way $\mathcal{M}$ is divided into disjoint subsets—equivalence classes. These subsets are called *simultaneity surfaces* and they indeed form hypersurfaces in $\mathcal{M}$. They represent the well-known absolute simultaneity of Newtonian physics.

Postulate 1.2 *Every simultaneity surface R carries the structure of Euclidean space. Let V^3 be the three-dimensional vector space over the real numbers with the scalar product $(\cdot, \cdot)$; then*

1. *for every R a map $E: R \times R \mapsto V^3$ is given, which we denote by $E(p,q) = q - p \in V^3$. We have*
2. $(p-q) = -(q-p)$,
3. $(p-q) + (q-r) = (p-r)$,
4. *let p be fixed in R. Then the map $q - p: R \mapsto V^3$, which maps q to V^3, is bijective for every p.*

This structure defines a distance $D(p,q)$ for any two simultaneous events p and q:

$$D(p,q) := \sqrt{(p-q, p-q)},$$

which can be measured by *measuring rods*.

A general *reference frame* in $\mathcal{M}$ is given by an event p_0 (initial time), one event $O(t)$ (the origin), and an orthonormal triad $\vec{e}_k(t)$ for each simultaneity surface corresponding to the value t of the time function determined by p_0. It is then possible to attribute to every event p four numbers (x^0, x^1, x^2, x^3), where $x^0 = \Delta T(p_0, p)$ and numbers x^1, x^2, x^3 are defined by:

$$p - O(t) = \sum_k x^k \vec{e}_k(t).$$

On the other hand, the four numbers (x^0, x^1, x^2, x^3), together with a general reference frame, determine exactly one event in $\mathcal{M}$. The relation between the coordinates

(x^0, x^1, x^2, x^3) and (y^0, y^1, y^2, y^3) of an event in two different reference frames is given by four invertible functions of four variables.

In Newtonian physics the so-called *free motion* plays a particular role: it serves to define an *inertial frame*. This is a reference frame with the property that every trajectory of a free motion admits, in the inertial frame, the form

$$x^\mu = v^\mu \lambda + b^\mu , \quad \mu = 0, 1, 2, 3 , \tag{1.2}$$

where λ is the curve parameter. The two 4-vectors v^μ and b^μ are defined by the free motion (up to affine reparameterizations).

Postulate 1.3 *There exists at least one inertial frame.*

1.4 Free Motion

Sometimes free motion is defined by the fact that it looks uniform and linear in an inertial frame. To avoid a circular definition we assume that a dynamical definition of free motion is possible. This means that all possible forces are known and can be eliminated. For instance, electromagnetic forces are switched off by allowing only test bodies whose electric and magnetic multipoles vanish. Contact forces such as friction, air resistance, etc., are eliminated by using contact-free test bodies.

Let us study (1.2) in more detail:

1. We see that one and the same free motion can have different representations of type (1.2). What is important is only the path, i.e., the set of points in $\mathcal{M}$, and not the values of the parameter λ. The choice of this parameter is arbitrary, restricted only by the condition that (1.2) has to be linear. In this way, λ is determined up to an affine transformation:

$$\lambda \mapsto \lambda' = \alpha \lambda + \beta , \quad \alpha \neq 0 .$$

λ is called an *affine parameter*. The arbitrariness we get is the prize for treating the time and the spatial coordinates in a symmetric way.

2. Equation (1.2) is equivalent to four differential equations:

$$\frac{d^2 x^\mu}{d\lambda^2} = 0 , \quad \forall \mu . \tag{1.3}$$

Every solution $x^\mu(\lambda)$ admits the form (1.2), and four arbitrary functions of the form (1.2) solve the system (1.3).

The importance of free motions is that it has a different description with respect to an inertial frame than with respect to a non-inertial frame. Hence we can distinguish these two classes of reference systems. How does the free motion look in an arbitrary reference frame?

Let us consider an inertial frame $\bar{K}$ with coordinates $\bar{x}^\mu$ and a non-inertial frame K with x^μ. The transformation between these two coordinate systems is given in general by

$$\bar{x}^0 = x^0 + \tau, \quad \bar{x}^k = \sum_l O_{kl}(t) \left[x^k - r^k(t) \right],$$

where τ is a real number, $O_{kl}(t)$ is a time-independent orthogonal matrix, and $r^k(t)$ is a time-dependent 3-tuple. This follows from the definition of a general reference system given in the previous section. We want to write the transformations in shorthand notation:

$$\bar{x}^\mu = \bar{x}^\mu \left(x^0, \ldots, x^3 \right).$$

For the curve $x^\mu(\lambda)$ of a free motion we have

$$\bar{x}^\mu(\lambda) = \bar{x}^\mu \left(x^0(\lambda), \ldots, x^3(\lambda) \right).$$

If we insert this relation for $\bar{x}^\mu(\lambda)$ in (1.3), we obtain

$$\ddot{x}^\mu + \sum_{\rho=0}^{3} \sum_{\sigma=0}^{3} \Gamma^\mu_{\rho\sigma} \dot{x}^\rho \dot{x}^\sigma = 0, \tag{1.4}$$

where

$$\Gamma^\mu_{\rho\sigma} = \sum_{v=0}^{3} \frac{\partial x^\mu}{\partial \bar{x}^v} \frac{\partial^2 \bar{x}^v}{\partial x^\rho \partial x^\sigma}.$$

It is easy to show that the components of $\Gamma^\mu_{\rho\sigma}$ are independent of the inertial frame $\bar{K}$. The coefficients $\Gamma^\mu_{\rho\sigma}$ will play an important role in the following. It is possible to show that they are connected with the so-called apparent forces (exercise). Summarizing: for inertial frame the $\Gamma^\mu_{\rho\sigma}$ are zero, not for non-inertial frame.

The differential equation (1.4) of free motions also has a formal mathematical aspect. Its coefficients Γ define a geometric object, a so-called *affine connection* [5]. The affine connection plays an important role in many fields of mathematics and modern theoretical physics. For instance the so-called gauge fields or Yang–Mills fields [6] are affine connections.

As geometrical structures of Newtonian space–time, we have all together

1. absolute time distance in space–time (determined by the ideal clock),
2. distances and angles in the simultaneity surfaces (determined by the measuring rods),
3. affine connection in space–time (determined by the free motions).

In the next section we want to look at the affine connection in more detail.

1.5 Affine Connection

1.5.1 Curvilinear Coordinates

The general reference frames of Newton's theory are examples of curvilinear coordinates since the coordinate lines $x^k = $ const, $k = 1, 2, 3$ are not straight lines. The transformation between two systems of curvilinear coordinates is not linear. The non-linearity of the allowed coordinate transformations is the most important new property. So far we had Newton's theory based on Galilei transformations and the theory of relativity, which is controlled by Poincaré transformations. These theories are linear. At this point we start to leave behind the well-known area of linear theories.

Another example: the surface of a sphere, with the spheric coordinates ϑ and φ. If ϑ and φ are contained in the intervals

$$0 < \vartheta < \pi, \quad 0 < \varphi < 2\pi, \tag{1.5}$$

the whole surface of the sphere is covered, except for the closed segment

$$0 \le \vartheta \le \pi, \quad \varphi = 0. \tag{1.6}$$

As soon as we try to cover more than this, some points on the surface of the sphere get several pairs of coordinates: the pole $\vartheta = 0$, for instance, allows any value for φ, etc. It is known that the surface of the sphere cannot be fully covered using one coordinate system only, but it can be covered using two "coordinate charts" $\{\vartheta_1, \varphi_1\}$ and $\{\vartheta_2, \varphi_2\}$, where the corresponding segments (1.6) are not allowed to overlap. We are thus naturally led to the concept of manifolds.

Definition 1 A differentiable n-manifold is a topological space $\mathcal{M}$ and a family $\{U_i\}$ of open subsets which cover $\mathcal{M}$:

$$\mathcal{M} = \bigcup_i U_i.$$

For each subset U_i, there exists a homeomorphism (a continuous bijection with continuous inverse) $h_i : U_i \mapsto \mathbb{R}^n$. The pair (U_i, h_i) is called a coordinate chart. If two coordinate charts overlap, $U_i \cap U_j \ne \emptyset$, then the map

$$h_i \circ h_j^{-1} : \mathbb{R}^n \mapsto \mathbb{R}^n$$

with domain $h_j(U_i \cap U_j)$ is differentiable (C^∞, i.e., all derivatives are continuous) and the Jacobi determinant of this map vanishes nowhere in $h_j(U_i \cap U_j)$.

We thus have two coordinate systems in the overlapping region $U_i \cap U_j$: $\{x^\mu\}$, which is pulled back by h_i from $\mathbb{R}^n$ to $\mathcal{M}$, and $\{\bar{x}^\mu\}$, which comes from h_j (Fig. 1.1). The map $h_i \circ h_j^{-1}$ is represented on $h_j(U_i \cap U_j) \subset \mathbb{R}^n$ by the function $x^\mu(\bar{x}^1, \ldots, \bar{x}^n)$, $\mu = 1, \ldots, n$, and the Jacobi determinant is

$$\frac{\partial (x^1, \ldots, x^n)}{\partial (\bar{x}^1, \ldots, \bar{x}^n)}.$$

We want to write the derivatives of the transformation functions $x^\mu(\bar{x}^1, \ldots, \bar{x}^n)$ in shorthand notation:

$$\frac{\partial x^\mu}{\partial \bar{x}^\nu} = X^\mu_\nu.$$

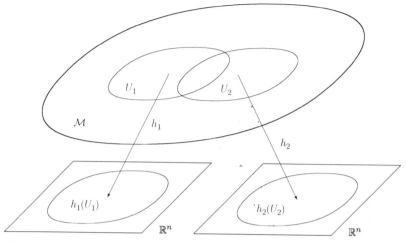

Fig. 1.1 Some coordinate charts of manifold $\mathcal{M}$

Note that the bar on the coordinate becomes a bar on the index! In the following we will calculate a lot with these symbols. The following relations are often used:

$$\sum_{\nu=1}^{n} X_{\bar{\nu}}^{\mu} X_{\rho}^{\bar{\nu}} = \delta_{\rho}^{\mu}, \quad \sum_{\rho=1}^{n} X_{\rho}^{\bar{\nu}} X_{\bar{\mu}}^{\rho} = \delta_{\bar{\mu}}^{\bar{\nu}}.$$

They follow from the chain rule for derivatives of composite functions. More about differentiable manifolds, see [5].

1.5.2 Curves and Tangential Vectors

In the first half of these Notes we want to treat mainly the dynamics of mass points. The curves, which serve as trajectories, become a basic concept. First we define what a curve is.

Definition 2 Let $\mathcal{M}$ be a n-manifold. A curve C is a map

$$C: \mathbb{R} \longmapsto \mathcal{M},$$

which is piecewise differentiable in the following sense: in the coordinates $\{x^{\mu}\}$ the map is represented by n functions $x^{\mu}(\lambda)$ and these functions are piecewise C^{∞}.

From the definition, it follows that the functions $\bar{x}^{\mu}(\lambda)$, which represent the curve C in other coordinates $\{\bar{x}^{\mu}\}$, are determined by the functions $x^{\mu}(\lambda)$:

$$\bar{x}^{\mu}(\lambda) = \bar{x}^{\mu}\left(x^1(\lambda),\ldots,x^n(\lambda)\right). \tag{1.7}$$

Here the $\bar{x}^\mu(x^1,\ldots,x^n)$ are the transformation functions from $\{x^\mu\}$ to $\{\bar{x}^\mu\}$.

The *tangential vector* t^μ to C at a point p is determined by its *components* with respect to $\{x^\mu\}$. If the point p corresponds to the value λ_0 of the parameter λ we have

$$t^\mu = \dot{x}^\mu(\lambda_0), \quad \mu = 1,\ldots,n.$$

In differential geometry a very important question is how a quantity transforms if the coordinates are changed. Equation (1.7) gives:

$$\bar{t}^\mu = \dot{\bar{x}}^\mu(\lambda_0) = \sum_{v=1}^n \frac{\partial \bar{x}^\mu}{\partial x^v} \dot{x}^v(\lambda_0) = \sum_{v=1}^n X_v^{\bar\mu} t^v.$$

Hence, the tangential vector is an example of a quantity with the following properties:

1. It is always connected to a particular point p of $\mathcal{M}$.
2. In the coordinates $\{x^\mu\}$ around p, it is represented by n components $(t^1,\ldots,t^n)$.
3. These components transform as follows:

$$\bar{t}^\mu = \sum_{v=1}^n X_v^{\bar\mu}(p) t^v.$$

Such a quantity is called a *vector*.

What is new in this definition is the fact that every vector is defined at a particular point of the manifold (the vector "lives" at this point). In special relativity, for instance, this is not demanded. In differential geometry, we must always connect a vector to a point of $\mathcal{M}$. Otherwise we do not know how the vector transforms: the matrix $X_v^{\bar\mu}$ is not constant over the whole manifold $\mathcal{M}$, as is the case in special relativity, since the transformation of the coordinates is in general non-linear.

1.5.3 Affine Connection of a Manifold

Affine connection can be defined in many different ways. For us, the most advantageous definition is the following.

Definition 3 Let $\mathcal{M}$ be a n-manifold. For physical or geometric reasons a class of curves may be selected in a way that the coordinate representation $x^\mu(\lambda)$ of any such curve satisfies the following differential equations:

$$\ddot{x}^\mu + \sum_{\rho=1}^n \sum_{\sigma=1}^n \Gamma_{\rho\sigma}^\mu \dot{x}^\rho \dot{x}^\sigma = 0, \quad \forall \mu, \tag{1.8}$$

and that any such solution of this system defines a curve of this class. Here, the functions $\Gamma_{\rho\sigma}^\mu(x)$ are assumed to be C^∞-functions of x^μ, and we demand

$$\Gamma_{\rho\sigma}^\mu(x) = \Gamma_{\sigma\rho}^\mu(x), \quad \forall x,\mu,\rho,\sigma.$$

The curves then define an affine connection on $\mathcal{M}$. The $\Gamma^\mu_{\rho\sigma}$ are called the *components* of the affine connection and the curves the *autoparallels* of the affine connection, and the manifold is called *affine connected*.

The system (1.8) consists of n coupled, ordinary, non-linear differential equations of second order. They are all solved for second-order derivatives; thus for every point p with coordinates x^μ_p and each vector v^μ we get exactly one autoparallel in this point.[2] It satisfies (1.8) with the initial conditions

$$x^\mu(0) = x^\mu_p ,$$
$$\dot{x}^\mu(0) = v^\mu .$$

Moreover the differential equation (1.8) is invariant with respect to affine transformations of the parameter λ. The parameterization of the autoparallel is therefore, by (1.8), given up to an affine transformation.

The components of the affine connection are only determined with respect to the chosen coordinates. Now we go back to the question of how the quantities transform under a change of coordinates. The answer is contained in the definition of the affine connection. To find it we need more technical knowledge. It is worthwhile to introduce this technique now, since it will be used many times later.

We define an *index-carrying quantity* (IQ) as a multidimensional table of numbers, the so-called *components* of the IQ, which are labeled by indices. For instance, t^μ is a one-dimensional table of numbers with n elements $(t^1, \ldots, t^n)$, $X^{\bar\mu}_\nu$ is a two-dimensional IQ with n^2 elements and $\Gamma^\mu_{\rho\sigma}$ a three-dimensional IQ. Every index runs from 1 to n. The order of the indices is important. There are two types of indices: upper and lower ones. The number p of upper and the number q of lower indices determine the so-called *type* (p,q) of the IQ. The IQ t^μ is of type $(1,0)$ and $\Gamma^\mu_{\rho\sigma}$ of type $(1,2)$. We have the following calculation rules for the IQ.

Equality Two IQs are equal if they are of the same type and if every component of the first IQ is equal to the corresponding component of the second. The equality can be expressed by demanding equality for all components with arbitrary indices, e.g.,

$$A^\mu_{\rho\sigma} = B^\mu_{\rho\sigma} .$$

The indices that appear on the left- and on the right-hand side must be given the same letters. Equal indices must appear on both sides in the same position, e.g., both indices upper or both lower. An equation like

$$A^\mu_{\rho\sigma} = B^\gamma_{\alpha\beta} ,$$

for instance, can only make sense if μ, ρ, σ, and α take special values. Then this equation means that only the corresponding components are equal, but not necessarily that the whole IQs are equal.

---

[2] We use an existence theorem from the theory of ordinary differential equations, see for example [7], Chap. 1.

Summation Two IQs of the same type can be added to define a new IQ, which is again of the same type. For instance:

$$Z^{\mu}_{\rho\sigma} = X^{\mu}_{\rho\sigma} + Y^{\mu}_{\rho\sigma} \ .$$

This equation means nothing else than adding the corresponding components.

Product Two arbitrary IQs of type (p_1, q_1) and (p_2, q_2) can be multiplied. This gives an IQ of type $(p_1 + p_2, q_1 + q_2)$. For instance:

$$Z^{\rho\sigma}_{\mu\nu\kappa} = X^{\rho}_{\kappa} B^{\sigma}_{\mu\nu} \ .$$

This equation gives a prescription of how to get a component of $Z^{\rho\sigma}_{\mu\nu\kappa}$ by multiplying the components of X^{ρ}_{κ} and $B^{\sigma}_{\mu\nu}$. Note that the product is only well defined if the order of the indices is fixed. We can define for example

$$W^{\rho\sigma}_{\kappa\mu\nu} = X^{\rho}_{\kappa} B^{\sigma}_{\mu\nu} \ ;$$

W is then an IQ other than Z!

Algebra An important observation is that calculating with index-carrying quantities means *not* calculating with the whole table, as is the case when we calculate with matrices. Calculating with IQs always means calculating with the single components. The two operations that we introduced above are therefore nothing but the common operations known from calculating with numbers. This is why the standard rules apply here also: two commutative, two associative, and one distributive law.

Contraction Given two IQs of type (p, q), where $p > 0$ and $q > 0$, we can construct an IQ of type $(p - 1, q - 1)$ by choosing one of the upper and one of the lower indices, then adding the components with the same values for those two indices. For instance

$$W^{\alpha}_{\beta} = \sum_{\mu=1}^{n} Z^{\mu\alpha}_{\beta\mu} \ .$$

We can obtain different IQs depending on the choice of the upper and the lower index. Therefore it is important to write all indices explicitly. The summation indices are called *dummy* whereas the others are called *free* indices. Einstein proposed the following convention for dummy indices: the summation symbol is omitted. For example one writes the IQ W^{α}_{β} as $Z^{\mu\alpha}_{\beta\mu}$. This is known as the *Einstein convention* (see also [8], p. 9). It shortens calculations and makes expressions easier to read.

We want to apply the above rules and conventions to calculations with IQs in order to derive the transformation law for the components of the affine connection. We have two coordinate systems $\{x^{\mu}\}$ and $\{\bar{x}^{\mu}\}$. Let an autoparallel C be represented in the coordinates $\{x^{\mu}\}$ by functions $x^{\mu}(\lambda)$. These functions satisfy (1.8). With respect to $\{\bar{x}^{\mu}\}$ this autoparallel is given by the functions $\bar{x}^{\mu}(\lambda)$. Which equations satisfy these functions? We know that

$$x^{\mu}(\lambda) = x^{\mu}\left(\bar{x}^{1}(\lambda), \ldots, \bar{x}^{n}(\lambda)\right) \ .$$

Let us calculate the derivatives

$$\dot{x}^\mu = \sum_{\rho=1}^{n} \frac{\partial x^\mu}{\partial \bar{x}^\rho} \dot{\bar{x}}^\rho = X^\mu_{\bar{\rho}} \dot{\bar{x}}^\rho \,,$$

$$\ddot{x}^\mu = \sum_{\rho=1}^{n} \frac{\partial x^\mu}{\partial \bar{x}^\rho} \ddot{\bar{x}}^\rho + \sum_{\rho=1}^{n} \sum_{\sigma=1}^{n} \frac{\partial^2 x^\mu}{\partial \bar{x}^\rho \bar{x}^\sigma} \dot{\bar{x}}^\rho \dot{\bar{x}}^\sigma = X^\mu_{\bar{\rho}} \ddot{\bar{x}}^\rho + X^\mu_{\bar{\rho}\bar{\sigma}} \dot{\bar{x}}^\rho \dot{\bar{x}}^\sigma \,,$$

and insert them in (1.8). Then, we obtain

$$X^\mu_{\bar{\rho}} \ddot{\bar{x}}^\rho + X^\mu_{\bar{\rho}\bar{\sigma}} \dot{\bar{x}}^\rho \dot{\bar{x}}^\sigma + \Gamma^\mu_{\alpha\beta} X^\alpha_{\bar{\rho}} \dot{\bar{x}}^\rho X^\beta_{\bar{\sigma}} \dot{\bar{x}}^\sigma = 0 \,.$$

Using the commutative law and the distributive law we find

$$X^\mu_{\bar{\rho}} \ddot{\bar{x}}^\rho + \left(X^\mu_{\bar{\rho}\bar{\sigma}} + \Gamma^\mu_{\alpha\beta} X^\alpha_{\bar{\rho}} X^\beta_{\bar{\sigma}} \right) \dot{\bar{x}}^\rho \dot{\bar{x}}^\sigma = 0 \,.$$

The left-hand side is an IQ of type $(1,0)$. If we multiply it with the IQ $X^{\bar{\nu}}_\kappa$ and contract the indices μ and κ, we end up with

$$\ddot{\bar{x}}^\nu + \left(X^{\bar{\nu}}_\mu X^\mu_{\bar{\rho}\bar{\sigma}} + \Gamma^\mu_{\alpha\beta} X^{\bar{\nu}}_\mu X^\alpha_{\bar{\rho}} X^\beta_{\bar{\sigma}} \right) \dot{\bar{x}}^\rho \dot{\bar{x}}^\sigma = 0 \,.$$

This equation already has the same structure as (1.8). This yields

$$\bar{\Gamma}^\nu_{\bar{\rho}\bar{\sigma}} = \Gamma^\mu_{\alpha\beta} X^{\bar{\nu}}_\mu X^\alpha_{\bar{\rho}} X^\beta_{\bar{\sigma}} + X^{\bar{\nu}}_\mu X^\mu_{\bar{\rho}\bar{\sigma}} \,, \tag{1.9}$$

which is the desired *transformation law*. It is an inhomogeneous law. This was to be expected: if we had, in the transformation law, only the first term on the right-hand side, the components of the affine connection would vanish in all coordinate systems as soon as they vanish in one. The affine connection of Newtonian space–time gives a counterexample: the components vanish in inertial frames but not in arbitrary reference frames.

1.5.4 The Metric Affine Connection

An important example for an affine connection is the following. Let us consider the manifold given by the surface of a sphere. A special class of curves on the sphere consists of the great circles. These are distinguished from all other curves by their geometric properties. Do they define an affine connection? To answer this question, we have to derive the differential equations that define great circles.

1.5.4.1 The Length of Curves

Let p and q be two arbitrary points on the sphere. Every curve connecting p and q (connecting curve) has a proper length. Great circles are defined as curves whose segments give rise to the shortest connecting curves. We can therefore deduce the differential equations for great circles as Euler–Lagrange equations coming from a variation principle. Here the length of the connecting curve plays the role of the action.

The length of an arbitrary curve C, which is given in the coordinates ϑ and φ by the parameter representation

$$\vartheta = \vartheta(\lambda), \quad \varphi = \varphi(\lambda), \quad a \leq \lambda \leq b,$$

can be calculated as follows. The relation between the coordinates ϑ and φ on the one hand, and the coordinates y^k, $k = 1, 2, 3$, of the Euclidean space $\mathbb{E}^3$, in which the sphere is embedded, on the other read

$$y^1 = r \sin \vartheta \cos \varphi, \tag{1.10}$$

$$y^2 = r \sin \vartheta \sin \varphi, \tag{1.11}$$

$$y^3 = r \cos \vartheta. \tag{1.12}$$

Here r is the radius of the sphere; C therefore admits in $\mathbb{E}^3$ the following representation:

$$y^1(\lambda) = r \sin \vartheta(\lambda) \cos \varphi(\lambda),$$
$$y^2(\lambda) = r \sin \vartheta(\lambda) \sin \varphi(\lambda),$$
$$y^3(\lambda) = r \cos \vartheta(\lambda).$$

Its length L is given by

$$L = \int_a^b d\lambda \sqrt{(\dot{y}^1)^2 + (\dot{y}^2)^2 + (\dot{y}^3)^2}.$$

Inserting the functions $y^k(\lambda)$ results in

$$L = \int_a^b d\lambda \sqrt{r^2 \dot{\vartheta}^2 + r^2 \sin^2 \vartheta \dot{\varphi}^2}.$$

An arbitrary n-dimensional surface F in $\mathbb{E}^m$ can be defined by its *embedding equations* (in analogy to (1.10), (1.11) and (1.12)):

$$y^k = y^k(x^\mu), \quad k = 1, \ldots, m,$$

and a curve is represented by

$$x^\mu = x^\mu(\lambda), \quad \mu = 1, \ldots, n.$$

The expression under the square root then becomes

$$(\dot{y}^1)^2 + \cdots + (\dot{y}^m)^2 = \sum_{k=1}^m \frac{\partial y^k}{\partial x^\mu} \frac{\partial y^k}{\partial x^\nu} \dot{x}^\mu \dot{x}^\nu.$$

1.5.4.2 The Metric

It is worthwhile to study this expression. It is a quadratic form in the components $\dot{x}^{\mu}$ of the tangential vector to the curve. The coefficients of this quadratic form are denoted by $g_{\mu\nu}$, i.e.,

$$g_{\mu\nu} = \sum_{k=1}^{3} \frac{\partial y^k}{\partial x^{\mu}} \frac{\partial y^k}{\partial x^{\nu}} . \qquad (1.13)$$

This equation determines an IQ $g_{\mu\nu}$ at any point x^{μ} of the surface and not only along the curve. The field $g_{\mu\nu}x^{\rho}$ is called *metric* of the surface F. It is induced by the embedding of F in $\mathbb{E}^3$. Actually, the metric does not depend on the curve and can be used to calculate the length of any curve. We can hence store this information in the metric, which has the advantage that it is not necessary to know how the manifold is embedded in the Euclidean space. In particular it is not necessary to embed at all. An example is given by the metric of the Euclidean space $\mathbb{E}^n$.

The components of the metric depend on the chosen coordinate system. Let us calculate this dependence. We choose new coordinates x' on the surface F. The transformation to the old coordinates is given by

$$x^{\mu} = x^{\mu}(x') .$$

The chain rule leads to

$$\frac{\partial y^k}{\partial x'^{\mu}} = \frac{\partial y^k}{\partial x^{\rho}} X_{\mu'}^{\rho} .$$

If we insert this equation in the definition of $g_{\mu\nu}$, we obtain

$$g'_{\mu\nu} = X_{\mu'}^{\rho} X_{\nu'}^{\sigma} g_{\rho\sigma} .$$

This is the transformation property of the components of the metric. It is linear and homogeneous as the one for vectors, but a little more complicated.

1.5.4.3 Tensors

Such IQs, which are connected to the coordinate system and whose components transform linearly and homogeneously, and where the coefficients of the transformation law are given by products of the matrix elements $X_{\mu'}^{\rho}$ or $X_{\rho}^{\mu'}$, represent the so-called *tensors*. More precisely, a tensor $\mathbf{A}$ of type (p,g) is a quantity that is in any coordinate system represented by an IQ of type (p,g). The transformation law of the representation $A_{\sigma...}^{\rho...}$, with respect to $\{x^{\mu}\}$, to $A_{\nu...}^{'\mu...}$, with respect to $\{x'^{\mu}\}$, is given by

$$A_{\nu...}^{'\mu...} = X_{\rho}^{\mu'} \dots X_{\nu'}^{\sigma} \dots A_{\sigma...}^{\rho...} ,$$

where the matrix $X_{\rho}^{\mu'}$ appears p times and its inverse $X_{\mu'}^{\rho}$ q times, with corresponding indices on the right-hand side. Such a tensor is also called of type (p,q) or p times *contravariant* and q times *covariant*. Thus a vector is simply contravariant and the metric is a two times covariant tensor. The sum $p+q$ is referred to as the rank of the tensor. This terminology expresses nothing but certain transformation properties of physical and geometric quantities: very different physical and geometric quantities can be described by tensors of the same type. Examples of quantities that are not tensors: Γ, X.

From the transformation law it follows that every tensor is connected with a point on the manifold, exactly as we saw it already for vectors. A *tensor field* assigns a tensor to every point on the manifold. This tensor is connected with that point. A tensor field is smooth if the components depend smoothly on the coordinates in one of the tensor's coordinate representations.

We have defined tensors as objects that are represented by IQs with respect to coordinate systems and these IQs have to transform from one coordinate system to another in a precise way. We can speak about a *tensorial property* if this property is independent of the representation (i.e., is valid in all representations). Some examples for tensorial properties follow. We have introduced some operations for the IQs: equality, summation, product, contraction. If we combine the IQs that represent tensors with such operations in any coordinate system, will the results be tensors again, i.e., will the corresponding transformation properties hold for such combinations? We want to study this question for each of the above-mentioned operations.

Equality Consider two tensors S and T, of the same type, which are connected to a point. In a fixed coordinate system they are represented by two IQs, $S^{\mu\cdots}_{\nu\cdots}$ and $T^{\mu\cdots}_{\nu\cdots}$, and these IQs are assumed to be equal. Then the IQs that represent the tensors are equal in every coordinate system. This follows from the fact that the transformation properties of the components are equal. We call such tensors equal. In this way the equality of the IQs coincides with the equality of the tensors.

Addition Let us study an example. Consider two tensors S and T of the same type $(1,2)$ connected to a point. In the coordinate system $\{x^{\mu}\}$ they are represented by the IQs $S^{\mu}_{\rho\sigma}$ and $T^{\mu}_{\rho\sigma}$. These IQs are of the same type and can be added:

$$W^{\mu}_{\rho\sigma} = S^{\mu}_{\rho\sigma} + T^{\mu}_{\rho\sigma} .$$

In another coordinate system, $\{x'^{\mu}\}$, we have two other IQs, $S'^{\mu}_{\rho\sigma}$ and $T'^{\mu}_{\rho\sigma}$, and their sum can be denoted by $W'^{\mu}_{\rho\sigma}$. Is the representation of the tensor W by such a sum unique? This means: Do the IQs $W^{\mu}_{\rho\sigma}$ and $W'^{\mu}_{\rho\sigma}$ satisfy the correct transformation rule? We calculate:

$$W'^{\mu}_{\rho\sigma} = S'^{\mu}_{\rho\sigma} + T'^{\mu}_{\rho\sigma} = X^{\mu'}_{\alpha} X^{\beta}_{\rho'} X^{\gamma}_{\sigma'} S^{\alpha}_{\beta\gamma} + X^{\mu'}_{\alpha} X^{\beta}_{\rho'} X^{\gamma}_{\sigma'} T^{\alpha}_{\beta\gamma} .$$

This follows since S and T are tensors. Using the calculation rules for IQs we can now factor out the transformation matrices:

$$X^{\mu'}_{\alpha} X^{\beta}_{\rho'} X^{\gamma}_{\sigma'} S^{\alpha}_{\beta\gamma} + X^{\mu'}_{\alpha} X^{\beta}_{\rho'} X^{\gamma}_{\sigma'} T^{\alpha}_{\beta\gamma} = X^{\mu'}_{\alpha} X^{\beta}_{\rho'} X^{\gamma}_{\sigma'} \left(S^{\alpha}_{\beta\gamma} + T^{\alpha}_{\beta\gamma} \right)$$

$$= X^{\mu'}_{\alpha} X^{\beta}_{\rho'} X^{\gamma}_{\sigma'} W^{\alpha}_{\beta\gamma} .$$

This is the right transformation law. We now see why summation is a tensorial operation. It is possible to add tensors of the same type (p,q) and the sum is again a tensor of this type.

Product Let S and T be tensors connected to a point and of type $(2,1)$ respectively $(1,1)$. Let us take the product of their IQ representations in the two coordinate systems from above:

$$V^{\mu\nu\kappa}_{\rho\sigma} = S^{\mu\nu}_{\rho}T^{\kappa}_{\sigma}, \quad V'^{\mu\nu\kappa}_{\rho\sigma} = S'^{\mu\nu}_{\rho}T'^{\kappa}_{\sigma}.$$

Then we obtain

$$V'^{\mu\nu\kappa}_{\rho\sigma} = S'^{\mu\nu}_{\rho}T'^{\kappa}_{\sigma} = X^{\mu'}_{\alpha}X^{\nu'}_{\beta}X^{\gamma}_{\rho'}S^{\alpha\beta}_{\gamma}X^{\kappa'}_{\delta}X^{\varepsilon}_{\sigma'}T^{\delta}_{\varepsilon}$$
$$= X^{\mu'}_{\alpha}X^{\nu'}_{\beta}X^{\gamma}_{\rho'}X^{\kappa'}_{\delta}X^{\varepsilon}_{\sigma'}S^{\alpha\beta}_{\gamma}T^{\delta}_{\varepsilon} = X^{\mu'}_{\alpha}X^{\nu'}_{\beta}X^{\gamma}_{\rho'}X^{\kappa'}_{\delta}X^{\varepsilon}_{\sigma'}V^{\alpha\beta\delta}_{\gamma\varepsilon}.$$

Hence the product is also a tensorial operation. The product of tensors of types (p_1,q_1) and (p_2,q_2) leads to a tensor of type (p_1+p_2,q_1+q_2).

Contraction Let us take the tensor S of the previous example and calculate the contraction of the IQs corresponding to the coordinate systems $\{x^\mu\}$ and $\{x'^\mu\}$:

$$U^\mu = S^{\mu\nu}_\nu, \quad U'^\mu = S'^{\mu\nu}_\nu.$$

We then obtain

$$U'^\mu = S'^{\mu\nu}_\nu = X^{\mu'}_\alpha X^{\nu'}_\beta X^{\gamma}_{\nu'}S^{\alpha\beta}_\gamma = (X^{\nu'}_\beta X^{\gamma}_{\nu'})X^{\mu'}_\alpha S^{\alpha\beta}_\gamma.$$

Using

$$X^{\nu'}_\beta X^{\gamma}_{\nu'} = \delta^{\gamma}_\beta$$

we find

$$U'^\mu = X^{\mu'}_\alpha S^{\alpha\gamma}_\gamma = X^{\mu'}_\alpha U^\alpha.$$

We see that contraction is a tensorial operation that transforms a tensor of type (p,q) into a tensor of type $(p-1,q-1)$.

1.5.4.4 Symmetry

The metric in every point is a so-called *symmetric* tensor. Namely, we see from the definition in (1.13) that

$$g_{\mu\nu}(x) = g_{\nu\mu}(x), \quad \forall x,\mu,\nu.$$

This property is invariant with respect to coordinate transformations:

$$g'_{\mu\nu} = X^{\rho}_{\mu'}X^{\sigma}_{\nu'}g_{\rho\sigma} = X^{\rho}_{\mu'}X^{\sigma}_{\nu'}g_{\sigma\rho} = X^{\sigma}_{\nu'}X^{\rho}_{\mu'}g_{\sigma\rho} = X^{\rho}_{\nu'}X^{\sigma}_{\mu'}g_{\rho\sigma} = g'_{\nu\mu}.$$

In analogy to symmetric tensors we can also define *antisymmetric* tensors: a tensor $A_{\mu\nu}$ is antisymmetric if $A_{\mu\nu} = -A_{\nu\mu}$.

Every IQ can be split uniquely, with respect to two indices of the same type
(i.e., both indices are upper or lower indices), into the sum of its symmetric and its
antisymmetric part. For example let $B_\mu^{\rho\sigma}$ be an arbitrary IQ of type $(2,1)$. Then we
can write

$$B_\mu^{\rho\sigma} = S_\mu^{\rho\sigma} + A_\mu^{\rho\sigma} \,,$$

where

$$S_\mu^{\rho\sigma} = S_\mu^{\sigma\rho} \,, \quad A_\mu^{\rho\sigma} = -A_\mu^{\sigma\rho} \,.$$

Both $S_\mu^{\rho\sigma}$ and $A_\mu^{\rho\sigma}$ are uniquely defined by the components of $B_\mu^{\rho\sigma}$:

$$S_\mu^{\rho\sigma} = \frac{1}{2}\left(B_\mu^{\rho\sigma} + B_\mu^{\sigma\rho}\right) \,, \quad A_\mu^{\rho\sigma} = \frac{1}{2}\left(B_\mu^{\rho\sigma} - B_\mu^{\sigma\rho}\right) \,.$$

Symmetric and antisymmetric IQs have the following property, which is often
used. Let for example $S^{\rho\sigma}$ be symmetric and $A_{\rho\sigma}$ antisymmetric. Then the equation

$$S^{\rho\sigma} A_{\rho\sigma} = 0$$

holds. It is true in general that the contraction of two symmetric with two antisymmetric indices always gives zero. The proof in the above case works as follows:

$$S^{\rho\sigma} A_{\rho\sigma} = -S^{\rho\sigma} A_{\sigma\rho} = -S^{\sigma\rho} A_{\sigma\rho} = -S^{\rho\sigma} A_{\rho\sigma} \,.$$

If IQs represent tensors in a given coordinate system, then all relations hold in an
arbitrary coordinate system.

Contravariant metric

The tensor field $g_{\mu\nu}(x)$ defines the so-called *contravariant metric*. In the following
we want to explain what this means.

The right-hand side of the defining equation (1.13) for the metric on a surface F
can be interpreted as the scalar product of two vectors in $\mathbf{E}^3$: we have (Fig. 1.2)

$$\sum_{k=1}^{3} \frac{\partial y^k}{\partial x^\mu} \frac{\partial y^k}{\partial x^\nu} = \left(\vec{Y}_\mu, \vec{Y}_\nu\right) \,,$$

where the vectors $\vec{Y}_\mu$, $\mu = 1,2$, are tangent vectors to the curves $x^1 = \lambda$, $x^2 = $ const,
respectively $x^1 = $ const, $x^2 = \lambda$ on the surface F. If the coordinates $\{x^\mu\}$ form an
allowed system (i.e., if they are independent), then also these two vectors are linearly
independent. The determinant $\det g_{\mu\nu}$ of the metric $g_{\mu\nu}$ is nothing but the Gram
determinant of these vectors and hence must not vanish.

In general the following holds for symmetric, contravariant tensors $T_{\mu\nu}$ of second
rank on an arbitrary n-manifold: if the determinant does not vanish with respect
to a given coordinate system $\{x'^\mu\}$, $\det T_{\mu\nu} \neq 0$, then we have $\det T'_{\mu\nu} \neq 0$ in any
coordinate system $\{x'^\mu\}$. Indeed, we have

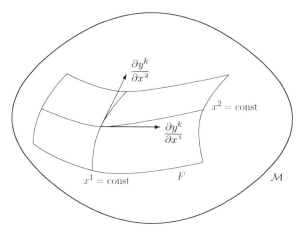

Fig. 1.2 Partial derivatives as components of tangential vectors to coordinate curves on the submanifold F

$$T'_{\mu\nu} = X^\rho_{\mu'} X^\sigma_{\nu'} T_{\rho\sigma} \, .$$

This can be written as the matrix product of three $n \times n$ matrices if we define the matrices **T** and **X** as:

$$\mathbf{T} := \begin{pmatrix} T_{11} & \cdots & T_{1n} \\ \vdots & & \vdots \\ T_{n1} & \cdots & T_{nn} \end{pmatrix} ,$$

and

$$\mathbf{X} := \begin{pmatrix} X^1_1 & \cdots & X^1_n \\ \vdots & & \vdots \\ X^n_1 & \cdots & X^n_n \end{pmatrix} .$$

Then it obviously holds that

$$\mathbf{T}' = \mathbf{X}^\top \mathbf{T} \mathbf{X} \, .$$

We therefore get[3]

$$\det T'_{\mu\nu} = \det{}^2 X^\rho_{\mu'} \det T_{\rho\sigma} \, . \tag{1.14}$$

The matrix **X** is always regular, $\det{}^2 X^\rho_{\mu'} > 0$, and therefore even the sign of $\det T_{\rho\sigma}$ is independent of the coordinate system.

The metric $g_{\mu\nu}(x)$ at the point x, considered as matrix, has therefore always an inverse. We denote the components of this inverse by $g^{\mu\nu}(x)$, i.e.,

$$g_{\mu\nu}(x) g^{\nu\rho}(x) = \delta^\rho_\mu \, .$$

This way we define a new IQ $g^{\mu\nu}(x)$ in every point and with respect to every coordinate system. It is possible to show (exercise) that the new IQ transforms as a tensor of type $(2,0)$. It is known as the *contravariant metric*.

---

[3] We use the theorem on the determinant of the product of matrices, see for instance [9], p. 133.

1.5.4.5 General Definition of the Metric

So far we defined the metric only on surfaces embedded in $\mathbb{E}^m$. A general definition for the metric can also be given.

Definition 4 Let $\mathcal{M}$ be a n-manifold and $g_{\mu\nu}(x)$ a symmetric non-degenerate $(\det g_{\mu\nu} \neq 0)$ tensor field of type $(0,2)$, which is well defined and smooth everywhere on $\mathcal{M}$. Then the pair $(\mathcal{M}, g_{\mu\nu})$ is called a *metric manifold* and $g_{\mu\nu}(x)$ is called the metric on $\mathcal{M}$.

The metric defines a scalar product in the space of vectors at a given point. Consider for example two vectors V^μ and U^μ at the point p. Then the expression $g_{\mu\nu}(p)V^\mu U^\nu$ defines a scalar that depends linearly on each of the vectors and is symmetric under the exchange of the two vectors. However, this expression is not necessarily positive-definite. We want to allow here explicitly indefinite metrics also. The positive-definite metrics are called Riemannian [5].

Definition 5 Let $(\mathcal{M}, g_{\mu\nu})$ be a metric n-manifold and $C : [\lambda_1, \lambda_2] \mapsto \mathcal{M}$ an arbitrary piecewise smooth curve in $\mathcal{M}$. Then the length $L(C)$ of this curve is defined by

$$L(C) = \int_{\lambda_1}^{\lambda_2} d\lambda \sqrt{g_{\mu\nu}(x(\lambda))\dot{x}^\mu(\lambda)\dot{x}^\nu(\lambda)}. \tag{1.15}$$

It is independent of the choice of parameterization $x^\mu(\lambda)$. (Note that the length is only defined for non-negative expressions under the square root.)

We see that the manifold under consideration does not necessarily need to be a surface in $\mathbb{E}^3$ to define the length of curves. However, such surfaces are examples for metric manifolds.

1.5.4.6 Geodesics

We can now derive the differential equations satisfied by connecting curves of extremal length on an arbitrary metric manifold. These connecting curves are called *geodesics* and the differential equations are called *geodesic equations*. The differential equation defining a great circle on the surface of a sphere is a special case of such geodesic equations.

Let p and q be two points on a metric n-manifold $(\mathcal{M}, g)$. Moreover let $\{x^\mu\}$ be a coordinate system and $C : [\lambda_1, \lambda_2] \mapsto \mathcal{M}$ an arbitrary curve connecting the points p and q, $C(\lambda_1) = p$, $C(\lambda_2) = q$. Let $x^\mu(\lambda)$ be the representation of the curve C with respect to $\{x^\mu\}$. Then the length of C is given by the integral (1.15). To find the extrema of the length amounts to a typical variation problem where the endpoints are fixed. The Lagrange function $\mathcal{L}$ reads

$$\mathcal{L}(x(\lambda), \dot{x}(\lambda)) = \sqrt{g_{\mu\nu}(x(\lambda))\dot{x}^\mu(\lambda)\dot{x}^\nu(\lambda)}.$$

The solution $x(\lambda)$ must satisfy the Euler–Lagrange equations:

$$\frac{\partial \mathscr{L}}{\partial x^{\kappa}} - \frac{d}{d\lambda}\frac{\partial \mathscr{L}}{\partial \dot{x}^{\kappa}} = 0.$$

We calculate

$$\frac{\partial \mathscr{L}}{\partial x^{\kappa}} = \frac{1}{2\mathscr{L}}\frac{\partial g_{\mu\nu}}{\partial x^{\kappa}}\dot{x}^{\mu}\dot{x}^{\nu},$$

$$\frac{\partial \mathscr{L}}{\partial \dot{x}^{\kappa}} = \frac{1}{2\mathscr{L}}\left(g_{\kappa\nu}\dot{x}^{\nu} + g_{\mu\kappa}\dot{x}^{\mu}\right) = \frac{1}{\mathscr{L}}g_{\kappa\nu}\dot{x}^{\nu},$$

$$\frac{d}{d\lambda}\frac{\partial \mathscr{L}}{\partial \dot{x}^{\kappa}} = \frac{1}{\mathscr{L}}g_{\kappa\nu}\ddot{x}^{\nu} + \frac{1}{\mathscr{L}}\frac{\partial g_{\kappa\nu}}{\partial x^{\mu}}\dot{x}^{\nu}\dot{x}^{\mu} - \frac{1}{\mathscr{L}^{2}}\dot{\mathscr{L}}g_{\kappa\nu}\dot{x}^{\nu}$$

(the above holds under the assumption that $g_{\mu\nu}(x)\dot{x}^{\mu}\dot{x}^{\nu} > 0$). Inserting this in the Euler–Lagrange equations yields

$$\frac{1}{\mathscr{L}}g_{\kappa\nu}\ddot{x}^{\nu} + \frac{1}{\mathscr{L}}\frac{\partial g_{\kappa\nu}}{\partial x^{\mu}}\dot{x}^{\nu}\dot{x}^{\mu} - \frac{1}{2\mathscr{L}}\frac{\partial g_{\mu\nu}}{\partial x^{\kappa}}\dot{x}^{\nu}\dot{x}^{\mu} = \frac{1}{\mathscr{L}^{2}}\dot{\mathscr{L}}g_{\kappa\nu}\dot{x}^{\nu}.$$

We can solve this equation for the second-order derivatives $\ddot{x}^{\nu}$ by multiplying it by $\mathscr{L}g^{\rho\kappa}$:

$$\ddot{x}^{\rho} + \frac{1}{2}g^{\rho\kappa}\left(2\frac{\partial g_{\kappa\nu}}{\partial x^{\mu}} - \frac{\partial g_{\mu\nu}}{\partial x^{\kappa}}\right)\dot{x}^{\mu}\dot{x}^{\nu} = \frac{d\ln\mathscr{L}}{d\lambda}\dot{x}^{\rho}.$$

The second term on the left-hand side can be written differently. The expression in the brackets is not symmetric in the indices μ and ν, but the expression $\dot{x}^{\mu}\dot{x}^{\nu}$ is. Hence, only the symmetric part of the expression in the brackets contributes. We want to denote the coefficient of $\dot{x}^{\mu}\dot{x}^{\nu}$, which arises this way, by $\{^{\rho}_{\mu\nu}\}$, the so-called *Christoffel symbol*. Then we have

$$\{^{\rho}_{\mu\nu}\} = \frac{1}{2}g^{\rho\kappa}\left(\frac{\partial g_{\kappa\nu}}{\partial x^{\mu}} + \frac{\partial g_{\kappa\mu}}{\partial x^{\nu}} - \frac{\partial g_{\mu\nu}}{\partial x^{\kappa}}\right), \tag{1.16}$$

and

$$\ddot{x}^{\rho} + \{^{\rho}_{\mu\nu}\}\dot{x}^{\mu}\dot{x}^{\nu} = \frac{d\ln\mathscr{L}}{d\lambda}\dot{x}^{\rho}. \tag{1.17}$$

This is the geodesic equation.

If the right-hand side of the equation was zero, then this equation would be formally identical to (1.8) of an autoparallel of an affine connection. Why is the right-hand side not zero? We varied a functional which defines the length of a curve and which is therefore invariant under arbitrary reparameterizations of the curve. A reparameterization is a transformation of the parameters, for instance $\lambda = \lambda(\kappa)$. The derived equation possesses this symmetry, too. On the other hand, the equation of an autoparallel does not have this symmetry; it determines a class of curve parameters, the so-called affine parameters.

Is there a particular class of parameters for which the right-hand side of the geodesic equation vanishes? These have to be exactly those parameters for which $\mathscr{L} = 0$, hence $\mathscr{L} = \text{const}$, holds. If we plug this in the expression for the length given in (1.15), we obtain

$$s(\lambda) = \int_\beta^\lambda d\xi \mathscr{L} = \mathscr{L}(\lambda - \beta) .$$

Hence

$$\lambda = \alpha s + \beta ,$$

where $\alpha = \mathscr{L}^{-1}$ and s is the length of the curve.

We therefore get the following result. The geodesics on a metric manifold are autoparallels of an affine connection if they are parameterized by their length or by an affine transformation of it. The components of the affine connection are given by the Christoffel symbols. This affine connection is called *metric* affine connection.

It is important to note the following: an affine connection can exist even without having a metric. The manifolds with affine connections are more general than metric manifolds.

1.6 Cartan–Friedrichs Space–Time

We now have enough mathematical tools at hand to finally express the idea that the gravitational force is an apparent force in a mathematical language. To do so, we first rewrite Newton's second law in an arbitrary reference frame, such that all apparent forces as well as all physical ones appear in this equation. Then we try to make a different distinction between these forces from what was done in Newton's theory.

The motion of a mass point with mass μ under the influence of a total force $\vec{f}$ is determined, with respect to an inertial frame $\{\bar{x}^\mu\}$, by the following equations:

$$\ddot{\bar{x}}^0 = 0, \quad \ddot{\bar{x}}^k = \mu^{-1} \bar{f}^k \left(\dot{\bar{x}}^0\right)^2 . \tag{1.18}$$

The term $(\dot{\bar{x}}^0)^2$ on the right-hand side is a correction term stemming from the fact that λ is not the time and hence $\ddot{\bar{x}}^k$ is not the acceleration (a dot denotes the derivative with respect to λ). The first equation in (1.18) defines a special class of parameters λ: they are given by a time function up to affine transformations.

The transformation of the left-hand side of (1.18) to a general reference frame (with Cartesian axes!) is the same as the transformation of a free motion. We differentiate twice the equations

$$x^0 = \bar{x}^0 - T , \quad x^k = x^k \left(\bar{x}^0, \bar{x}^1, \bar{x}^2, \bar{x}^3\right)$$

and then substitute (1.18). We obtain

$$\ddot{x}^0 = 0 ,$$
$$\ddot{x}^k + \Gamma^k_{N\rho\sigma} \dot{x}^\rho \dot{x}^\sigma = \mu^{-1} f^k \left(\dot{x}^0\right)^2 , \tag{1.19}$$

where

$$f^k = X_i^k \tilde{f}^l .$$

Here, $\Gamma^\mu_{N\rho\sigma}$ are the components of the Newton affine connection; X_i^k is a (time-dependent) orthogonal matrix. In (1.19) the apparent forces stand on the left, the physical forces on the right-hand side.

The total force $\vec{f}$ can be split into the gravitational force and the rest:

$$f^k = -\mu \partial_k \Phi + F^k , \tag{1.20}$$

where Φ is the Newton potential. Let us insert (1.20) in (1.19):

$$\ddot{x}^k + \Gamma^k_{N\rho\sigma} \dot{x}^\rho \dot{x}^\sigma = -\partial_k \Phi \left(\dot{x}^0\right)^2 + \mu^{-1} F^k \left(\dot{x}^0\right)^2 .$$

As expected, the gravitational part has the same form as the geometric terms on the left-hand side. In particular, it does not contain any information about the test particle. If we add this part to the left-hand side, we obtain

$$\ddot{x}^k + \left(\Gamma^k_{N\rho\sigma} + \delta^0_\rho \delta^0_\sigma \partial_k \Phi\right) \dot{x}^\rho \dot{x}^\sigma = \mu^{-1} F^k \left(\dot{x}^0\right)^2 . \tag{1.21}$$

The coefficients of the quadratic form on the left-hand side of (1.21) define a new affine connection, which we call the *Einstein affine connection*. Its components $\Gamma^\mu_{E\rho\sigma}$ are then given by

$$\Gamma^0_{E\rho\sigma} = \Gamma^0_{N\rho\sigma} = 0 , \quad \Gamma^k_{Emn} = \Gamma^k_{Nmn} = 0 ,$$
$$\Gamma^k_{E00} = \Gamma^k_{N00} + \partial_k \Phi , \quad \Gamma^k_{E0l} = \Gamma^k_{N0l} . \tag{1.22}$$

The autoparallels of the Einstein affine connection are the *free falls of Newton's theory* ($F^k = 0$).

This way we actually obtained a new space–time. It possesses the following structures: an absolute time, as is the case for Newton's space–time, the $\mathbb{E}^3$-structure in the simultaneity hypersurfaces as for Newton's space–time, but an Einstein affine connection instead of a Newton affine connection. The space–time with this geometry is called *Cartan–Friedrichs space–time* [10], see also [8], p. 287. In this space–time, gravitation is not considered as a field of force but as a part of the geometry of space–time.

Equations (1.21) and (1.19) are mathematically "equivalent": their solutions describe the same motion of the mass points. The measurable predictions did not change. Nevertheless the interpretation is changed significantly. What in Newton's theory is described as a motion under the influence of a gravitational force (free fall), becomes a free motion in Cartan–Friedrichs theory. Whereas the geometry of space–time in Newton's theory is rigid and does not change, it depends on the distribution of the masses in Cartan–Friedrichs theory.

Let us consider for example the motion of a man who stands on the floor of a lecture hall. The viewpoint of Newton's theory is that this motion is free of forces: the gravitational force of the Earth is exactly compensated by the contact force of the Earth's surface. On the other hand, in the perspective of Cartan–Friedrichs theory, the man's motion is not free of forces: the contact force acts on the man and accelerates him upward.

The acceleration is now defined directly by an affine connection (left-hand side of (1.21)) and not by an inertial frame. This resolves the paradox of the expanding Earth. Do we have in the Cartan–Friedrichs theory an analogue for the Newtonian inertial frames? We want to answer this question now. The formal definition of the inertial frame is based on the affine connection and reads: the inertial frame of an affine connection is a reference frame in which all components of the affine connection vanish everywhere:

$$\Gamma^\rho_{\mu\nu}(x) = 0, \quad \forall \rho, \mu, \nu, x . \tag{1.23}$$

This question is important for the solution of the problem of an expanding Earth. That is to say, if there was a Cartesian frame $\bar{x}^\mu$ where $\bar{\Gamma}_E$ vanishes, then (1.21) would read

$$\ddot{\bar{x}}^k = \mu^{-1} \bar{F}^k \left(\dot{\bar{x}}^0 \right)^2 .$$

The force $\bar{F}^k$ is now a contact force, which points everywhere away from the center of the Earth. Then the coordinates $\bar{x}^k$ of the surface must change exactly in this direction and hence the Earth has to expand. Therefore such inertial frames are not supposed to exist.

A coordinate system that satisfies (1.23) is called in differential geometry a *global geodesic system*. Our question can then be formulated mathematically: Does the Einstein affine connection allow global geodesic systems? To answer this question we first have to study again some more mathematics.

1.7 Curvature Tensor

The problem mentioned in the previous section can in general be stated precisely as follows. Let $\mathcal{M}$ be an affine-connected n-manifold. Let $\Gamma^\mu_{\rho\sigma}$ be the components of the affine connection with respect to a reference frame $\{x^\mu\}$. Do there exist coordinates $\{\bar{x}^\mu\}$ such that the corresponding components of the affine connection vanish everywhere? That is to say

$$\bar{\Gamma}^\mu_{\rho\sigma}(\bar{x}) = 0 \quad \forall \bar{x}, \mu, \rho, \sigma .$$

We can formulate it differently: we need to find the transformation functions $\bar{x}^\mu(x)$. If we use the transformation law for the components of Γ, we obtain

$$0 = \bar{\Gamma}^\mu_{\rho\sigma}(\bar{x}) = \Gamma^\nu_{\lambda\kappa} X^{\bar\mu}_\nu X^\lambda_{\bar\rho} X^\kappa_{\bar\sigma} + X^{\bar\mu}_\nu X^\nu_{\bar\rho\bar\sigma} .$$

This seems to be a difficult differential equation for the transformation functions $\bar{x}^\mu(x)$. We can, however, simplify this equation. We can transform the last term as

$$X_\nu^{\bar{\mu}} X_{\bar{\rho}\bar{\sigma}}^\nu = X_\lambda^{\bar{\mu}} \left(\partial_{\bar{\sigma}} X_{\bar{\rho}}^\lambda \right) = \partial_{\bar{\sigma}} \left(X_\lambda^{\bar{\mu}} X_{\bar{\rho}}^\lambda \right) - \left(\partial_{\bar{\sigma}} X_\lambda^{\bar{\mu}} \right) X_{\bar{\rho}}^\lambda$$

$$= \partial_{\bar{\sigma}} \left(\delta_{\bar{\rho}}^\mu \right) - \left(X_{\lambda\kappa}^{\bar{\mu}} X_{\bar{\sigma}}^\kappa \right) X_{\bar{\rho}}^\lambda = -X_{\lambda\kappa}^{\bar{\mu}} X_{\bar{\sigma}}^\kappa X_{\bar{\rho}}^\lambda .$$

This yields

$$\Gamma_{\lambda\kappa}^\nu X_\nu^{\bar{\mu}} X_{\bar{\rho}}^\lambda X_{\bar{\sigma}}^\kappa + X_\nu^{\bar{\mu}} X_{\bar{\rho}\bar{\sigma}}^\nu = \left(\Gamma_{\lambda\kappa}^\nu X_\nu^{\bar{\mu}} - X_{\lambda\kappa}^{\bar{\mu}} \right) X_{\bar{\sigma}}^\kappa X_{\bar{\rho}}^\lambda .$$

Hence, we obtain the equation

$$\frac{\partial^2 \bar{x}^\mu}{\partial x^\lambda \partial x^\kappa} = \Gamma_{\lambda\kappa}^\nu(x) \frac{\partial \bar{x}^\mu}{\partial x^\nu} . \tag{1.24}$$

This is a system of linear differential equations for the functions $\bar{x}^\mu(x)$.

A necessary condition for the solvability of this system can be deduced by differentiating both sides with respect to x^α and demanding symmetry in the indices α and κ. Let us start with the differentiation:

$$\frac{\partial^3 \bar{x}^\mu}{\partial x^\lambda \partial x^\kappa \partial x^\alpha} = \partial_\alpha \Gamma_{\lambda\kappa}^\nu(x) X_\nu^{\bar{\mu}} + \Gamma_{\lambda\kappa}^\nu(x) X_{\nu\alpha}^{\bar{\mu}}$$

$$= \partial_\alpha \Gamma_{\lambda\kappa}^\nu(x) X_\nu^{\bar{\mu}} + \Gamma_{\lambda\kappa}^\nu(x) \Gamma_{\nu\alpha}^\rho(x) X_{\bar{\rho}}^{\bar{\mu}}$$

$$= [\partial_\alpha \Gamma_{\kappa\lambda}^\rho(x) + \Gamma_{\lambda\kappa}^\nu(x) \Gamma_{\nu\alpha}^\rho(x)] X_{\bar{\rho}}^{\bar{\mu}} .$$

Here we substituted the expression given in (1.24) for $X_{\nu\alpha}^{\bar{\mu}}$. The symmetry condition leads to

$$\partial_\alpha \Gamma_{\lambda\kappa}^\rho(x) + \Gamma_{\lambda\kappa}^\nu(x) \Gamma_{\nu\alpha}^\rho(x) - \partial_\kappa \Gamma_{\lambda\alpha}^\rho(x) - \Gamma_{\lambda\alpha}^\nu(x) \Gamma_{\nu\kappa}^\rho(x) = 0 .$$

The expression on the left-hand side, which we want to write in a shorthand notation as $R_{\lambda\alpha\kappa}^\rho$, plays an important role in differential geometry. We have two theorems:

Theorem 1 *Let $\mathcal{M}$ be a n-manifold with affine connection and $\Gamma_{\nu\kappa}^\rho$ the components of the affine connection with respect to an arbitrary coordinate system $\{x^\mu\}$. An IQ $R_{\lambda\alpha\kappa}^\rho$ is assumed to be defined in any such coordinate system as follows:*

$$R_{\lambda\alpha\kappa}^\rho(x) = \partial_\alpha \Gamma_{\lambda\kappa}^\rho(x) - \partial_\kappa \Gamma_{\lambda\alpha}^\rho(x) + \Gamma_{\lambda\kappa}^\nu(x) \Gamma_{\nu\alpha}^\rho(x) - \Gamma_{\lambda\alpha}^\nu(x) \Gamma_{\nu\kappa}^\rho(x) . \tag{1.25}$$

Then $R_{\lambda\alpha\kappa}^\rho$ is a tensor of type $(1,3)$, which is antisymmetric in the last two indices. $R_{\lambda\alpha\kappa}^\rho$ is called the curvature tensor *of the affine connection $\Gamma_{\lambda\kappa}^\rho$.*

The proof is direct, even though a bit tedious: one has to use the transformation law (1.9) for $\Gamma_{\lambda\kappa}^\rho$ and those for the derivatives with respect to the coordinates. All terms that contain second- and third-order derivatives of the transformation functions cancel.

Theorem 2 *The necessary and sufficient condition for having at least one solution of the differential equation (1.24) in the neighborhood U of p reads*

$$R^{\rho}_{\lambda\alpha\kappa}(x) = 0 \qquad \forall x \in U, \rho, \lambda, \alpha, \text{and } \kappa .$$

We have shown only the necessary condition. The opposite direction is difficult [11].

The affine connection, whose components can be transformed to zero, is called *integrable* or *flat*. We note that $R^{\rho}_{\lambda\alpha\kappa}$ as tensor vanishes either in every coordinate system or in none. The integrability of an affine connection is therefore a coordinate-independent property.

Let us calculate the curvature of the Cartan–Friedrichs space–time. For the affine connection, we want to insert the relation (1.22) in (1.25) and want to use the fact that the curvature tensor vanishes for the Newton affine connection. We obtain for instance, for the components $R_{E}{}^{k}_{0l0}$:

$$
\begin{aligned}
R_{E}{}^{k}_{0l0} &= \partial_l \Gamma_{E}{}^{k}_{00} - \partial_0 \Gamma_{E}{}^{k}_{0l} + \Gamma_{E}{}^{k}_{l\rho}\Gamma_{E}{}^{\rho}_{00} - \Gamma_{E}{}^{k}_{0\rho}\Gamma_{E}{}^{\rho}_{0l} \\
&= \partial_l \left(\Gamma_{N}{}^{k}_{00} + \partial_k \Phi \right) - \partial_0 \Gamma_{N}{}^{k}_{0l} + \Gamma_{N}{}^{k}_{l0}\Gamma_{N}{}^{0}_{00} + \Gamma_{N}{}^{k}_{lr}\left(\Gamma_{N}{}^{r}_{00} + \partial_r \Phi \right) \\
&\quad - \left(\Gamma_{N}{}^{k}_{00} + \partial_k \Phi \right)\Gamma_{N}{}^{0}_{0l} - \Gamma_{N}{}^{k}_{0r}\Gamma_{N}{}^{r}_{0l} \\
&= \partial_l \partial_k \Phi + \left(\partial_l \Gamma_{N}{}^{k}_{00} - \partial_0 \Gamma_{N}{}^{k}_{0l} + \Gamma_{N}{}^{k}_{l\rho}\Gamma_{N}{}^{\rho}_{00} - \Gamma_{N}{}^{k}_{0\rho}\Gamma_{N}{}^{\rho}_{0l} \right) \\
&\quad + \Gamma_{N}{}^{k}_{lr}\partial_r \Phi - \partial_k \Phi \Gamma_{N}{}^{0}_{0l} .
\end{aligned}
\tag{1.26}
$$

The last two terms are zero and the four terms in brackets are equal to the component $R_{N}{}^{k}_{0l0}$ of the Newton curvature tensor, which of course vanishes. This way we obtain

$$R_{E}{}^{k}_{0l0} = \partial_l \partial_k \Phi . \tag{1.27}$$

Analogously, we can calculate all the other components of the curvature tensor with the simple result

$$
\begin{aligned}
R_{E}{}^{k}_{mln} = R_{E}{}^{0}_{lmn} = R_{E}{}^{k}_{0mn} = R_{E}{}^{k}_{l0m} \\
= R_{E}{}^{0}_{0mn} = R_{E}{}^{0}_{l0m} = R_{E}{}^{0}_{00l} = 0 .
\end{aligned}
\tag{1.28}
$$

Hence, as expected, the Einstein affine connection is curved for every inhomogeneous gravitational field; in this case, no global inertial frames can exist.

It is surprising that the curvature exists, so to speak, only in the time direction. This is implied by the fact that all purely space-like components $R_{E}{}^{k}_{mln}$ of the tensor vanish. If we use the property of the components of the Newton affine connection that only Γ^{k}_{N00} and Γ^{k}_{N0l} are different from zero (exercise) and that the same is true for the Einstein affine connection, we easily see: the straight lines in every simultaneity surface (which is the three-dimensional Euclidean space) are autoparallels of the Einstein affine connection, exactly as is the case for the Newton affine connection. This property, however, will not survive in the relativistic theory, since space and time are then not separable anymore. But in those cases where the Newton theory is a good approximation, the curvature of the space will be smaller than the curvature in the time direction by some powers of the speed of light.

What is the meaning of the individual components of the curvature tensor? So far we just know that the vanishing of the curvature tensor guarantees the existence of inertial frames. This is just a very rough understanding of the components. We can gain more insight in the following way.

What is the physical meaning of the matrix $R_{\mathrm{E}\,0l0}^{\ \ k}$? Let us consider two points, one with coordinates x^m and the other with coordinates $x^m + \delta x^m$. The relative acceleration a^k of two bodies free falling through these points is

$$a^k = -\partial_k \Phi(x^m + \delta x^m) + \partial_k \Phi(x^m) = -\partial_k \partial_l \Phi(x^m)\,\delta x^l = -R_{\mathrm{E}\,0l0}^{\ \ k}\delta x^l\ .$$

The components of the curvature tensor thus denote a difference in the acceleration of the freely falling system in different points. On the Earth we perceive such relative accelerations, for instance in the field of the Moon. They evoke forces which produce tides. This is why it is sometimes said that the curvature tensor of general relativity describes *tidal forces*. It is also true in general that the components of the curvature tensor of an arbitrary affine connection contain information about the relative acceleration of its autoparallels.

1.8 The Equivalence Principle

1.8.1 Galilei Equivalence Principle

We were able to write Newtonian mechanics in Cartan–Friedrichs form and to make gravitation become the geometry of space–time. But this was only possible under the following assumption:

> **Galilei equivalence principle:** The motion of an arbitrary free-falling test particle is independent of its consistence and of its structure.

For this it is sufficient that (a) the test particle is electrically neutral, (b) its gravitational binding energy is negligible compared with the mass, (c) its angular momentum is negligible, (d) its radius is small enough for the inhomogeneities of the gravitational field not to affect the motion. This is the so-called *Galilei equivalence principle* (or *weak equivalence principle*). The first experiments for it (these were pendular experiments, the Pisa tower was just an Gedankenexperiment) were done by Galilei (see [2]).

We now face two problems. First, if the gravitational effects on all systems, not only on the dynamical trajectories, simply result from reference frames that are not moving correctly, then it should be possible to generalize the Galilei principle. But how? Second, a complete abolishment of the inertial frames is not convincing. The inertial frames play far too important a role, in particular in special relativity, and special relativity is experimentally very well established. They should not be absolutely wrong but survive maybe as an approximation. The answers to both the questions are related. We first need some mathematics.

1.8.2 Geodesic Systems

Equation (1.27) shows that in Cartan–Friedrichs space–time no exact analogy to
inertial frames exists. Can we weaken the condition (1.24) and obtain nevertheless
a reasonable class of inertial frames? The following definition is suitable:

Definition 6 Let $\mathcal{M}$ be a n-manifold with affine connection Γ and p an arbitrary
point of $\mathcal{M}$. The coordinate system $\{x^\mu\}$ is called *geodesic* in p if

$$\Gamma^\mu{}_{\rho\sigma}(p) = 0 .$$

Theorem 3 *For every point p of any affine-connected manifold $\mathcal{M}$ there exists at
least one coordinate system, which is geodesic in p.*

Proof Let us choose coordinates $\{x^\mu\}$ around the point p; the components of the
affine connection with respect to $\{x^\mu\}$ may be denoted by $\Gamma^\mu{}_{\rho\sigma}$, and let $\Gamma^\mu{}_{\rho\sigma}(p) \neq$
0. We want to show that there exist new coordinates $\{\bar{x}^\mu\}$ such that $\bar{\Gamma}^\mu{}_{\rho\sigma}(p) = 0$.

The condition $\bar{\Gamma}^\mu{}_{\rho\sigma}(p) = 0$ is equivalent to (1.24), restricted to a point p:

$$\frac{\partial^2 \bar{x}^\mu}{\partial x^\rho \partial x^\sigma}(p) = \frac{\partial \bar{x}^\mu}{\partial x^\nu}(p)\Gamma^\nu{}_{\rho\sigma}(p) .$$

This condition is satisfied by the following transformation functions:

$$\bar{x}^\mu = x^\mu - x^\mu(p) + \frac{1}{2}\Gamma^\mu{}_{\rho\sigma}(p)\left(x^\rho - x^\rho(p)\right)\left(x^\sigma - x^\sigma(p)\right) \tag{1.29}$$

and hence the functions $\{\bar{x}^\mu\}$, which are defined by these transformations, are the
coordinates we were looking for, qed.

1.8.3 Local Inertial Frames

What is the physical meaning of the geodesic coordinates in Cartan–Friedrichs
space–time? Let us consider such a space–time with a given gravitational field Φ
and let us choose a point p. We can choose the coordinates in an intelligent way: let
$\{x^\mu\}$ be an inertial frame in the corresponding Newton space–time. Then we have

$$\Gamma_{\text{N}}{}^\mu{}_{\rho\sigma} = 0 .$$

Equation (1.22) then yields

$$\Gamma_{\text{E}}{}^k{}_{00} = \partial_k\Phi ,$$

where all other components of Γ_{E} vanish. If we insert this in the transformation
equation (1.29), we obtain

$$\bar{x}^0 = x^0 - x^0(p),$$

$$\bar{x}^k = x^k - x^k(p) + \frac{1}{2}\partial_k\Phi(p)\left(x^0 - x^0(p)\right)^2 .$$

The trajectory of the origin $\bar{x}^\kappa = 0$ is

$$x^k = x^k(p) - \frac{1}{2}\partial_k\Phi(p)\left(x^0 - x^0(p)\right)^2 .$$

This is the trajectory of the free fall through the point p around $x^0 = x^0(p)$, where the two coordinate systems coincide. The relative rotation of the spatial axes is given by a constant matrix

$$X_l^{\bar{k}} = \frac{\partial \bar{x}^k}{\partial x^l} = \delta_l^k , \qquad \frac{\partial^2 \bar{x}^k}{\partial x^l \partial x^0} = 0 .$$

This is because the coordinate x^k is only contained in the first part of the transformation function. Hence, the locally geodesic frame is nothing but a freely falling, non-rotating system through point p. We call such systems *local inertial frames*.

1.8.4 Formulation of the Principle

Which meaning can such systems have for physics? All quantities entering our considerations are smooth. If such a quantity vanishes at a point p, it is immeasurably small in a whole neighborhood of p, too. This means: if we restrict ourselves to a sufficiently small neighborhood in an arbitrary affine-connected manifold, then the affine connection (or the geometry) is arbitrarily well approximated by a flat affine connection. This can be visualized on the sphere: the smaller the neighborhood of p is, the better it is approximated by the tangential plane. These considerations lead to the following hypothesis:

Strong equivalence principle. No physical experiment, which is conducted with an arbitrary but fixed precision within a freely falling, non-rotating box, can detect the gravitational field of outer sources if the box is sufficiently small and the total duration of the experiment is sufficiently short.

In other words, the influence of gravity on physical systems or processes can be locally transformed away or created with arbitrary precision by choosing an adequate coordinate system. However, the systems or processes must be such that they are sufficiently well localized in space as well as in time.

This principle is the precise way to state our claim that gravitation is an apparent force. It possesses moreover a big practical importance. With its help we can predict the influence of gravitation on arbitrary physical systems or processes, as soon as we know how these systems or processes are described without gravitation in curvilinear coordinates.

There are several weaker versions of this principle. If the "physical experiments" are restricted in such a way that no experiments with gravity are allowed (i.e., the gravitational field of the measuring device and of the tools used can be neglected in the experiment), we speak about the *Einstein equivalence principle*. If we restrict the experiments furthermore, so that we permit only the observation of trajectories of test particles in mechanics, then we call it the (*weak*) *Galilei equivalence principle*.

The Einstein equivalence principle will be the basis for the relativistic theory of gravitation.

The equivalence principle (in each of the three versions) is an experimentally extensively tested physical law [1, 2, 3] of all.

1.9 Parallel Transport

One of the most important tasks of the global inertial frame of Newton's theory was to make it possible to compare the values attributed to different physical quantities at points that are far away from each other. Let us consider for instance the trajectory $x^\mu = x^\mu(\lambda)$ of a mass point in an arbitrary reference frame in Newton space–time. Its velocity at the time $t_1 = t(\lambda_1)$ and at the time $t_2 = t(\lambda_2)$ is $\dot{x}^\mu(\lambda_1)$ and $\dot{x}^\mu(\lambda_2)$, respectively. Are these velocities equal or different? Similar questions are important when considering the balance of momenta, etc. The recipe known from Newton's theory is: transform the velocity vectors into one and the same inertial frame and compare the corresponding components. This kind of equality obviously is independent of the chosen inertial frame due to the linearity of the transformation between two inertial frames. Hence it is possible to identify vectors at different points in Minkowski space–time, so that one vector space emerges, which is common to all points; this way it is also possible to combine vectors linearly at different points, as we are used to from Minkowski space–time. This definition of equality can be generalized to arbitrary tensors.

In differential geometry, tensors at different points, whose components with respect to global geodesic coordinates are equal, are not called equal but *parallel*. The term "equal" is reserved for equal tensors at equal points. This is why we want to talk about parallel tensors in the following.

Whether or not two tensors at two different points of a flat affine-connected manifold are parallel is determined by a global geodesic system. These systems are in turn determined by the affine connection. But there is a direct way from the flat affine connection to parallel tensors: the so-called *parallel transport*. We define the parallel transport only for a vector. For other tensors the considerations are analogous.

Let thus $(\mathscr{M}, \Gamma)$ be a flat affine-connected manifold, p and q two points at this manifold, and $\{x^\mu\}$ an arbitrary coordinate system. Let us choose a curve C connecting the two points, i.e., $C : [\lambda_1, \lambda_2] \mapsto \mathscr{M}, C(\lambda_1) = p$ and $C(\lambda_2) = q$. Let C be defined by the functions $x^\mu(\lambda)$ with respect to the coordinates $\{x^\mu\}$. Let moreover V^μ be a vector in p. This vector determines at every point of C, including q, a vector that is parallel to it. The vector field obtained this way along C is represented by the n functions $V^\mu(\lambda)$: these are its components with respect to $\{x^\mu\}$ at the point $C(\lambda)$ for every value of λ. Vector field $V^\mu(\lambda)$ along C is called the *parallel transport of V^μ along C*.

We want to determine the parallel transport directly by means of $\Gamma^\mu{}_{\rho\sigma}$. For this purpose we introduce an arbitrary global geodesic system $\{\bar{x}^\mu\}$ around C. Since $\bar{\Gamma}^\mu{}_{\rho\sigma}$ vanishes, it follows from the transformation rule (1.9) for the affine connection

$$\Gamma^\mu{}_{\rho\sigma} = X^\mu_{\bar{\nu}} X^{\bar{\nu}}_{\rho\sigma} .$$

According to the definition of parallelism, the components $\bar{V}^\mu(\lambda)$ of the parallel transport with respect to $\{\bar{x}^\mu\}$ satisfy the following equations

$$\frac{d\bar{V}^\mu}{d\lambda} = 0 .$$

We have

$$\bar{V}^\mu(\lambda) = X_\nu^{\bar{\mu}} V^\nu(\lambda) .$$

Let us substitute the right-hand side of this equation for $\bar{V}^\mu(\lambda)$. After differentiation we obtain

$$\frac{dV^\nu}{d\lambda} X_\nu^{\bar{\mu}} + X_{\nu\rho}^{\bar{\mu}} \dot{x}^\rho V^\nu = 0 .$$

By multiplication with $X_{\bar{\mu}}^\kappa$ we find

$$\frac{dV^\kappa}{d\lambda} + \left(X_{\bar{\mu}}^\kappa X_{\nu\rho}^{\bar{\mu}} \right) \dot{x}^\rho V^\nu = 0 ,$$

hence

$$\frac{dV^\kappa}{d\lambda} + \Gamma^\kappa{}_{\nu\rho} \dot{x}^\rho V^\nu = 0 . \tag{1.30}$$

This is the desired relation. It is called *parallel transport equation* along C. Equation (1.30) represents a system of n ordinary linear differential equations of first order for $V^\nu(\lambda)$, which are solved for the derivatives. Its solution is hence uniquely determined by the initial value $V^\kappa(\lambda_1)$ and we can thus obtain parallel vectors in two arbitrary points directly from the values of $\Gamma^\mu{}_{\nu\rho}$.

Since the parallel transport is determined entirely by $\Gamma^\mu{}_{\nu\rho}$, it can be generalized to arbitrary affine-connected manifolds.

Definition 7 Let $(\mathcal{M}, \Gamma)$ be a n-manifold with affine connection. Let $\{x^\mu\}$ be arbitrary coordinates, and $\Gamma^\mu{}_{\nu\rho}$ the components of the affine connection with respect to $\{x^\mu\}$. Moreover, a curve C shall be given by the functions $x^\mu(\lambda)$, and a vector V^μ at the point $x^\mu(\lambda_1)$. The vector field $V^\mu(\lambda)$ along C, which is uniquely determined by the differential equation (1.30) and the initial value V^μ, is called *parallel transport of V^μ along C*.

An example for such a parallel transport is the vector field, which is tangential to an autoparallel. Indeed, the equation of an autoparallel is a particular case of the parallel transport equation. The vector that is transported is, in this case, the tangential vector of the curve. This explains also the name "autoparallel".

Is it now possible to define, using this parallel transport, parallel vectors at different points even on curved manifolds? A difficulty could be that two points can be connected by different curves. Indeed, on curved manifolds the parallel transport depends on the path:

Theorem 4 *Let $(\mathcal{M}, \Gamma)$ be a n-manifold with affine connection. Let an infinitesimal quadrangle be given by its corners $p = (x^\mu)$, $q_1 = (x^\mu + \delta x_1^\mu)$, $q_2 = (x^\mu + \delta x_2^\mu)$, and $r = (x^\mu + \delta x_1^\mu + \delta x_2^\mu)$ in the coordinates $\{x^\mu\}$. The components of the curvature* *tensor in the point p with respect to $\{x^{\mu}\}$ are denoted by $R^{\mu}{}_{\nu\rho\sigma}$. Let the parallel transport of an arbitrary vector V^{μ} from p to r via q_1 be given by $V_1^{\mu}(r)$, via q_2 be given by $V_2^{\mu}(r)$. Then the equation*

$$V_2^{\mu} - V_1^{\mu} = R^{\mu}{}_{\nu\rho\sigma}V^{\nu}\delta x_1^{\rho}\delta x_2^{\sigma} \tag{1.31}$$

holds (up to second order in $\delta x_{1,2}$).

Proof Let V^{μ} be a vector at (x^{μ}) and $V^{\mu} + dV^{\mu}$ a parallel vector in an arbitrary neighboring point $(x^{\mu} + dx^{\mu})$. For dV^{μ} we find from the equation of the parallel transport

$$dV^{\mu} = -\Gamma^{\mu}{}_{\rho\sigma}V^{\rho}dx^{\sigma} \, .$$

The parallel transport via q_1 is assumed to be given by $V^{\mu}(p)$, $V_1^{\mu}(q_1)$, $V_1^{\mu}(r)$, the parallel transport via q_2 by $V^{\mu}(p)\, V_2^{\mu}(q_2)$, $V_2^{\mu}(r)$ (Fig. 1.3). Then we have (actually we should expand up to second order, but the end results are equal):

$$V_1^{\mu}(q_1) = V^{\mu}(p) - \Gamma^{\mu}{}_{\rho\sigma}(p)V^{\rho}(p)\delta x_1^{\sigma}$$

and

$$\begin{aligned}
V_1^{\mu}(r) &= V_1^{\mu}(q_1) - \Gamma^{\mu}{}_{\rho\sigma}(q_1)V_1^{\rho}(q_1)\delta x_2^{\sigma} = V^{\mu}(p) - \Gamma^{\mu}{}_{\rho\sigma}(p)V^{\rho}(p)\delta x_1^{\sigma} \\
&\quad - \left(\Gamma^{\mu}{}_{\rho\sigma}(p) + \partial_{\tau}\Gamma^{\mu}{}_{\rho\sigma}(p)\delta x_1^{\tau}\right)\left(V^{\rho}(p) - \Gamma^{\rho}{}_{\alpha\beta}(p)V^{\alpha}(p)\delta x_1^{\beta}\right)\delta x_2^{\sigma} \\
&= V^{\mu}(p) - \Gamma^{\mu}{}_{\rho\sigma}(p)V^{\rho}(p)\delta x_1^{\sigma} - \Gamma^{\mu}{}_{\rho\sigma}(p)V^{\rho}(p)\delta x_2^{\sigma} \\
&\quad - \partial_{\tau}\Gamma^{\mu}{}_{\rho\sigma}(p)\delta x_1^{\tau}V^{\rho}(p)\delta x_2^{\sigma} + \Gamma^{\mu}{}_{\rho\sigma}(p)\Gamma^{\rho}{}_{\alpha\beta}(p)V^{\alpha}(p)\delta x_1^{\beta}\delta x_2^{\sigma} + O^3,
\end{aligned}$$

where O^3 stands for higher order terms in $\delta x_{1,2}^{\tau}$ (terms of order 3 and higher). For the path via q_2 we obtain the same expression, only the indices 1 and 2 have to be exchanged. This way we find

$$V_2^{\mu}(r) - V_1^{\mu}(r) = (\partial_{\tau}\Gamma^{\mu}{}_{\rho\sigma} - \partial_{\sigma}\Gamma^{\mu}{}_{\rho\tau} + \Gamma^{\mu}{}_{\alpha\tau}\Gamma^{\alpha}{}_{\rho\sigma} - \Gamma^{\mu}{}_{\alpha\sigma}\Gamma^{\alpha}{}_{\rho\tau})\delta x_1^{\tau}V^{\rho}(p)\delta x_2^{\sigma} \, ,$$

which is what we were looking for, qed.

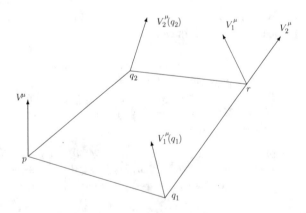

Fig. 1.3 Parallel transport of vector V along two different ways

From the proof it also follows that the parallel vectors in neighboring points are uniquely defined and that they have equal components with respect to a locally geodesic system. Thanks to the affine connection we can hence compare tensors in neighboring points. This is one of the main features of the affine connection: it establishes a connection between neighboring points. It is now also clear why the acceleration can be defined by means of the affine connection.

To compare tensors at points of a space–time that are far away from each other, we need special conditions; we shall come back to this later.

1.10 Exercises

1. A row of weights runs over two wheels with a difference in height given by l. The weights, which are coming from below, each emit a photon with frequency v_1 in the upward direction, thus lose rest mass. They get lighter and ascend further upward where they absorb a photon (emitted by another weight and with frequency v_2). This way they get heavier and pull the weights system down. *Question:* Which redshift

$$z = \frac{v_1 - v_2}{v_2}$$

is necessary to just avoid obtaining a perpetuum mobile?

2. A Gedankenexperiment: connect a mass point with *negative* mass $-m$ through a firm, massless bar with a second mass point of mass $m > 0$. The length of the bar is assumed to be l. Calculate the motion of the system by means of the standard formulas known from Newton mechanics and gravity theory. What can be concluded from the result?

3. Consider a box of height l moving upward with constant acceleration a. A source at the bottom of the box is sending a light signal to the top of the box with the frequency v_1. *Question:* With which frequency v_2 is the signal absorbed at the top? (To simplify things, please use all formulas, i.e., the equations of motion and the Doppler effect, in non-relativistic approximation!)

4. Find the concrete form of the transformation between two inertial frames expressed by the displacement T_0 of the origin of the time, the time-dependent displacements $r_k(x^0)$ of the origins of the equal-time planes, and the time-dependent rotation $O_{kl}(x^0)$ of the spatial axes.

5. Use the results of the previous exercise and calculate the coefficients $\Gamma^\mu_{\rho\sigma}$ in the equation

$$\ddot{x}^\mu + \sum_{\rho=0}^{3} \sum_{\sigma=0}^{3} \Gamma^\mu_{\rho\sigma} \dot{x}^\rho \dot{x}^\sigma = 0 \,,$$

which describes free motion with respect to a non-inertial frame K, and determine their relation to the known apparent forces. Prove the correctness of the following definition for the angular velocity $\omega_k(x^0)$:

$$\sum_r O_{kr}(x^0)\dot{O}_{lr}(x^0) = \sum_r \varepsilon_{rkl}\omega_r(x^0) .$$

6. Identify the free and the dummy indices in the following expressions

$$C^\alpha D_\alpha = 5 ; \quad A^\rho_\sigma A^\sigma = B^\rho ; \quad T^{\alpha\mu\lambda} A_\mu C^\gamma_\lambda = Q^{\alpha\gamma} .$$

How many different equations does each expression represent?
7. Given are the following numeric values:

$$A^\mu = (5,0,-1,-6) ,$$
$$B_\mu = (1,-2,4,0) ,$$

$$C_{\rho\sigma} = \begin{pmatrix} 1 & 0 & 2 & 3 \\ 5 & -2 & -2 & 0 \\ 4 & 5 & 2 & 2 \\ -1 & -1 & -3 & 0 \end{pmatrix} ;$$

Compute: $A^\rho B_\rho$, $A^\rho C_{\rho\sigma}$ for all σ, $A^\alpha C_{\alpha\beta}$ for all β, and $A^\lambda B_\kappa$ for all λ and κ.
8. Which of the following relations satisfy the conditions of a coordinate transformation?

$$(x,y) \mapsto (\xi,\eta)?$$

(a)
$$\xi = x, \quad \eta = 1 ,$$

(b)
$$\xi = \sqrt{x^2 + y^2}, \quad \eta = \arctan\left(\frac{y}{x}\right) ,$$

(c)
$$\xi = \log x, \quad \eta = y .$$

9. Determine the metric on the following surfaces in $\mathbb{E}^3$:

(a) $x^2 + y^2 = 1$,
(b) $x^2 + y^2 - z^2 = -1$;

(x,y,z) are Cartesian coordinates.
10. How does a tensor of type $(0,0)$ transform?
11. Let $D_{\mu\nu}$ be a tensor of type $(0,2)$. Show that $\sum_{\mu=1}^n D_{\mu\mu}$ is in general *not* a tensor.
12. Prove that the quantity whose components form the following table with respect to an *arbitrary* coordinate system is a tensor:

$$\begin{pmatrix} 1 & 0 & \cdots & 0 & 0 \\ 0 & 1 & \cdots & 0 & 0 \\ & \vdots & & & \vdots \\ 0 & 0 & \cdots & 0 & 1 \end{pmatrix}$$

(n-dimensional unit matrix). Of which type is this tensor?

13. Let $T_{\mu\nu}$ be a tensor of type $(0,2)$. Its components form an $n \times n$ matrix:

$$\begin{pmatrix} T_{11}, & \cdots, & T_{1n} \\ \vdots & & \vdots \\ T_{n1}, & \cdots, & T_{nn} \end{pmatrix}$$

and let

$$\begin{pmatrix} T^{11}, & \cdots, & T^{1n} \\ \vdots & & \vdots \\ T^{n1}, & \cdots, & T^{nn} \end{pmatrix}$$

be its inverse matrix. Show that the quantity $T^{\mu\nu}$ defined this way is also a tensor. Of which type is it?

14. How does $\det(T^{\mu}_{\nu})$ transform, where T^{μ}_{ν} is a tensor of type $(1,1)$?

15. Compute the affine connection $\Gamma^{\mu}_{\rho\sigma}$ on the surface of a sphere of radius $r = 1$. Show that the great circles are solutions for the geodesic equations. Hint: the function $\vartheta(\lambda)$ can be directly computed from the definition of a great circle, if the function $\varphi(\lambda)$ is chosen arbitrarily (this choice is arbitrary and this fact can be used to obtain simplifications).

16. Let $W^{\mu\nu}_{\rho\sigma\tau}$ be a tensor of type $(2,3)$ and let the quantities U, V, and X be defined as follows:

$$U^{\mu\nu}_{\rho\sigma\tau} = W^{\mu\nu}_{\sigma\tau\rho} \; ;$$

$$V^{\mu\nu}_{\rho\sigma\tau} = W^{\rho\nu}_{\mu\sigma\tau} \; ;$$

$$X^{\mu\nu}_{\rho\sigma\tau} = W^{\nu\mu}_{\rho\sigma\tau} \; .$$

This is assumed to hold in arbitrary coordinates. Which of the quantities U, V, and X is a tensor? Under which conditions does one of the new quantities equal W?

17. Compute the curvature tensor for the two-dimensional metric with components

$$g_{11} = \pm 1 , \quad g_{12} = g_{21} = 0 , \quad g_{22} = f(x^1) ,$$

where $f(x^1)$ is an arbitrary function.

(a) For which f does the curvature vanish?
(b) For which f does the curvature admit the form

$$R^{\mu}_{\nu\rho\sigma} = K(\delta^{\mu}_{\rho} g_{\nu\sigma} - \delta^{\mu}_{\sigma} g_{\nu\rho}) ,$$

where $K = K(x^1, x^2)$ is a function of the coordinates (depending on f)? K is called "Gauss curvature".

18. Is it possible to find coordinates on the well-known surface of a sphere such that the metric admits the form as given in Exercise 17? Which property has the Gauss curvature?

19. The three-dimensional Minkowski space–time is the metric manifold $(\mathbf{R}^3, \eta_{kl})$, where

$$\eta = \begin{pmatrix} -1 & 0 & 0 \\ 0 & 1 & 0 \\ 0 & 0 & 1 \end{pmatrix}.$$

Study the surfaces

(a) $(x^0)^2 - (x^1)^2 - (x^2)^2 = R^2$,

(b) $(x^0)^2 - (x^1)^2 - (x^2)^2 = -R^2$.

Find coordinates such that the metric on the surfaces admits the form as given in Exercise 17. Compute the Gauss curvature. Hint: try "spherical coordinates" with hyperbolic functions ($\sinh\theta$ or $\cosh\theta$).

20. Compute the numerical value of the curvature of the Cartan–Friedrichs space–time on the Earth's surface (in the "spherically symmetric approximation").

21. Determine the concrete form of the parallel transport equation for the vector $V^\mu = (1,0)$ along the curve $\vartheta(\lambda), \varphi(\lambda)$ given by

$$\vartheta = a(= \text{const}), \quad \varphi = \lambda,$$

in a metric manifold with metric

$$ds^2 = d\vartheta^2 + f(\vartheta)d\varphi^2.$$

22. Assume that the manifold in Exercise 21 is such that the points $(\vartheta, 0)$ and $(\vartheta, 2\pi)$ coincide (similar to what happens on a sphere). Calculate the parallel transport along the now closed curve $\lambda \in [0, 2\pi]$ from Exercise 21 for the following values for the function $f(\vartheta)$

(a) $f = R^2\vartheta^2$,

(b) $f = R^2 \sin^2(\vartheta/R)$,

(c) $f = R^2 \sinh^2(\vartheta/R)$.

Which form do the three surfaces have?
How does the transport depend on the parameter a?

23. Let $\mathscr{M}$ be an arbitrary n-manifold, $\{x^\mu\}$ coordinates on it, and C_1 and C_2 two curves given by

$$C_1: \quad x^\mu = u^\mu(\lambda), \quad \lambda \in [0, 1]$$

$$C_2: \quad x^\mu = v^\mu(\lambda), \quad \lambda \in [0, 1],$$

where $u^\mu(\lambda)$ and $v^\mu(\lambda)$ are $2n$ arbitrary functions, which satisfy

$$u^\mu(1) = v^\mu(0), \quad \forall \mu;$$

moreover, let

$V^\mu \ldots$ be a vector in $u^\mu(0)$;
$V_1^\mu \ldots$ be parallel to V^μ in $u^\mu(1)$ along C_1;
$V_2^\mu \ldots$ be parallel to V_1^μ in $v^\mu(1)$ along C_2.

Show that:

(a) V_2^μ is parallel to V^μ in $v^\mu(1)$ along the composite, piecewise smooth curve C given by

$$x^\mu = z^\mu(\lambda) , \quad \lambda \in [0,2] ,$$

where

$$z^\mu(\lambda) = u^\mu(\lambda) , \quad \lambda \in [0,1] ,$$
$$z^\mu(\lambda) = v^\mu(\lambda - 1), \quad \lambda \in [1,2] .$$

(b) V^μ is parallel to V_1^μ along the curve C^{-1} defined by

$$x^\mu = u^\mu(1 - \lambda) , \quad \lambda \in [0,1] .$$

24. Use an inertial frame of the Newton theory as coordinate system in a Cartan–Friedrichs space–time with a fixed, arbitrary gravitational potential $\Phi(\vec{x})$. Study the parallel transport in the coordinates

(a) along arbitrary spatial curves, i.e., $x^0 = $ const, and of arbitrary vectors;
(b) along a closed curve, which consists of two free falls and two spatial curves; consider in particular the 4-velocity of the free falls.

References

1. C. M. Will, *Theory and Experiment in Gravitational Physics*. Cambridge University Press, Cambridge, 1993.
2. I. Ciufolini and J. A. Wheeler, *Gravitation and Inertia*. Princeton University Press, Princeton, NJ, 1995.
3. B. Schutz, *Gravity from the Ground up*, Cambridge University Press, Cambridge, UK, 2003.
4. W. Pauli, *Theory of Relativity*, Dover Publications, New York, 1981.
5. S. Kobayashi and K. Nomizu, *Foundations of Differential Geometry*, Wiley, New York, Vol. I 1963, Vol. II 1969.
6. S. Weinberg, *The Quantum Theory of Fields* Vol. I, Cambridge University Press, Cambridge UK, 1995.
7. E. A. Coddington and N. Levinson, *Theory of Ordinary Differential Equations*, McGraw-Hill, New York, 1955.
8. C. W. Misner, K. S. Thorne and J. A. Wheeler, em Gravitation, Freeman, San Francisco, CA, 1973.
9. I. M. Gel'fand, *Lectures on Linear Algebra*. Interscience Publishers, New York, 1961.
10. E. Cartan, Ann. Ecole Norm. Sup. **40** (1923) 325; **41** (1924) 1.
11. L. P. Eisenhart, *Riemannian Geometry*. Princeton University Press, Princeton, NJ, 1949.

Chapter 2
Relativistic Particle Dynamics in Gravitational Fields

2.1 Relativistic Gravity

In Chap. 1 we learned that Newtonian gravitation can be considered as a geometry of space–time. In addition we learned the fundamentals of the corresponding mathematical apparatus, differential geometry. In this chapter we want to apply those methods to relativistic particle dynamics. In contrast to the non-relativistic case, this will lead to a new theory, the so-called general relativity, which will modify the theory of special relativity as well as the theory of Newtonian gravitation. This means that general relativity predicts new physical phenomena.

2.2 Geometry of Minkowski Space–Time

In this section we want to formulate well-known properties of the relativistic space–time in terms of differential geometry. Special relativity is assumed to be known (for an introduction, see [1]). We choose the units such that $c = 1$.

Special relativity is based on inertial frames, exactly as is the case in Newton's theory. These privileged reference frames are again determined by free motion. Physically they can in principle be realized by the following devices:

1. a freely floating radar equipment,
2. an ideal clock,
3. non-rotating orthonormal axes of coordinates (measuring rods, gyroscopes, etc.).

The coordinates $\{\bar{x}^\mu\}$, which correspond to such an inertial frame and describe an event in this inertial frame, can be measured by means of this equipment. A transformation between the coordinates, which correspond to two different inertial frames, determines an element of the Poincaré group

$$x'^\mu = \Lambda^\mu_\nu x^\nu + a^\mu ,$$

Hájíček, P.: *Relativistic Particle Dynamics in Gravitational Fields.* Lect. Notes Phys. **750**, 39–92 (2008)
DOI 10.1007/978-3-540-78659-7_2

where Λ_ν^μ is a Lorentz matrix, i.e.,

$$\eta_{\mu\nu}\Lambda_\rho^\mu \Lambda_\sigma^\nu = \eta_{\rho\sigma} \, .$$

Vice versa, every element of this group gives rise to an inertial frame if we apply this transformation to an arbitrary inertial frame.

The time intervals (in particular the simultaneity) and the distances of events in Minkowski space–time are relative: they depend on the choice of the inertial frame. The only absolute quantity (i.e., the only quantity that does not depend on the chosen reference frame) that is related to time intervals and distances is the so-called *interval*. The interval I between two events with coordinates $\{\bar{x}^\mu\}$ and $\{\bar{y}^\mu\}$ with respect to an inertial frame is

$$I = \left(\bar{x}^0 - \bar{y}^0\right)^2 - \left(\bar{x}^1 - \bar{y}^1\right)^2 - \left(\bar{x}^2 - \bar{y}^2\right)^2 - \left(\bar{x}^3 - \bar{y}^3\right)^2 \, .$$

The geometric meaning of the interval is the following. If the interval between two events is positive $(I > 0)$, then there exists an inertial frame where the differences of the spatial coordinates vanish

$$\bar{x}^1 - \bar{y}^1 = \bar{x}^2 - \bar{y}^2 = \bar{x}^3 - \bar{y}^3 = 0$$

and $I = T^2$, where T equals the time difference between these two events happening at the same position. If $I < 0$, there exists an inertial frame where the two events occur at the same time, $\bar{x}^0 - \bar{y}^0 = 0$, such that $I = -d^2$, where d is the distance of these two simultaneous events. Finally, if $I = 0$, the two events can be connected by a light signal.

Another geometric structure is determined by free motions; they play a crucial role for the definition of inertial frame. We already know the following properties of free motions:

1. with respect to an inertial frame, every free motion has the form

$$\bar{x}^\mu = \bar{a}^\mu \lambda + \bar{b}^\mu,$$

where $\bar{a}^\mu$ and $\bar{b}^\mu$ are constant 4-tuples and λ is a parameter;
2. the 4-tuple $\bar{a}^\mu$ satisfies the inequality:

$$\left(\bar{a}^0\right)^2 - \sum_k \left(\bar{a}^k\right)^2 \geq 0.$$

We now want to replace these non-local geometric objects by local differential geometric objects such that all inertial frames can be reconstructed out of those new objects (and this way the whole geometry of Minkowski space–time). All quantities and equations are to be written in a way that suggests their transformation properties with respect to arbitrary (non-linear) coordinate transformations.

We can consider the Minkowski space–time $\mathcal{M}$ as a manifold; there are global coordinates, e.g., those of an inertial frame, which map $\mathcal{M}$ to $\mathbb{R}^4$. This way the manifold is defined!

Then the interval determines a Lorentzian metric $g_{\mu\nu}(x)$ in every point of $\mathcal{M}$ by its components with respect to the inertial frame:

$$\bar{g}_{\mu\nu}(\bar{x}) = \begin{pmatrix} 1 & 0 & 0 & 0 \\ 0 & -1 & 0 & 0 \\ 0 & 0 & -1 & 0 \\ 0 & 0 & 0 & -1 \end{pmatrix} \qquad \forall \bar{x} \in \mathcal{M}. \tag{2.1}$$

(The elements of the matrix on the right-hand side are often denoted by $\eta_{\mu\nu}$.) Note that this equation is valid at every point of $\mathcal{M}$, so that the metric has constant components with respect to the inertial frame. This metric is clearly non-degenerate and independent of the inertial frame chosen to define it. This follows from the fact that the Poincaré transformations leave this tensor field invariant. The metric (2.1) is called *Minkowski metric*.

We want to emphasize that the Minkowski metric and the inertial frame connected with it are determined only by observation and measurements. Why? Are there other Minkowski metrics on $\mathbb{R}^4$? The surprising answer is "Yes"! Let us construct two non-equivalent Minkowski metrics on $\mathbb{R}^4$.

As a first step we assume that the metric $g_{\mu\nu}$ is given by its components $\eta_{\mu\nu}$ with respect to some coordinates x^μ. These coordinates thus define an inertial frame for the metric $g_{\mu\nu}$. In the second step we choose a coordinate system x'^μ, such that this one is *not* connected to x^μ by a Poincaré transformation. For instance,

$$x'^\mu = A^\mu_\nu x^\nu , \tag{2.2}$$

where A^μ_ν is a constant, invertible matrix, which satisfies

$$\eta_{\rho\sigma} A^\rho_\mu A^\sigma_\nu \neq \eta_{\mu\nu} ,$$

but it is arbitrary apart from this condition. By assumption we have

$$g'_{\mu\nu} \neq \eta_{\mu\nu} .$$

In a third step we introduce a new tensor field $h_{\mu\nu}$ by defining its components with respect to the coordinates x'^μ:

$$h'_{\mu\nu} = \eta_{\mu\nu} . \tag{2.3}$$

Obviously $h_{\mu\nu}$ is a Minkowski metric on $\mathbb{R}^4$, since there are coordinates, namely x'^μ, in which (2.3) holds. But $g_{\mu\nu} \neq h_{\mu\nu}$, since the components of these two tensor fields do not agree with respect to the coordinates x'^μ. Hence, the construction is finished.

Let us remark that the transformation (2.2) can be chosen to be much more general and the construction would still work: it only has to be an invertible transformation (non-linear), which maps $\mathbb{R}^4$ to $\mathbb{R}^4$. There are hence many non-equivalent Minkowski metrics on $\mathbb{R}^4$.

The Minkowski metric can be used to classify vectors X^μ at a point x. Namely, depending on the sign of the expression $g_{\mu\nu}(x)X^\mu X^\nu$, they can be separated into time-like, light-like, and space-like vectors. The "length" of a curve in Minkowski space–time is only well defined using this metric, if the curve is not space-like, i.e., its tangent vectors in every point are not space-like. Then the coordinate-independent integral

$$\tau_{12} = \int_{\lambda_1}^{\lambda_2} d\lambda \sqrt{g_{\mu\nu}\dot{x}^\mu \dot{x}^\nu} \tag{2.4}$$

is real. Its physical meaning is the *proper time* along the curve passing from λ_1 to λ_2; it is the time measured by an ideal clock, which moves along this curve.

The free motions can again (as in Newton's theory) be described by a differential equation; with respect to inertial frame $\{\bar{x}^\mu\}$, their trajectories are given by

$$\ddot{\bar{x}} = 0. \tag{2.5}$$

We have however the additional condition $\bar{g}_{\mu\nu}\dot{\bar{x}}^\mu \dot{\bar{x}}^\nu \geq 0$. The differential equation (2.5) determines an affine connection. With respect to the inertial frame, the corresponding components $\bar{\Gamma}^\mu_{\rho\sigma}$ are equal to zero. But not all autoparallels of this affine connection are free motions! Nevertheless we can determine this affine connection by studying only the free motions (exercise). The affine connection determined this way is *globally flat*; it admits global geodesic coordinates (example: an inertial frame).

We have now identified two geometrical objects: the Minkowski metric $g_{\mu\nu}(x)$ and the free-motion affine connection $\Gamma^\mu_{\rho\sigma}(x)$. It can be shown that these objects are not independent: $\Gamma^\mu_{\rho\sigma}$ is the metric affine connection corresponding to the metric $g_{\mu\nu}(x)$. Proof: In an arbitrary inertial frame, we have

$$\bar{\Gamma}^\mu_{\rho\sigma} = 0$$

and

$$\{^{\mu}_{\rho\sigma}\} = \frac{1}{2}\bar{g}^{\mu\nu}(\bar{\partial}_\rho \bar{g}_{\sigma\nu} + \bar{\partial}_\sigma \bar{g}_{\rho\nu} - \bar{\partial}_\nu \bar{g}_{\rho\sigma}) = 0,$$

since $\bar{g}_{\rho\sigma}$ is constant for all ρ and σ. The components of the metric affine connection and the affine connection, which is defined by the free motions, are therefore equal in *one* coordinate system. They obey, in addition, the same transformation law and therefore have to be equal in every coordinate system:

$$\Gamma^\mu_{\rho\sigma} = \{^{\mu}_{\rho\sigma}\}. \tag{2.6}$$

Equation (2.6) means that the whole geometry of Minkowski space–time is contained in a single object—in the Minkowski metric. Let us summarize: the geometry of Minkowski space–time is completely determined by a pair $(\mathcal{M}, g)$, where $\mathcal{M}$ is a 4-manifold and g is a metric on $\mathcal{M}$. The metric g on $\mathcal{M}$ is called *Minkowski metric*, if there exists a coordinate chart $h : \mathcal{M} \mapsto \mathbb{R}^4$ such that (1) h has as domain the whole manifold $\mathcal{M}$ (global chart) and (2) the components $g_{\mu\nu}(x)$ of the metric satisfy with respect to this chart: $g_{\mu\nu}(x) = \eta_{\mu\nu}$.

In particular, also the inertial frames are defined by the metric. Obviously h itself defines an inertial frame. Let us denote its coordinates by x^μ. All other inertial frames must be global geodesic systems and therefore they can be obtained from x^μ by a linear transformation

$$x^\mu = A^\mu_\nu x'^\nu + B^\mu \,,$$

where A^μ_ν and B^μ are constant matrices. The corresponding transformation of the metric reads

$$g'_{\mu\nu} = A^\rho_\mu A^\sigma_\nu \eta_{\rho\sigma}.$$

But the system x'^μ is only an inertial frame, even if $g'_{\mu\nu} = \eta_{\mu\nu}$ holds. Then A^μ_ν must be a Lorentz matrix, since the defining equation for a Lorentz matrix is

$$\eta_{\mu\nu} = A^\rho_\mu A^\sigma_\nu \eta_{\rho\sigma} \,.$$

This way we obtain all inertial frames out of the local object $g_{\mu\nu}$, and thus the interval too.

At the end of this section, we want to clarify some technical issues. The free motions are geodesics of the Minkowski metric. When deriving the geodesic equation, we have to vary the integral (2.4). Undertaking the variation, we see that a square root appears in the denominator. As long as we are dealing with a time-like curve, there is no problem—the square root is positive. For light-like curves, the variation cannot be done. One way out, which is often used, is to consider a different variation principle of the dynamics, namely the one with the action

$$S = \frac{1}{2} \int d\lambda \, g_{\mu\nu} \dot{x}^\mu \dot{x}^\nu \,.$$

Indeed, the corresponding Euler–Lagrange equation is

$$\partial_\rho \left(\frac{1}{2} g_{\mu\nu} \dot{x}^\mu \dot{x}^\nu \right) - \frac{d}{d\lambda} (g_{\mu\rho} \dot{x}^\mu) = 0 \,,$$

and this yields directly the equation of the autoparallels:

$$\ddot{x}^\rho + \{^{\rho}_{\mu\nu}\} \dot{x}^\mu \dot{x}^\nu = 0 \,. \tag{2.7}$$

We therefore obtain the right dynamics, including the affine parameter.

The Euler–Lagrange equations corresponding to the above Lagrangian offer the fastest way to compute the Christoffel symbols. We just have to solve them for the second derivatives of the coordinates and read off the Γs.

We note that the canonical momentum [2]

$$p_\mu = \frac{\partial L}{\partial \dot{x}^\mu} = g_{\mu\rho} \dot{x}^\rho$$

is closely related to the tangent vector; we can even write

$$p^\mu = \dot{x}^\mu \; . \tag{2.8}$$

Here we used a convention: in Lorentzian geometry, where the metric is at our disposal, we can *raise* and *lower* the indices—in doing so, the quantity is always denoted by the same letter. So the covariant momentum p_μ and the contravariant momentum p^μ are connected by the relations $p_\mu = g_{\mu\nu} p^\nu$ and $p^\mu = g^{\mu\nu} p_\nu$.

It is possible to obtain for the canonical momentum the value of the 4-momentum p^μ, if the curve is parameterized by the so-called *physical parameter*. For massive particles, we have for example

$$p^\mu = \mu \frac{\mathrm{d}x^\mu}{\mathrm{d}s} \; ,$$

where s is the proper time and $\mathrm{d}x^\mu / \mathrm{d}s$ is the so-called 4-velocity. In this case we obtain for the physical parameter λ the equation

$$\frac{\mathrm{d}s}{\mathrm{d}\lambda} = \mu \; . \tag{2.9}$$

For light-like particles, the parameter λ has to be chosen in such a way that

$$\dot{x}^\mu = h k^\mu \; ,$$

where h is the Planck constant and k^μ is the wave vector.

2.3 Particle Dynamics in General Relativity

General relativity is a relativistic theory of gravitation, which we will construct in the following. We use the same methods that led us in the non-relativistic case from the Newton to the Cartan–Friedrichs theory: we modified the geometric structure of Newton space–time without gravity—the affine connection—in such a way that the free fall in a gravitational field could be represented as autoparallel of the new affine connection. Because of the relative acceleration between two freely falling particles, we lost the existence of global inertial frame. The properties of inertial frames could only be recovered approximately in small, freely falling, non-rotating boxes, and only in the case where the measurement times were not too long.

Therefore we now want to modify the geometric structure of Minkowski space–time without gravity to obtain gravitational effects. For simplicity we want to try to make this modification in such a way that the new affine connection is also metric.

If we want to modify an affine connection this way, we cannot leave the metric unchanged. This time, however, the modification to be made is not determined in advance: the "relativistic" free falls are not given by any formula known so far.

One point we can start from is the existence of local inertial systems. The space–time is a 4-manifold $\mathcal{M}$ with metric $g_{\mu\nu}$, which has the following properties: for every point $p \in \mathcal{M}$, there exists a coordinate system $\{x^\mu\}$ such that the corresponding components of the metric and the affine connection satisfy:

$$g_{\mu\nu}(p) = \eta_{\mu\nu}, \tag{2.10}$$

$$\left\{ {}^\rho_{\mu\nu} \right\}(p) = 0. \tag{2.11}$$

We call such systems again *local inertial frames*. In a small neighborhood of p, the metric and the affine connection then have nearly the same form as in an inertial frame of special relativity. We abandon the requirement that there exist coordinates such that (2.10) and (2.11) hold everywhere on the whole manifold as in special relativity. As we will see later, the new ansatz allows for a wide class of space–times, which is rich enough to describe all known gravitational fields.

The gravitational field can thus be interpreted as the tensor field $g_{\mu\nu}(x)$ on a manifold $\mathcal{M}$; different metric fields therefore imply different geometries; the manifolds are only restricted by the condition that they have to admit such a metric. This allows for a wide class of topologies, not only the trivial topology of $\mathbb{R}^4$ [3]. The metric $g_{\mu\nu}$ therefore gets two different functions: first to describe the geometry of space–time as in special relativity and second to describe the gravitational field.

We want to formulate our ansatz also in a mathematically clean way. For this purpose, we first have to introduce the notion of the *signature* of a metric.

Definition 8 Let $\mathcal{M}$ be an n-manifold with metric $g_{\mu\nu}$ and p a point in $\mathcal{M}$. If there exist coordinates $\{x^\mu\}$ such that

$$g_{\mu\nu}(p) = \mathrm{Diag}(+1, \ldots, +1, -1, \ldots, -1)$$

with n_+-times $+1$ and n_--times -1, where $\mathrm{Diag}(a_1, \ldots, a_n)$ is an abbreviation for the diagonal matrix with the elements $a_1, \ldots, a_n$ sitting on the diagonal, then the metric has the *signature* $\sigma = n_+ - n_-$ in p. (Zeros are not allowed because the metric has to be regular in p.)

The signature does not depend on the way the metric is transformed to the above diagonal form. This is because the transformation of the metric at a point p is a real general linear transformation:

$$g'_{\mu\nu}(p) = X^\rho_{\mu'}(p) X^\sigma_{\nu'}(p) g_{\rho\sigma}(p) .$$

It is known from linear algebra [4] that every symmetric matrix can be brought by such a transformation to the form

$$\mathrm{Diag}(+1, \ldots, +1, -1, \ldots, -1, 0, \ldots, 0) ,$$

and that the number of $+1$, of -1 and of 0 is an invariant of the matrix (see quadratic forms in [4]). For metrics that are regular everywhere, σ is even independent from the point! (Exercise.)

As soon as we know the signature σ of the metric and the dimension n of the manifold, we can determine the numbers n_+ and n_- from

$$n_+ + n_- = n, \quad n_+ - n_- = \sigma .$$

We can hence formulate the requirement (2.10) as follows: the metric has the signature -2 at every point of $\mathscr{M}$. The requirement (2.11) taken alone is always satisfied. The corresponding theorem was proved in Chap. 1.

We now want to prove that both requirements (2.10) and (2.11) are satisfied if the signature of the metric is -2. The first step is to find geodesic coordinates $\{\tilde{x}^\mu\}$ for the point p. Then we have

$$\widetilde{\left\{ {}^\rho_{\mu\nu} \right\}}(p) = 0 .$$

In these coordinates, $\tilde{g}_{\mu\nu}(p)$ does not necessarily have to be equal to $\eta_{\mu\nu}$. The second step relies on the observation that a linear transformation of the coordinates does not affect the validity of (2.11). Indeed, the second term on the right-hand side of the transformation formula (1.9) for the components of the affine connection vanishes if the transformation is linear, i.e., for every transformation of the form

$$\tilde{x}^\mu = A^\mu_\rho \bar{x}^\rho$$

where A^μ_ρ is a constant matrix, we obtain

$$\overline{\left\{ {}^\rho_{\mu\nu} \right\}}(p) = 0 .$$

The third step shows the existence of an appropriate matrix A^μ_ρ. The fact that the metric $\tilde{g}_{\mu\nu}$ has, at the point p, the signature -2 and that the dimension of the manifold is 4 means that there is a real regular matrix A^μ_ρ such that

$$A^\rho_\mu A^\sigma_\nu \tilde{g}_{\rho\sigma} = \eta_{\mu\nu} ;$$

therefore the existence is shown.

It follows that we can formulate our ansatz as follows.

Postulate 2.1 *The space–times of general relativity are differentiable 4-manifolds with metrics $g_{\mu\nu}$, which have at every point the signature -2.*

We can consider the above as the first fundamental postulate of general relativity.

The local inertial frame will play an important role in what follows. We can characterize it directly by properties of the metric because the following identity holds (exercise)

$$\partial_\rho g_{\mu\nu} = g_{\mu\sigma} \left\{ {}^\sigma_{\rho\nu} \right\} + g_{\nu\sigma} \left\{ {}^\sigma_{\rho\mu} \right\} . \tag{2.12}$$

It follows that a local inertial frame $\bar{x}^\mu$ in $p \in \mathcal{M}$ can be characterized in an equivalent way by the relation

$$\bar{g}_{\mu\nu}(p) = \eta_{\mu\nu} , \quad \frac{\partial \bar{g}_{\mu\nu}}{\partial \bar{x}^\rho}(p) = 0 .$$

In the given event p, the metric has the Minkowski form and its first derivatives vanish. The local inertial frames whose existence is guaranteed by Postulate 1 are physically interpreted as freely falling non-rotating reference frames, exactly as in the non-relativistic theory.

We then use the Einstein equivalence principle to determine the effect of gravitation on arbitrary physical phenomena. The equivalence principle can be adopted without any change from the first chapter. It is considered as a heuristic principle. We apply the equivalence principle often in the form of the so-called *principle of the general covariance*. This works in two steps.

1. We assume that a description of special relativity for a situation without gravity exists in form of a *local* (i.e., concerning neighboring points) equation. We transform this equation into general curvilinear coordinates such that it admits a *covariant* form, i.e., the equation holds in this form in all coordinate systems (also in an inertial frame). We assume in addition that the equation in this new, covariant form contains no higher derivatives of the metric than those of the first order.

2. We postulate that the equation holds in this form in the curved space–times of general relativity. This way we obtain a very definite information about how the phenomena under consideration are influenced by an arbitrary gravitational field. It is also clear that the Einstein equivalence principle is satisfied: in a local inertial frame, the first derivatives of the metric vanish and its components have the same form as in an inertial frame of special relativity.

We note that the expression "covariant" has two very different meanings in mathematics: covariant indices, tensors, etc., are lower indices. Covariant equations are equations that are satisfied in arbitrary coordinate systems. A covariant equation can also contain contravariant tensors.

If the coupling of other systems to gravity is defined in this way, then it is called *minimal coupling*. We now want to transfer different equalities and inequalities from special relativity to general relativity.

The classification of vectors can be adopted without changes; the sign of the expression

$$g_{\mu\nu} \dot{x}^\mu \dot{x}^\nu$$

determines whether a curve is time-, light-, or space-like. The meaning of this classification is shown by the following postulate:

Postulate 2.2 *Massive (massless) test particles always move-also under the influence of forces-along time-like (light-like) curves.*

The causal structure of space–time, which determines which events can have causal influence on a given event, is also given by the metric. Hence it is a part of the space–time geometry.

For time measurements in the absence of gravitation, (2.4) holds; it is already written in the covariant form and does not contain derivatives of the metric. In the absence of gravity, we therefore have (minimal coupling of the clock mechanism) the following:

Postulate 2.3 *The proper time* ds *between* λ *and* $\lambda + d\lambda$ *along the time-like curve* $x^\mu = x^\mu(\lambda)$ *of an ideal clock, which is measuring it, reads*

$$ds = d\lambda \sqrt{g_{\mu\nu}\dot{x}^\mu \dot{x}^\nu} \ . \tag{2.13}$$

We will see that this postulate reproduces in the correct way the redshift in a gravitational field.

For the free motion (no forces act) "in the absence of gravitation", (2.7) holds. It already has a covariant form and contains only components of the metric and its first-order derivatives. It can be derived from the variation principle of the same form as the one in flat space–time. In the presence of gravity, we therefore have the following:

Postulate 2.4 *The trajectory* $x^\mu(\lambda)$ *of a massive or massless test particle, on which no forces (free fall) act, satisfies* (2.7):

$$\ddot{x}^\rho + \left\{{\rho \atop \mu\nu}\right\} \dot{x}^\mu \dot{x}^\nu = 0 \ .$$

Its 4-momentum $p^\mu(\lambda)$ *is proportional to the velocity*

$$p^\mu = k\dot{x}^\mu \ , \tag{2.14}$$

where k is a constant number, depending on the choice of parameter of the autoparallel.

First it follows that the curve extremizes the action:

$$S = \int_{\lambda_1}^{\lambda_2} d\lambda L \ ,$$

where

$$L = \frac{1}{2} g_{\mu\nu}(x(\lambda)) \dot{x}^\mu(\lambda) \dot{x}^\nu(\lambda) \ . \tag{2.15}$$

Second, $\dot{k} = 0$ along the trajectory of a freely falling test particle. A solution to (2.7) is uniquely determined by the initial data

$$x^\mu(\lambda_0) = x_0^\mu, \quad \dot{x}^\mu(\lambda_0) = p_0^\mu \ , \tag{2.16}$$

where p_0^μ is the 4-momentum of the test particle at the initial point x_0^μ and the trajectory is then parameterized by the physical parameter. The tangent vector and thus also the canonical momentum undergo a parallel transport. This is only true for the free fall. All test particles therefore move along the geodesics of the metric $g_{\mu\nu}$.

For photons we will study only free motions or scattering. For massive particles, we want to consider in addition accelerated motions. For this purpose, we still need the following two postulates.

Postulate 2.5 *The equation of motion of a massive particle is generalized to*

$$\frac{dp^\mu}{ds} + \{^\mu_{\rho\sigma}\} p^\rho \frac{dx^\sigma}{ds} = f^\mu , \tag{2.17}$$

where f^μ is the sum of all acting 4-forces (except for gravitation), p^μ is the 4-momentum of the particle,

$$p^\rho := \mu \frac{dx^\rho}{ds} ,$$

s is the proper time along the trajectory of the particle, and μ is the mass.

It follows that (1) $g_{\mu\nu} p^\mu p^\nu = \mu^2$; (2) If the mass is constant, $\mu = $ const (this is not self-evident, e.g., for rockets this is not true!), then the 4-force satisfies the equation

$$g_{\mu\nu} f^\mu p^\nu = 0 , \tag{2.18}$$

and vice versa. We can prove this by multiplying (2.17) with $g_{\nu\mu} p^\nu$ and by summing over μ :

$$\begin{aligned}
g_{\mu\nu} f^\mu p^\nu &= g_{\nu\mu} p^\nu \frac{dp^\mu}{ds} + g_{\nu\mu} p^\nu \{^\mu_{\rho\sigma}\} p^\rho \dot{x}^\sigma \\
&= g_{\mu\nu} p^\nu \dot{p}^\mu + \frac{1}{2} \left(g_{\mu\nu} \{^\mu_{\rho\sigma}\} + g_{\mu\rho} \{^\mu_{\nu\sigma}\} \right) p^\nu p^\rho \dot{x}^\sigma \\
&= \frac{1}{2} \left(g_{\mu\nu} \dot{p}^\mu p^\nu + g_{\mu\nu} \dot{p}^\nu p^\mu + \partial_\sigma g_{\nu\rho} p^\nu p^\rho \dot{x}^\sigma \right) \\
&= \frac{1}{2} \frac{d}{ds} \left(g_{\mu\nu} p^\mu p^\nu \right) = \frac{1}{2} \frac{d}{ds} \mu^2 .
\end{aligned} \tag{2.19}$$

We then have, in particular for the autoparallels

$$g_{\mu\nu} \dot{x}^\mu \dot{x}^\nu = \text{const} ; \tag{2.20}$$

if the initial condition (2.16) is satisfied then

$$g_{\mu\nu} \dot{x}^\mu \dot{x}^\nu = \mu^2 \tag{2.21}$$

holds, where μ stands for the mass of the test particle. The mass is then conserved throughout the motion. It also follows from the above that the autoparallels remain time-like (light-like) if they started time-like (light-like). Equation (2.20) is an important conservation law; we will use it often.

For scattering processes we have the following:

Postulate 2.6 *If r particles with 4-momenta $\mathbf{p}_1, \ldots, \mathbf{p}_r$ are scattered at the point x, such that s particles with 4-momenta $q_1^\mu, \ldots, q_s^\mu$ result, then*

Fig. 2.1 Scattering of test
particles

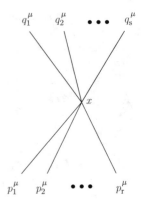

$$p_1^\mu + \ldots + p_r^\mu = q_1^\mu + \ldots + q_s^\mu$$

holds in x.

These first six postulates and the interpretation of the local inertial frame as a freely falling, non-rotating system with orthogonal axes shall be sufficient for us at the moment. They suffice to explore the structure of the curved space–time and to gain a first understanding of it. They concern, however, only the particle dynamics in the gravitational field. We still have to postulate the dynamics of the fields in the gravitational fields as well as the dynamics of the gravitational field itself, to make the theory complete (Fig. 2.1).

2.4 Local Measurements

The measurements that an observer is making in his close space–time neighborhood form the basis for all observations. If we observe for instance an object that is far away, we then analyze the light signals that reach our observatory. The whole observation can hence be traced back to local measurements.

In this section we will get to know an often used approximation, which is to consider "small" neighborhoods as "infinitesimal". Such a method is very suggestive for physicists. Its strict mathematical justification in differential geometry is, however, tedious. We want to do without this justification here; it can be found in [5].

2.4.1 General Reference Frame

From our experience with Newton's theory and special relativity, we are used to considering a coordinate system simply as a physical reference frame, i.e., to relate it to some measurement apparatus for determining the coordinates. In general relativity, the situation is more complicated. In general the coordinates carry information only about the topological and differential structure of space–time, and have *nothing* to do with any physical reference frame! The choice of the coordinates in general relativity is comparable to choosing a gauge in electrodynamics rather than to the choice of a reference frame. This will be explained more precisely in Sect. 4.3. Components of tensors with respect to coordinates have no measurable meaning; measurable is only what is invariant with respect to transformations of the coordinates. This fact is also reflected in the dimension of the coordinates: in general, they are dimensionless, and the metric carries the dimension of length squared.

In a similar way as the gauge, the coordinates can be chosen such that they simplify the problems under consideration. For example they can be adapted to a real reference frame. In such cases, the coordinates can get a non-trivial dimension, and the dimensions of different components of the metric can differ. For instance, coordinates of a local inertial frame have dimension of length ($c = 1$!), and all components of the metric are dimensionless, just as is the case in special relativity and in inertial frames.

Let $\{x^\mu\}$ be a coordinate system and let $g_{\mu\nu}$ be the corresponding components of the metric. In this section we restrict ourselves to systems whose x^0 trajectories are time-like. The x^0 curves are those with the parameter representation

$$x^0 = \lambda, \quad x^k = x_0^k, \quad k = 1, 2, 3 ,$$

where x_0^k are arbitrary constants. The tangent vector $\dot{x}^\mu$ has the components

$$\dot{x}^\mu = (1, 0, 0, 0) . \tag{2.22}$$

The curves are time-like:

$$g_{\mu\nu} \dot{x}^\mu \dot{x}^\nu > 0 ,$$

which implies

$$g_{00} > 0 . \tag{2.23}$$

If we put on each of these curves an observer or a mass point, these observers then fill the space–time in a dense way. Sometimes people speak about "observer dust" or "reference fluid". This reference fluid should be sufficiently "thin" for its influence on the geometry of space–time to be neglected. The three numbers x^k can be considered as the name of an observer or of a mass point; a choice for the coordinate x^0 then fixes an event for every observer. Such a coordinate system is called *general reference frame*. The values of the coordinates corresponding to different events are conventions within the family of observers; for instance one observer can emit periodically a radio signal, which determines everywhere particular values for x^0. We leave these coordinates arbitrary.

If a "puritan" was asked to describe such a reference frame, he would have to leave the coordinates arbitrary. He then would have to represent our reference frame by a family of curves (*congruence*, see [5]), which are parameterized by a time parameter t; such a structure would then be coordinate independent.

2.4.2 Proper Time

With an ideal clock, the observer measures the time interval ds between the two nearby values x^0 and $x^0 + dx^0$. Equation (2.13) yields

$$ds = \sqrt{g_{\mu\nu}\dot{x}^\mu \dot{x}^\nu}\, d\lambda = \sqrt{g_{00}}\, dx^0 .$$

This way the component g_{00} of the metric can in principle be measured:

$$g_{00} = \left(\frac{ds}{dx^0}\right)^2 . \tag{2.24}$$

2.4.3 Radar Measurement

The observer (x^k) sends a radar signal at the time x^0 (event p with coordinates (x^0, x^k)) to his neighbor $(x^k + dx^k)$, who reflects this signal and communicates the time value $(x^0 + dx^0)$, which is measured in his coordinate system and indicates the moment the signal arrived (event q with coordinates $(x^0 + dx^0, x^k + dx^k)$) there. The echo reaches the observer (x^k), which is described by the event r with coordinates $(x^0 + \Delta x^0, x^k)$. The (infinitesimal) coordinate differences read (Fig. 2.2)

$$x^\mu(q) - x^\mu(p) = \left(dx^0, dx^k\right),$$

$$x^\mu(r) - x^\mu(q) = \left(\Delta x^0 - dx^0, -dx^k\right),$$

and hence form light-like vectors. The following two equations are therefore satisfied

$$g_{00}(dx^0)^2 + 2g_{0k}dx^0 dx^k + g_{kl}dx^k dx^l = 0,$$

$$g_{00}(dx^0 - \Delta x^0)^2 + 2g_{0k}(dx^0 - \Delta x^0)dx^k + g_{kl}dx^k dx^l = 0 . \tag{2.25}$$

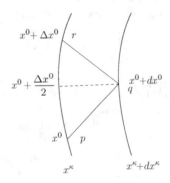

Fig. 2.2 Radar measurement

Subtracting the second equation from the first leads to

$$g_{0k}dx^k = g_{00}\left(\frac{1}{2}\Delta x^0 - dx^0\right). \tag{2.26}$$

If we calculate dx^0 from (2.26) and insert it back in (2.25), we obtain

$$-\hat{g}_{kl}dx^k dx^l = g_{00}\left(\frac{1}{2}\Delta x^0\right)^2, \tag{2.27}$$

where we used the abbreviation:

$$\hat{g}_{kl} := g_{kl} - \frac{g_{0k}g_{0l}}{g_{00}}. \tag{2.28}$$

Equations (2.26), (2.27), and (2.28) can be used to determine the components g_{0k} and g_{kl} of the metric from the measured values for g_{00}, dx^0, and Δx^0, and from the known values for dx^k. In order to do so, it is necessary to carry out the measurement simultaneously with six appropriately chosen neighbors. All components of the metric with respect to these general reference frames can hence be measured in principle. Next we want to find out the geometric meaning of g_{0k} and $\hat{g}_{kl}$.

2.4.4 Simultaneity

We define in general relativity: two events with coordinates

$$\left(x^0 + \frac{1}{2}\Delta x^0, x^k\right), \quad \left(x^0 + dx^0, x^k + dx^k\right)$$

are *simultaneous* for the observer (x^k). This definition is as in special relativity but is now only true for nearby events. Hence, if $dx^0 = \frac{1}{2}\Delta x^0$, then both time coordinates are at the instant x^0 *synchronous*. Equation (2.26) then yields $g_{0k}dx^k = 0$. Hence the quantity $g_{0k}dx^k$ is a measure of how asynchronous the time coordinates of the two observers (x^k) and $(x^k + dx^k)$ are.

The condition of simultaneity of nearby events for the observer (x^k) can equivalently be formulated by noting that the infinitesimal vector dl^μ,

$$dl^\mu = \left(dx^0 - \frac{1}{2}\Delta x^0, dx^k\right), \tag{2.29}$$

which connects the two simultaneous events, is orthogonal to the 4-velocity of the observer. This is because (2.26) says that

$$g_{00}\left(dx^0 - \frac{1}{2}\Delta x^0\right) + g_{0k}dx^k = 0.$$

But this is because of (2.29) being equivalent to

$$g_{0\mu}dl^\mu = 0 ,$$

and (2.22) then gives

$$g_{\mu\nu}\dot{x}^\mu dl^\nu = 0 .$$

The vectors, which are orthogonal to their 4-velocity, are hence spatial for the observer.

2.4.5 Distances

From the above radar measurement, we also obtain the distance da between the observers (x^k) and $(x^k + dx^k)$. In our case da is half of the proper time between the instants x^0 and $x^0 + \Delta x^0$ (multiplied by the speed of light):

$$da = \frac{1}{2}\sqrt{g_{00}}\Delta x^0 .$$

Then we obtain from (2.27) that

$$da^2 = -\hat{g}_{kl}dx^k dx^l .$$

The metric $\hat{g}_{kl}$ hence measures the momentary distance of the observers (x^k) and $(x^k + dx^k)$. We can calculate this distance also directly from $g_{\mu\nu}$; namely, it is equal to the length of the vector dl^μ (exercise).

Summarizing, we can therefore say that the metric in general relativity is not a purely formal object but is determined by measurements; this is in fact, locally, exactly like the Minkowski metric.

2.4.6 Spectra and Directions

Let us consider an observer with trajectory $x^\mu(\lambda)$. Its 4-velocity $e^\mu_{(0)}$ at the point λ_0 of its trajectory satisfies

$$e^\mu_{(0)} = N\dot{x}^\mu(\lambda_0) ,$$

where the normalization constant N is determined by

$$g_{\mu\nu}e^\mu_{(0)}e^\nu_{(0)} = 1 .$$

The observer perceives those directions orthogonal to $e^\mu_{(0)}$ as spatial. The observer can choose three axes out of those, which are mutually orthogonal, and can represent them by the unit vectors $e^\mu_{(k)}, k = 1, 2, 3$. The four vectors $e^\mu_{(\alpha)}, \alpha = 0, 1, 2, 3$, then

Fig. 2.3 Frequency and
directions

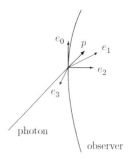

satisfy the equations

$$g_{\mu\nu} e^{\mu}_{(\alpha)} e^{\nu}_{(\beta)} = \eta_{\alpha\beta} \ . \tag{2.30}$$

The 4-tuple of vectors $e^{\mu}_{(\alpha)}$ constructed this way is called *vierbein*. This vierbein represents a *rest frame* of the observer, who can relate the components of different tensors in the event $x^{\mu}(\lambda_0)$ to this vierbein (Fig. 2.3).

Let us consider for instance a light signal with 4-momentum p^{μ}, which reaches the observer at the time $x^0(\lambda_0)$. How are the frequency ν, which is measured by the observer, and the unit vector n^k, which denotes the direction of the photon, related to the 4-momentum? The observer decomposes p^{μ} in the vierbein elements:

$$p^{\mu} = h\nu(e^{\mu}_{(0)} + n^k e^{\mu}_{(k)}) \ . \tag{2.31}$$

From (2.30) we obtain

$$h\nu = g_{\mu\nu} p^{\mu} e^{\nu}_{(0)} \ , \tag{2.32}$$

$$n^k = -(h\nu)^{-1} g_{\mu\nu} p^{\mu} e^{\nu}_{(k)} \ . \tag{2.33}$$

It is true in principle that the components of tensors with respect to a vierbein are related to directly measurable numbers. This is why they are sometimes called *physical components* of the tensor.

2.5 Stationary Space–Time

To begin with, we want to consider some particular, symmetric space–times and study their physical properties. In this section we are going to treat the mathematical framework connected with symmetries without going into details; we simply postulate a certain form of the metric, try to convince ourselves that it possesses a symmetry, and then we pass directly over to the observable properties. Our aim is to understand how to obtain, for instance, a ballistic curve in space out of a corresponding autoparallel in space–time.

In the definition of stationary space–time, we must distinguish between *stationary* and *static* fields. "Stationary" means "time independent". Also a motion can be stationary, e.g., a stationary circular motion of an electric current in a superconducting ring, or the rotation of the stars. "Static" means even more symmetry, such that stationary motions are also excluded. For the definition of a stationary situation, we can use the fact that all stationary motions reverse when time is reversed.

Definition 9 A space–time is called stationary if it admits a coordinate system $\{x^\mu\}$ with the following properties:

1. the x^0 curves are time-like,
2. the components $g_{\mu\nu}$ of the metric are independent of the time coordinate x^0

$$\partial_0 g_{\mu\nu} = 0 . \tag{2.34}$$

It is called *static* if the components are, in addition, invariant with respect to time reversal $x^0 \rightarrow -x^0$.

The time reversal is considered here as a coordinate transformation:

$$x'^0 = -x^0, \quad x'^k = x^k . \tag{2.35}$$

The non-vanishing transformation coefficients for tensors are

$$\frac{\partial x^0}{\partial x'^0} = -1, \quad \frac{\partial x^k}{\partial x'^k} = 1 . \tag{2.36}$$

The components of the metric with respect to $\{x'^\mu\}$ hence satisfy

$$g'_{00} = g_{00}, \quad g'_{0k} = -g_{0k}, \quad g'_{kl} = g_{kl} .$$

The invariance of the metric with respect to time reversal then means that

$$g_{0k}(p) = 0 \quad \forall p . \tag{2.37}$$

Summarizing, we can write that there exist, in a stationary space–time, coordinates (t, x^k) such that the metric in these coordinates takes the form

$$ds^2 = \alpha^2 (dt + \beta_k dx^k)^2 + \hat{g}_{kl} dx^k dx^l , \tag{2.38}$$

where

$$\alpha = \sqrt{g_{00}(x^1, x^2, x^3)} ,$$

$$\beta_k = \frac{g_{0k}(x^1, x^2, x^3)}{g_{00}(x^1, x^2, x^3)} ,$$

$$\hat{g}_{kl} = g_{kl}(x^1, x^2, x^3) - \frac{g_{0k}(x^1, x^2, x^3) g_{0l}(x^1, x^2, x^3)}{g_{00}(x^1, x^2, x^3)} .$$

Here, the $\{x^\mu\}$ are called *stationary coordinates*, and the observers along the x^0 curves are called *stationary observers*. If the coordinates are chosen such that $\beta_k = 0$, then the metric, the coordinates, and the observer are called *static*.

If we find coordinates such that a given metric admits the form (2.38) with $\beta_k \neq 0$, this does not imply that there do not exist coordinates for which $\beta_k = 0$. Let us study this.

The stationary coordinates, it they exist, are only unique up to a certain transformation. First, a general coordinate transformation of the spatial coordinates is allowed

$$x'^k = x'^k(x^1, x^2, x^3) . \tag{2.39}$$

Thereby, $\alpha(x^1, x^2, x^3)$ transforms as a scalar field, $\beta_k(x^1, x^2, x^3)$ as a covector field, and $\hat{g}_{kl}(x^1, x^2, x^3)$ as a tensor field of type $(0,2)$. Furthermore, we can shift the time coordinate in a way that depends on the space coordinates

$$t' = t + T(x^1, x^2, x^3) , \tag{2.40}$$

where $T(x^1, x^2, x^3)$ is an arbitrary function. Then the quantities transform as follows:

$$\alpha' = \alpha , \quad \beta'_k = \beta_k - \partial_k T , \quad \hat{g}'_{kl} = \hat{g}_{kl} . \tag{2.41}$$

We observe that β_k changes in the same way as an electromagnetic vector potential A_k does under a gauge transformation. The geometric meaning of transformation (2.41) is an arbitrary shift $T(x^1, x^2, x^3)$ of all $t = \text{const}$ hypersurfaces into the family of $t' = \text{const}$ hypersurfaces. Finally we can change the t-scale

$$t' = \kappa t , \tag{2.42}$$

where κ is a positive constant; then

$$\alpha' = \kappa^{-1}\alpha , \quad \beta'_k = \kappa\beta_k , \quad \hat{g}'_{kl} = \hat{g}_{kl} . \tag{2.43}$$

These three transformations and their combinations leave the form (2.38) invariant, and there does not exist a more general transformation with the same property. Strictly speaking, there are rare exceptions if space–time is time independent in more than one direction, as is the case, for instance, for Minkowski space–time: the Minkowski metric has the form (2.38); it is invariant with respect to all Lorentz transformations, and these do not belong to the above class of transformations.

Equation (2.41) implies the following theorem.

Theorem 5 *The metric (2.38) is static if and only if β_k is the gradient of a function.*

Then β_k is either zero or it can be transformed away.

2.5.1 The 3-Space

The stationary space–time is a special case that makes physically meaningful split-ting of space–time in space and time possible. We imagine that spatial points are defined by stationary observers: every observer sits in a fixed point of space and has a well-defined, fixed distance from its neighbors. This means that these spatial points are not points of a three-dimensional hypersurface anymore, as for instance the simultaneity surfaces in Newton theory, but every spatial point is identified with the orbit of a stationary observer. This orbit has three coordinates x^1, x^2, and x^3, and the distances between neighboring orbits, which are given by $\hat{g}_{kl}dx^k dx^l$, are now considered as distances between the spatial points. The coordinates x^1, x^2, and x^3 cover a manifold Σ, and the 3-space is defined as the metric manifold $(\Sigma, \hat{g}_{kl})$.

The space Σ can be obtained from the space–time $\mathcal{M}$ by the projection π : $\mathcal{M} \mapsto \Sigma$, defined as follows: Let $p \in \mathcal{M}$ be a point with stationary coordinates (t, x^1, x^2, x^3). Then $\pi(p) \in \Sigma$ is determined by the coordinates (x^1, x^2, x^3).

2.5.2 Free Falls

The equation defining autoparallels arises in a stationary space–time from the Lagrangian

$$L = \frac{1}{2}\alpha^2 \left(x^1, x^2, x^3\right) \left[\dot{t} + \beta_k \left(x^1, x^2, x^3\right) \dot{x}^k\right]^2 + \frac{1}{2}\hat{g}_{kl} \left(x^1, x^2, x^3\right) \dot{x}^k \dot{x}^l .$$

The variation with respect to t yields

$$\alpha^2 \left(\dot{t} + \beta_k \dot{x}^k\right) = e , \tag{2.44}$$

where e is a constant. Another conservation law is

$$\alpha^2 (\dot{t} + \beta_k \dot{x}^k)^2 + \hat{g}_{kl} \dot{x}^k \dot{x}^l = \sigma , \tag{2.45}$$

where σ is the signature of the particle orbit, $\sigma = 1$ for massive test particles, and $\sigma = 0$ for massless ones; this means that we parameterize massive particles with the proper time but massless particles with an arbitrary parameter. Then follows:

$$\dot{t} + \beta_k \dot{x}^k = \frac{1}{\alpha}\sqrt{\sigma - \hat{g}_{kl} \dot{x}^k \dot{x}^l} . \tag{2.46}$$

The variation of the action with respect to x^k leads, after a simple calculation, to

$$\alpha^2 \beta_m \left(\ddot{t} + \beta_k \ddot{x}^k\right) + \hat{g}_{mk}\ddot{x}^k$$
$$- \alpha\alpha_{,m} \left(\dot{t} + \beta_k \dot{x}^k\right)^2 + (\alpha^2 \beta_m)_{,l}\dot{x}^l (\dot{t} + \beta_k \dot{x}^k) - \alpha^2 \beta_{l,m}\dot{x}^l \left(\dot{t} + \beta_k \dot{x}^k\right)$$
$$+ \alpha^2 \beta_m \beta_{k,l}\dot{x}^k \dot{x}^l + \frac{1}{2}\left(\hat{g}_{mk,l} + \hat{g}_{ml,k} - \hat{g}_{kl,m}\right)\dot{x}^k \dot{x}^l = 0 . \tag{2.47}$$

Writing this formula we used a convenient and popular abbreviation: a comma for the derivative. For instance,

$$\alpha_{,m} = \frac{\partial \alpha}{\partial x^m} \, ,$$

or

$$g_{kl,m} = \frac{\partial g_{kl}}{\partial x^m} \, .$$

We want to reformulate (2.47) in such a way that no $\ddot{t}$ and $\dot{t}$ are left. Thereto we first have to differentiate (2.44) with respect to λ,

$$\alpha^2 \left(\ddot{t} + \beta_k \ddot{x}^k \right) + 2\alpha\alpha_{,l}\dot{x}^l \left(\dot{t} + \beta_k \dot{x}^k \right) + \alpha^2 \beta_{k,l}\dot{x}^k \dot{x}^l = 0 \, ,$$

multiply this equation by β_m and subtract it from (2.47). Then we insert for $\dot{t} + \beta_k \dot{x}^k$ the expression given in (2.46), and we thus obtain

$$\ddot{x}^m + \hat{\Gamma}^m_{kl}\dot{x}^k \dot{x}^l = \hat{g}^{mk} \sqrt{\sigma - \hat{g}_{rs}\dot{x}^r \dot{x}^s} \left[\frac{\alpha_{,k}}{\alpha} \sqrt{\sigma - \hat{g}_{rs}\dot{x}^r \dot{x}^s} + \alpha(\beta_{l,k} - \beta_{k,l})\dot{x}^l \right] , \quad (2.48)$$

where

$$\hat{\Gamma}^k_{rs} = \frac{1}{2} g^{kl} \left(\partial_r \hat{g}_{ls} + \partial_s \hat{g}_{lr} - \partial_l \hat{g}_{rs} \right)$$

are the Christoffel symbols of the spatial metric.

Consequently the equation of autoparallels is split in a time (2.44) and three spatial equations (2.48). We can then see which role every component of the metric plays in the dynamics of test particles: the components g_{kl} and g_{0k} curve the 3-space; g_{00} and g_{0k} bend the orbits so that they deviate from space geodesics. This way we can also realize better what the four-component description of gravity in the relativistic theory means and understand some of the phenomena that appear new with respect to the Newton theory.

2.5.3 The Gravitoelectric and Gravitomagnetic Force

Let us now provide an interpretation for (2.48). Every curve $x^\mu = x^\mu(\lambda)$ in the four-dimensional space–time $\mathcal{M}$ is mapped by the projection π on a curve $x^k = x^k(\lambda)$ in the three-dimensional space $(\Sigma, \hat{g}_{kl})$. The left-hand side of (2.48) is the λ acceleration of the projected curve. If the right-hand side vanished, (2.48) would describe an autoparallel in $(\Sigma, \hat{g}_{kl})$. Otherwise the projection of the curve of a free-falling particle in this space does not look like an autoparallel. For instance, for the spatial metric $\hat{g}_{kl} = -\delta_{kl}$, every autoparallel is a straight line. But an autoparallel of the full space–time metric (2.38) leads in this case to a curved line, if projected on the space: we "see" an acceleration, which is described by the right-hand side of (2.38).

Let us study the acceleration for the massive case, i.e., $\lambda = s$ and $\sigma = 1$. What is the meaning of the square root? If we introduce the rest frame (u^μ, e^μ_k) of a stationary observer in the point p of its motion with the particle, we have $u^\mu = (1/\alpha, 0, 0, 0)$ and $\dot{x}^\mu$ is the (normalized) 4-velocity of the particle. The scalar product of these two unit vectors is

$$g_{\mu\nu} u^\mu \dot{x}^\nu = \alpha(\dot{t} + \beta_k \dot{x}^k) \,.$$

Together with (2.46), this yields

$$\sqrt{1 - \hat{g}_{rs} \dot{x}^r \dot{x}^s} = g_{\mu\nu} u^\mu \dot{x}^\nu \,. \tag{2.49}$$

Let us introduce now new stationary coordinates $\tilde{x}^\mu$, which are adapted to the observer: first we have $\tilde{x}^0 = \alpha(\bar{p})t$, where $\bar{p} = \pi(p)$; secondly the hypersurface $\tilde{x}^0 = $ const is orthogonal to the orbit of this observer (and of no other one); and finally $\tilde{x}^k$ is such that the vectors satisfy $\tilde{e}_k^\mu = \delta_{k\mu}$. It is easy to show that such stationary coordinates exist (exercise) but that they do not, in general, form geodetic systems in p. In this system we have $\tilde{g}_{\mu\nu}(p) = \eta_{\mu\nu}$, $\tilde{u}^\mu = \delta_0^\mu$, and we obtain

$$g_{\mu\nu}(p) u^\mu \dot{x}^\nu = \tilde{g}_{\mu\nu}(p) \tilde{u}^\mu \dot{\tilde{x}}^\nu = \dot{\tilde{x}}^0 \,. \tag{2.50}$$

The particle's velocity with respect to the observer's rest frame is

$$\left(\frac{d\tilde{x}^0}{d\tilde{x}^0}, \frac{d\tilde{x}^k}{d\tilde{x}^0} \right) = (1, \tilde{v}^k) \,,$$

such that the relation between the two velocities is given by

$$(\dot{\tilde{x}}^0, \dot{\tilde{x}}^k) = \frac{1}{\sqrt{1 - \tilde{v}^2}} (1, \tilde{v}^k) \,. \tag{2.51}$$

Equations (2.49), (2.50), and (2.51) yield

$$\sqrt{1 - \hat{g}_{rs} \dot{x}^r \dot{x}^s} = \frac{1}{\sqrt{1 - \tilde{v}^2}} \,. \tag{2.52}$$

In this way, we can write the right-hand side of (2.48) in the point $\bar{p}$ and in the adapted coordinates:

$$-\frac{\delta^{mk}}{\sqrt{1 - \tilde{v}^2}} \left[(\partial_k \ln \tilde{\alpha}) \dot{\tilde{x}}^0 + \left(\tilde{\partial}_k \tilde{\beta}_l - \tilde{\partial}_l \tilde{\beta}_k \right) \dot{\tilde{x}}^l \right] \,, \tag{2.53}$$

since $\tilde{\alpha}(\bar{p}) = 1$. Also the left-hand side can be written in adapted coordinates (where we simply would have to write everything with tildes).

The expression (2.53) describes the deviation of the spatial projection of the particle's orbit from the autoparallels of the space and therefore constitutes an effective force action on the particle. It reminds us of the expression for the Lorentz force of a stationary field potential A_μ acting on a particle with charge q in the special relativistic Maxwell theory:

$$-q\delta^{mk}\left[(\partial_k A_0)\dot{x}^0 + (\partial_k A_l - \partial_l A_k)\dot{x}^l\right],$$

where the first term in the square brackets stands for the electric force and the second for the magnetic one.

Correspondingly, the first term in expression (2.53) is called *gravitoelectric* if it is multiplied by the mass μ. The second is called *gravitomagnetic*. The gravitomagnetic effects of the Earth's gravitational field are weak and are being measured or observed (for a discussion, see [6]). The source of this gravitomagnetic field is the angular momentum of the Earth in analogy to an electricity loop in electromagnetism.

In the following we will restrict the discussion to static space–times and set $\beta_k = 0$ as well as $\ln \alpha = \Phi$. Then we have $\hat{g}_{kl} = g_{kl}$,

$$ds^2 = e^{2\Phi} dt^2 + g_{kl} dx^k dx^l.$$

The sign in expression (2.53) implies that the gravitoelectric force points from points with larger g_{00} to those with smaller g_{00}. For the metric, which Newtonian physics reproduces approximately, we obtain

$$ds^2 = e^{2\Phi}\left(dx^0\right)^2 - \delta_{kl} dx^k dx^l. \tag{2.54}$$

For instance, the only non-vanishing components of the curvature tensor for the metric (2.54) are

$$R^0_{k0l} = -\frac{\partial^2 \Phi}{\partial x^k \partial x^l} - \frac{\partial \Phi}{\partial x^k}\frac{\partial \Phi}{\partial x^l}.$$

This is different from (1.27), which we found in Newton's theory. The second term on the right-hand side can be neglected with respect to the first one if $\Phi \ll 1$, which is valid near the Earth.

2.5.4 Redshift

Consider two static observers (x_1^k) and (x_2^k) with 4-velocity u_1^μ and u_2^μ. Let (x_1^k) send a photon to (x_2^k). This leaves (x_1^k) with the 4-momentum p_1^μ, runs along the light-like autoparallel and reaches (x_2^k) with the 4-momentum p_2^μ. The frequency v_1 of the photon in the rest frame of (x_1^k) and the frequency v_2 in the rest frame of (x_2^k) are given by (2.32) (Fig. 2.4).

$$hv_1 = g_{\mu\nu} u_1^\mu p_1^\nu, \qquad hv_2 = g_{\mu\nu} u_2^\mu p_2^\nu.$$

If we insert the corresponding values for $g_{\mu\nu}$ and u^μ, we obtain

$$hv_i = \sqrt{g_{00}(x_i)} p_i^0, \quad i = 1, 2.$$

Fig. 2.4 Redshift

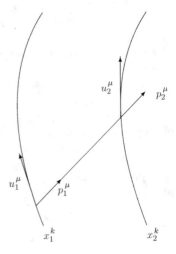

From this we would like to deduce the redshift

$$z = \frac{\lambda_2 - \lambda_1}{\lambda_1} = \frac{v_1 - v_2}{v_2} \, .$$

The light-like orbit of the photon can be parameterized

$$x^\mu = x^\mu(\kappa) \, .$$

The functions $x^\mu(\kappa)$ must satisfy (2.44) and (2.48) with $\beta_k = 0$. Let κ be a physical parameter:

$$p^\mu = \dot{x}^\mu \, .$$

Because of (2.44) with $\beta_k = 0$, we have

$$e = g_{00}\dot{x}^0 = \sqrt{g_{00}}\left(\sqrt{g_{00}}p^0\right) = h\sqrt{g_{00}}\,v \, ,$$

which yields, for the two frequencies,

$$\sqrt{g_{00}\left(x_1^k\right)}\,v_1 = \sqrt{g_{00}(x_2^k)}\,v_2 \, ,$$

and for the redshift we obtain

$$z = \frac{\sqrt{g_{00}\left(x_2^k\right)} - \sqrt{g_{00}\left(x_1^k\right)}}{\sqrt{g_{00}\left(x_1^k\right)}} \, . \tag{2.55}$$

The redshift therefore depends, in the static space–times, only on the relation of the metric's g_{00} components in the corresponding points, and it is positive if the signal climbs in the gravitational field (i.e., if it runs from points with small g_{00} to points with larger g_{00}). We have actually nowhere used the fact that geodesics are light-like. Equation (2.55) is therefore valid for massive particles, if we define

$$z = \frac{\varepsilon_1 - \varepsilon_2}{\varepsilon_2} \, ,$$

where ε is the particle's (kinetic) energy with respect to the rest frame of the static observer at the corresponding point, $\varepsilon = g_{\mu\nu} u^{\mu} p^{\nu}$. For a discussion of the experimental aspects, see [6].

2.5.5 Gravitational Time Dilation

Equation (2.54) shows that Newton's gravitational potential and the component g_{00} of the metric in static space–time are very closely connected. But this component is also related to time measurement. It seems therefore that even the ideal clocks are influenced by the gravitational field. In the following, we want to investigate this phenomenon further.

We would like to compare two clocks: one at a static observer $(x_{(1)}^k)$ and the other at $(x_{(2)}^k)$ (Fig. 2.5). The best way to do this is by using a light signal: the observer $(x_{(1)}^k)$ sends light signals to $(x_{(2)}^k)$, with an interval of $\Delta s_{(1)}$ seconds, which is determined by an ideal clock. The observer $(x_{(2)}^k)$ measures with his clock the interval $\Delta s_{(2)}$ between two arriving signals. The nth signal runs along the light-like geodesic $x^{\mu} = y_n^{\mu}(\lambda)$, which satisfies

$$y_n^{\mu}\left(\lambda_{(1)}\right) = \left(x_{(1)n}^0, x_{(1)}^k\right), \qquad y_n^{\mu}\left(\lambda_{(2)}\right) = \left(x_{(2)n}^0, x_{(2)}^k\right) ,$$

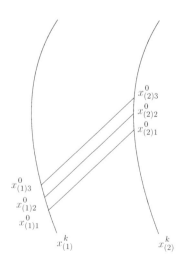

Fig. 2.5 Gravitational time dilation

where

$$x^0_{(1)n} - x^0_{(1)(n-1)} = \frac{\Delta s_{(1)}}{\sqrt{g_{00}\left(x^k_{(1)}\right)}} \, .$$

Let us assume that the orbit of the first signal is known. The orbit of the second can obviously be determined by the shift

$$y^0_2(\lambda) = y^0_1(\lambda) + \frac{\Delta s_{(1)}}{\sqrt{g_{00}\left(x^k_{(1)}\right)}}, \qquad y^k_2(\lambda) = y^k_1(\lambda) \, .$$

The shifted curve will indeed have the same values for $\dot{y}^\mu_2$ and $\ddot{y}^\mu_2$ as the first one. Also the coefficients of the geodesic equation (2.44) and (2.48) are independent of $y^0_2(\lambda)$, which is changed by the shift. Therefore the shifted curve $y^\mu_2(\lambda)$ solves the equation with the correct initial and final values.

The values of the time coordinate x^0 for the signal's arrival at (x^k_2) are then given by $x^0_{(2)n} = y^0_n(\lambda_{(2)})$, hence

$$x^0_{(2)n} = x^0_{(2)1} + (n-1)\frac{1}{\sqrt{g_{00}\left(x^k_{(1)}\right)}}\Delta s_{(1)} \, ,$$

and the time difference is

$$\frac{\Delta s_{(2)}}{\Delta s_{(1)}} = \frac{\sqrt{g_{00}\left(x^k_{(2)}\right)}}{\sqrt{g_{00}\left(x^k_{(1)}\right)}} \, . \tag{2.56}$$

This relation expresses the relative speed of the two clocks under consideration. We see that the clock runs slower, where the component g_{00} is smaller, i.e., in the direction of the gravitational acceleration. A discussion of the experiments can be found in [6].

This effect can also provide an explanation for the redshift, which is based on the wave theory of light. The light wave's maxima propagate along light-like geodesics. Hence we can use (2.56) also for the relation of the periods of the electromagnetic oscillations. This leads again to (2.55).

The fact that the gravitational field influences the speed of clocks has great significance. As we shall see, also the trajectories of light rays are influenced by gravity. The measurement of the proper time and the propagation of light determine the geometry of space–time according to the interpretations of special relativity theory. If now gravity is present—and this is always the case—then we are led to contradictions with the postulates of Minkowskian geometry. There are in principle two ways out of this dilemma.

1. We could keep Minkowskian geometry and could try to describe gravity as a field of Minkowskian space–time. The effect of gravity on the propagation of light and on the measurement of the proper time should then be explained by an appropriate interaction between the clocks and the electromagnetic field on the one hand and the gravitational field on the other hand. But then the geometry of Minkowskian space–time is not determined by measurements as it is demanded in special relativity theory. What else determines the geometry in this case? If gravity is weak, it turns out that a Minkowskian auxiliary metric can be introduced. This metric is however not unique (as a field on the manifold, compare the discussion in Sect. 2.2), and it is not measurable; the causal structure, distances, and time intervals are anyhow determined by the gravitational field.
2. Light propagation and measurement of proper time determine the geometry of the space–time, similarly as in special relativity theory. However, it is not a Minkowskian geometry (this is the way chosen by general relativity).

Notice that the measurable properties of the space–time geometry include the following structures:

1. the causal structure (which events influence which others),
2. the time intervals, and
3. the distances.

2.6 Isometry (A Mathematical Intermezzo)

In the previous section we dealt for the first time with symmetries of space–time. It is worthwhile to study symmetry more generally and in more detail. For this purpose, we first have to learn how objects such as tensors are transformed by a map. Let us start with an example.

2.6.1 Rotation in $\mathbb{E}^2$

Consider the Euclidean plane $\mathbb{E}^2$ with Cartesian coordinates $\{x^1, x^2\}$ and a rotation ϕ of this plane by an angle α. A point $p \in \mathbb{E}^2$ with coordinates $x^\mu(p)$ transforms to the point $q = \phi(p)$ with coordinates $x^\mu(q)$, where

$$(x^1(q), x^2(q)) = (x^1(p)\cos\alpha - x^2(p)\sin\alpha, x^1(p)\sin\alpha + x^2(p)\cos\alpha), \quad (2.57)$$

and

$$(x^1(p), x^2(p)) = (x^1(q)\cos\alpha + x^2(q)\sin\alpha, -x^1(q)\sin\alpha + x^2(q)\cos\alpha). \quad (2.58)$$

This equation describes a so-called *active* transformation; the points are moved but the coordinates are not changed; a *passive* transformation is nothing but a coordinate transformation where the points remain fixed and the coordinates are changed.

Let now V^μ be a vector in p given by its components with respect to $\{x^\mu\}$. How do we transform V^μ from p to $\phi(p)$ by ϕ? We have to make two steps.

(1) We rotate the coordinate system $\{x^\mu\}$ and obtain in this way its so-called ϕ-image $\{x'^\mu\}$:

$$x'^1 = x^1 \cos\alpha + x^2 \sin\alpha, \qquad x'^2 = -x^1 \sin\alpha + x^2 \cos\alpha, \tag{2.59}$$

$$x^1 = x'^1 \cos\alpha - x'^2 \sin\alpha, \qquad x^2 = x'^1 \sin\alpha + x'^2 \cos\alpha. \tag{2.60}$$

Then we have

$$x'^\mu(q) = x^\mu(p) . \tag{2.61}$$

(2) In the point $\phi(p)$ we define a vector U^μ by its components with respect to $\{x'^\mu\}$ determined by the equation

$$U'^\mu = V^\mu . \tag{2.62}$$

Therefore

$$U^\mu = X^\mu_{\nu'} U'^\nu = X^\mu_{\nu'} V^\nu ,$$

so that, according to (2.60),

$$(U^1, U^2) = (V^1 \cos\alpha - V^2 \sin\alpha, V^1 \sin\alpha + V^2 \cos\alpha) . \tag{2.63}$$

We denote this vector U^μ at the point $\phi(p)$ by $(\phi_* V)^\mu$. This way the mapping of tensors is defined by ϕ.

Let us consider a whole vector field $V^\mu(x)$ on $\mathbb{E}^2$. We say that $V^\mu(x)$ is invariant with respect to ϕ if, for every point $p \in \mathbb{E}^2$ the vector $(\phi_* V(p))^\mu$ equals the vector $V^\mu(\phi(p))$ at the point $\phi(p)$. If we define for example $V^\mu(x)$ by the equations:

$$V^1(x^1, x^2) = x^1, \qquad V^2(x^1, x^2) = x^2 ,$$

we obtain, according to (2.63) and (2.57), that

$$\begin{aligned}
&((\phi_* V(p))^1, (\phi_* V(p))^2) \\
&= (V^1(p)\cos\alpha - V^2(p)\sin\alpha, V^1(p)\sin\alpha + V^2(p)\cos\alpha) \\
&= (x^1(p)\cos\alpha - x^2(p)\sin\alpha, x^1(p)\sin\alpha + x^2(p)\cos\alpha) = (x^1(q), x^2(q)) \\
&= (V^1(q), V^2(q)).
\end{aligned}$$

Our vector field is therefore invariant.

2.6.2 Diffeomorphisms

We would like to extend these ideas to a broader class of maps, spaces, and objects. Thereby it is advantageous to consider the coordinate systems as maps, as we already did on p. 8. A coordinate system $\{x^\mu\}$ on an n-manifold $\mathcal{M}$ has a domain

$\mathcal{U}$, and there is a bijection $h : \mathcal{U} \to \mathbb{R}^n$, which assigns to every point $p \in \mathcal{U}$ its coordinates $x^\mu(p)$. Given two such charts, $(\mathcal{U}, h)$ with coordinates $\{x^\mu\}$ and $(\mathcal{V}, k)$ with coordinates $\{y^\mu\}$, then the transformations $y^\mu(x^1, \ldots, x^n)$ defined on $\mathcal{U} \cap \mathcal{V}$ are equal to the composed map $k \circ h^{-1}$.

Definition 10 Let $\mathcal{M}$ and $\mathcal{N}$ be two n-manifolds, $\phi : \mathcal{M} \to \mathcal{N}$ a bijection. ϕ is called a *diffeomorphism* (in short *diffeo*), if the following conditions are satisfied for every point $p \in \mathcal{M}$:

(1) If $(\mathcal{U}, h)$ is a chart around p and $(\mathcal{V}, k)$ is a chart around $\phi(p)$, then the functions $k \circ \phi \circ h^{-1} : \mathbb{R}^n \to \mathbb{R}^n$ are differentiable (C^∞).

(2) The matrix of the first derivatives of the functions $k \circ \phi \circ h^{-1}$ has a non-vanishing determinant.

The manifolds $\mathcal{M}$ and $\mathcal{N}$ as well as $(\mathcal{U}, h)$ and $(\mathcal{V}, k)$ are allowed to coincide. The right-hand side of (2.57) is an example of functions $k \circ \phi \circ h^{-1}$.

Let $\mathcal{M}$ and $\mathcal{N}$ be two n-manifolds, $\phi : \mathcal{M} \to \mathcal{N}$ a diffeo. Moreover let $p \in \mathcal{M}$ and $(\mathcal{U}, h)$ be charts around p in $\mathcal{M}$. Then $(\phi(\mathcal{U}), h \circ \phi^{-1})$ is a chart around $\phi(p)$ in $\mathcal{N}$. This follows immediately from the definition of diffeos, since the transformations between the charts $(\phi(\mathcal{U}), h \circ \phi^{-1})$ and $(\mathcal{V}, k)$ (generalization of (2.59)) coincide with the functions $k \circ \phi \circ h^{-1}$, and these satisfy all conditions for allowed coordinates. We call the chart $(\phi(\mathcal{U}), h \circ \phi^{-1})$ the ϕ-*image* of $(\mathcal{U}, h)$. If $\{x^\mu\}$ are the coordinates for the charts $(\mathcal{U}, h)$ and if $\{x'^\mu\}$ are the ones for $(\phi(\mathcal{U}), h \circ \phi^{-1})$, then (2.61) holds again.

Definition 11 Let $\mathcal{M}$ and $\mathcal{N}$ be two n-manifolds, $\phi : \mathcal{M} \to \mathcal{N}$ a diffeo. Let $\mathcal{O}$ be an object at the point $p \in \mathcal{M}$, which is represented by a well-defined index-carrying quantity (see Sect. 1.5.3) with respect to an arbitrary chart around p. Let $(\mathcal{U}, h)$ be a chart around p. Then the object $\phi_\star \mathcal{O}$ is defined by the following condition: the representation of $\phi_\star \mathcal{O}$ with respect to $(\phi(\mathcal{U}), h \circ \phi^{-1})$ coincides with the representation of $\mathcal{O}$ with respect to $(\mathcal{U}, h)$.

The condition that the object possesses well-defined components with respect to arbitrary coordinates is non-trivial. Although all objects that we have encountered so far (tensors and affine connections) satisfy this condition, there are objects that do not (namely spinors). Finally we come to the definition of a symmetry.

Definition 12 Let $\mathcal{M}$ be an n-manifold and $\phi : \mathcal{M} \to \mathcal{M}$ a diffeo. Furthermore let $\mathcal{O}(x)$ be a field of the object $\mathcal{O}$ on $\mathcal{M}$. We say that the field $\mathcal{O}(x)$ is invariant with respect to ϕ if

$$(\phi_\star \mathcal{O})(x) = \mathcal{O}(x) . \tag{2.64}$$

2.6.3 Lie Derivative

Let us study infinitesimal diffeos in more detail. Such a transformation can be denoted by $d\phi$. Let us choose a point $p \in \mathcal{M}$ and let $q = d\phi(p)$. We choose furthermore a fixed chart (U,h) around p; let $\{x^\mu\}$ be the corresponding coordinates, and let the functions $h \circ d\phi \circ h^{-1}$ be of the form

$$x^\mu(q) = x^\mu(p) + \xi^\mu(p)d\varepsilon . \tag{2.65}$$

The term $\xi^\mu(p)d\varepsilon$ represents the differences of the coordinates corresponding to the (arbitrary) point p and its $d\phi$-image; since $d\phi$ is infinitesimal, these differences form a vector at each point, and therefore $\xi^\mu(x)$ is a vector field on $\mathcal{M}$. Conversely we can relate to every smooth vector field on $\mathcal{M}$ an infinitesimal diffeomorphism according to (2.65). The inverse map (2.65) is

$$x^\mu(p) = x^\mu(q) - \xi^\mu(q)d\varepsilon . \tag{2.66}$$

In (2.65) and (2.66) it is not important, in which point, p or q, the vector field ξ^μ is taken, since the difference is again of the order of $d\varepsilon$; we calculate everything up to linear contributions in $d\varepsilon$. We denote the $d\phi$-image of $\{x^\mu\}$ by $\{x'^\mu\}$. From the definition, it follows that

$$x'^\mu(q) = x^\mu(p) .$$

In order to determine the transformation between the coordinate systems, we must compare the two coordinate systems in a given point. The relation (2.66) yields that the transformation rule and its inverse admit the form

$$x'^\mu = x^\mu - \xi^\mu(x)d\varepsilon, \qquad x^\mu = x'^\mu + \xi^\mu(x')d\varepsilon . \tag{2.67}$$

Taking derivatives of (2.67), we obtain

$$\frac{\partial x'^\mu}{\partial x^\nu}(p) = \delta_\nu^\mu - \xi_{,\nu}^\mu(p)d\varepsilon, \quad \frac{\partial x^\mu}{\partial x'^\nu}(p) = \delta_\nu^\mu + \xi_{,\nu}^\mu(p)d\varepsilon, \tag{2.68}$$

$$\frac{\partial^2 x'^\rho}{\partial x^\mu \partial x^\nu}(p) = -\xi_{,\mu\nu}^\rho(p)d\varepsilon, \tag{2.69}$$

where it is again unimportant whether we take the primed or unprimed components of the derivatives $\xi_{,\nu}^\mu$ and $\xi_{,\mu\nu}^\mu$, and whether we take their values at the point p or q.

By means of (2.65), (2.66), (2.68), and (2.69), we can calculate the $d\phi$-image of various fields. Let, for instance, $\chi(x)$ be a scalar field on $\mathcal{M}$ and $\psi(x)$ its $d\phi$-image. From the definition, it follows that

$$\psi(q) = \chi(p) .$$

We want to compare the fields at a given point. Let us therefore determine the value of χ at the point q using (2.66):

$$\psi(q) = \chi(q) - \chi_{,\mu}(q)\xi^\mu(q)d\varepsilon . \tag{2.70}$$

Consider now a vector field $U^\mu(x)$; its $d\phi$-image $V^\mu(x)$ satisfies by definition the equation

$$V'^\mu(q) = U^\mu(p),$$

where this time we have to write the components of the two fields in different coordinates. Let us transform everything in the coordinate system $\{x^\mu\}$ and at the point q.

$$V^\mu(q) = X^\mu_{\rho'}V'^\rho(q) = X^\mu_{\rho'}U^\rho(p)$$
$$= (\delta^\mu_\rho + \xi^\mu_{,\rho}(q)d\varepsilon)(U^\rho(q) - U^\rho_{,\sigma}(q)\xi^\sigma(q)d\varepsilon)$$
$$= U^\mu(q) - (U^\mu_{,\rho}(q)\xi^\rho(q) - U^\rho(q)\xi^\mu_{,\rho}(q))d\varepsilon.$$

The arguments do not have to be written out explicitly, since they coincide:

$$V^\mu = U^\mu - (U^\mu_{,\rho}\xi^\rho - U^\rho\xi^\mu_{,\rho})d\varepsilon. \tag{2.71}$$

Since affine connections have well-defined components with respect to coordinate systems, we can undertake an analogue calculation for those too. Let $\Gamma^\mu_{\rho\sigma}(x)$ be an affine connection and $\Delta^\mu_{\rho\sigma}(x)$ its $d\phi_\star$-image. By definition the following must hold:

$$\Delta'^\mu_{\rho\sigma}(q) = \Gamma^\mu_{\rho\sigma}(p).$$

In this case, the transformation to a particular point and coordinate system starts as follows:

$$\Delta^\mu_{\rho\sigma}(q) = X^\mu_{\alpha'}X^{\beta'}_\rho X^\gamma_\sigma \Delta'^\alpha_{\beta\gamma}(q) + X^\mu_{\alpha'}X^{\alpha'}_{\rho\sigma}$$
$$= X^\mu_{\alpha'}X^{\beta'}_\rho X^\gamma_\sigma \Gamma^\alpha_{\beta\gamma}(p) + X^\mu_{\alpha'}X^{\alpha'}_{\rho\sigma}.$$

We continue analogously to the previous procedure and obtain

$$\Delta^\mu_{\rho\sigma} = \Gamma^\mu_{\rho\sigma} - \left(\Gamma^\mu_{\rho\sigma,\tau}\xi^\tau - \Gamma^\tau_{\rho\sigma}\xi^\mu_{,\tau} + \Gamma^\mu_{\tau\sigma}\xi^\tau_{,\rho} + \Gamma^\mu_{\rho\tau}\xi^\tau_{,\sigma} + \xi^\mu_{,\rho\sigma}\right)d\varepsilon. \tag{2.72}$$

We still need an equation for the covariant tensor fields of second rank (type (0,2)). It is given by the following:

$$T_{\mu\nu} = S_{\mu\nu} - \left(S_{\mu\nu,\rho}\xi^\rho + S_{\rho\nu}\xi^\rho_{,\mu} + S_{\mu\rho}\xi^\rho_{,\nu}\right)d\varepsilon. \tag{2.73}$$

The difference between a field Φ and its $\phi_\star$-image Ψ defines the so-called *Lie derivative*. More precisely, the Lie derivative $\mathscr{L}_\xi\Phi$ of the field Φ with respect to the vector field ξ is defined by

$$\Phi(x) - \Psi(x) = (\mathscr{L}_\xi\Phi)(x)d\varepsilon. \tag{2.74}$$

Equations (2.70) (2.71), (2.72), and (2.73) lead for instance to the following:

$$\mathscr{L}_\xi\chi = \chi_{,\rho}\xi^\rho, \tag{2.75}$$

$$\mathscr{L}_\xi U^\mu = U^\mu_{,\rho}\xi^\rho - U^\rho\xi^\mu_{,\rho} \tag{2.76}$$

$$\mathscr{L}_\xi \Gamma^\mu_{\rho\sigma} = \Gamma^\mu_{\rho\sigma,\tau}\xi^\tau - \Gamma^\tau_{\rho\sigma}\xi^\mu_{,\tau} + \Gamma^\mu_{\tau\sigma}\xi^\tau_{,\rho} + \Gamma^\mu_{\rho\tau}\xi^\tau_{,\sigma} + \xi^\mu_{,\rho\sigma}, \qquad (2.77)$$

$$\mathscr{L}_\xi S_{\mu\nu} = S_{\mu\nu,\rho}\xi^\rho + S_{\rho\nu}\xi^\rho_{,\mu} + S_{\mu\rho}\xi^\rho_{,\nu}. \qquad (2.78)$$

From the definition of the Lie derivative, it follows that the Lie derivative of a tensor field of type (p,q) is again a tensor field of type (p,q) because its value at a point is the difference of two tensors at this point.

2.6.4 Killing Vector Field

In the section about the static space–times, we have learned about a special case of the metric, which was independent of one of the coordinates. We were able to derive two useful conclusions from this:

1. a first integral of the geodesic equations,
2. shifting this coordinate by a constant value, the metric properties are conserved (the shift of a geodesic is again a geodesic).

We would like to formulate the idea of symmetry in a covariant way. Let $(\mathscr{M},g)$ be a space–time; let us assume that there are coordinates $\{\bar{x}^\mu\}$ such that the components $\bar{g}_{\mu\nu}$ of the metric do not depend on one of the coordinates, $\bar{x}^\kappa$:

$$\bar{\partial}_\kappa \bar{g}_{\rho\sigma}(\bar{x}) = 0 \quad \forall \rho, \sigma, \bar{x}, \qquad (2.79)$$

where κ is fixed. The metric is then symmetric with respect to the one-dimensional group of transformations

$$\bar{x}^\mu \longrightarrow \bar{x}^\mu, \quad \mu \neq \kappa$$
$$\bar{x}^\kappa \longrightarrow \bar{x}^\kappa + t.$$

Each transformation of such a group is called *isometry*. (It is not difficult to show that this is a symmetry group in the sense of the previous part of this section.)

Let us define a vector field ξ^μ by the equation

$$\bar{\xi}^\mu = \delta^\mu_\kappa \, ;$$

i.e., the components of ξ with respect to $\{\bar{x}^\mu\}$ are given by the above equation: this defines a vector at every point. Then we have

$$\bar{\partial}_\kappa \bar{g}_{\rho\sigma}(\bar{x}) = \bar{\xi}^\mu \bar{\partial}_\mu \bar{g}_{\rho\sigma}(\bar{x}).$$

The corresponding transformation is generated by the vector field $\bar{\xi}^\mu$. The coordinates $\{\bar{x}^\mu\}$ are said to be *adapted to the symmetry*.

We now want to express the property (2.79) of the vector field ξ in general coordinates $\{x^\mu\}$. We obtain the following:

$$\xi^{\nu} = \delta^{\mu}_{\kappa} X^{\nu}_{\bar{\mu}} = X^{\nu}_{\bar{\kappa}},$$

and

$$\partial_{\bar{\kappa}} \bar{g}_{\rho\sigma} = \partial_{\bar{\kappa}} \left(X^{\alpha}_{\bar{\rho}} X^{\beta}_{\bar{\sigma}} g_{\alpha\beta} \right) = X^{\alpha}_{\bar{\rho}} X^{\beta}_{\bar{\sigma}} X^{\nu}_{\bar{\kappa}} g_{\alpha\beta,\nu} + X^{\alpha}_{\bar{\rho}\bar{\kappa}} X^{\beta}_{\bar{\sigma}} g_{\alpha\beta} + X^{\alpha}_{\bar{\rho}} X^{\beta}_{\bar{\sigma}\bar{\kappa}} g_{\alpha\beta}$$

$$= X^{\alpha}_{\bar{\rho}} X^{\beta}_{\bar{\sigma}} X^{\nu}_{\bar{\kappa}} g_{\alpha\beta,\nu} + X^{\gamma}_{\bar{\rho}\bar{\kappa}} X^{\beta}_{\bar{\sigma}} g_{\gamma\beta} + X^{\alpha}_{\bar{\rho}} X^{\gamma}_{\bar{\sigma}\bar{\kappa}} g_{\alpha\gamma}$$

$$= X^{\alpha}_{\bar{\rho}} X^{\beta}_{\bar{\sigma}} \left(X^{\nu}_{\bar{\kappa}} g_{\alpha\beta,\nu} + X^{\bar{\lambda}}_{\alpha} X^{\gamma}_{\bar{\kappa}\bar{\lambda}} g_{\gamma\beta} + X^{\bar{\lambda}}_{\beta} X^{\gamma}_{\bar{\kappa}\bar{\lambda}} g_{\alpha\gamma} \right)$$

$$= X^{\alpha}_{\bar{\rho}} X^{\beta}_{\bar{\sigma}} \left(g_{\alpha\beta,\nu} \xi^{\nu} + g_{\gamma\beta} \xi^{\gamma}_{,\alpha} + g_{\alpha\gamma} \xi^{\gamma}_{,\beta} \right).$$

From the above, it follows that (2.79) is equivalent to

$$g_{\alpha\beta,\nu} \xi^{\nu} + g_{\nu\beta} \xi^{\nu}_{,\alpha} + g_{\alpha\nu} \xi^{\nu}_{,\beta} = 0. \tag{2.80}$$

This equation is called *Killing equation*, and the vector field ξ^{μ} is called the *Killing vector field*. (Comparing with (2.78), we see that (2.80) is equivalent to $\mathscr{L}_{\xi} g = 0$.)

Isometries imply conservation laws of the dynamics. The importance of the Killing vector fields lies for us in their property to provide integrals of the geodesic equations.

Theorem 6 *Let $\xi^{\mu}(x)$ be a Killing vector field. Then the quantity*

$$P_{\xi} := g_{\mu\nu} \xi^{\mu} \dot{x}^{\nu}$$

is conserved along an autoparallel.

Proof Let us choose an autoparallel and describe it by the functions $x^{\mu}(\lambda)$. These functions must obey the Euler–Lagrange equation

$$\frac{d}{d\lambda} \left(g_{\rho\nu} \dot{x}^{\nu} \right) = \frac{1}{2} g_{\mu\nu,\rho} \dot{x}^{\mu} \dot{x}^{\nu},$$

which corresponds to the Lagrangian $\frac{1}{2} g_{\mu\nu} \dot{x}^{\mu} \dot{x}^{\nu}$. Let us calculate the λ derivative of P_{ξ}:

$$\frac{d}{d\lambda} \left(g_{\mu\nu} \xi^{\mu} \dot{x}^{\nu} \right) = g_{\mu\nu} \xi^{\mu}_{,\rho} \dot{x}^{\nu} \dot{x}^{\rho} + \frac{d}{d\lambda} \left(g_{\mu\nu} \dot{x}^{\nu} \right) \xi^{\mu}$$

$$= \frac{1}{2} \left(g_{\rho\nu} \xi^{\rho}_{,\mu} + g_{\mu\rho} \xi^{\rho}_{,\nu} \right) \dot{x}^{\mu} \dot{x}^{\nu} + \frac{d}{d\lambda} \left(g_{\rho\nu} \dot{x}^{\nu} \right) \xi^{\rho}.$$

If we insert for the last term from the Euler–Lagrange equation, we obtain

$$\frac{d}{d\lambda} \left(g_{\mu\nu} \xi^{\mu} \dot{x}^{\nu} \right) = \frac{1}{2} \left(g_{\rho\nu} \xi^{\rho}_{,\mu} + g_{\mu\rho} \xi^{\rho}_{,\nu} + g_{\mu\nu,\rho} \xi^{\rho} \right) \dot{x}^{\mu} \dot{x}^{\nu}.$$

The expression in brackets vanishes if ξ^{μ} is a Killing field and therefore the derivative vanishes, qed.

For static space–times in adapted coordinates, it holds that

$$g_{\mu\nu}\xi^\mu \dot{x}^\nu = g_{0\nu}\dot{x}^\nu.$$

We used the fact that this quantity is constant in order to calculate the redshift.

2.7 Rotationally Symmetric Space–Times

We can intuitively understand how the metric of a rotationally symmetric space–time looks like if we study the metric on two-dimensional rotational surfaces.

2.7.1 Rotation Surfaces

Let $\mathbb{E}^3$ be the three-dimensional Euclidean space and x, y, and z the orthogonal coordinates in $\mathbb{E}^3$. Let the *profile curve* of a rotation surface $\mathscr{F}$ with axis $x = 0$, $y = 0$ be given by the equation $z = z(r)$, where r is the distance to the axis. Furthermore, let ϑ be the angle of the rotation with value zero at the xz-plane. We choose the functions r and ϑ as coordinates on $\mathscr{F}$. The coordinates x, y, and z of a point r, ϑ at the surface are given by:

$$x = r\cos\vartheta, \qquad y = r\sin\vartheta, \qquad z = z(r). \tag{2.81}$$

The metric, which is induced on $\mathscr{F}$ by $\mathbb{E}^3$, can be determined by calculating the differentials dx, dy, and dz of the $\mathbb{E}^3$ metric from relation (2.81):

$$
\begin{aligned}
ds^2 &= dx^2 + dy^2 + dz^2 \\
&= (\cos\vartheta \, dr - r\sin\vartheta \, d\vartheta)^2 + (\sin\vartheta \, dr + r\cos\vartheta \, d\vartheta)^2 + (z' \, dr)^2 \\
&= \left[1 + z'^2(r)\right] dr^2 + r^2 d\vartheta^2.
\end{aligned}
$$

The metric therefore has the following form

$$ds^2 = A(r)dr^2 + r^2 d\vartheta^2, \tag{2.82}$$

where $A(r) = 1 + z'^2(r)$.

The rotations around the z-axis form a group, which transforms every point of the plane $\mathscr{F}$ into another point in $\mathscr{F}$. If we let the whole group act on a fixed point p in $\mathscr{F}$, we obtain a set of points in the plane $\mathscr{F}$, of which each one is the image of p with respect to a group element. We denote this set by $\mathscr{O}(p)$ and call it the *orbit* of the group through p. In our case, the orbits are circles $r = $ const. The function r is constant along the orbits. It is hence an invariant of the group. The function ϑ plays the role of a coordinate along the orbits.

The meaning of the elements of the metric in (2.82) can be described as follows. The element $r^2 d\vartheta^2$ measures the quantity $L(p)$, the length of the orbit. Indeed,

$$L(p) = \int d\vartheta \sqrt{g_{\vartheta\vartheta}}.$$

The element $A(r)dr^2$ measures the distance between the orbit with the value r and the one with the value $r + dr$. We observe that there are no elements of the form $g_{r\vartheta} dr\, d\vartheta$. This means geometrically: ϑ could be chosen along different orbits such that the curves $\vartheta = \text{const}$ are orthogonal to the orbits.

What is the geometric meaning of the invariant r? In the embedding space $\mathbb{E}^3$, $r(p)$ is for instance the distance of the point p to the z-axis. In order to measure this distance, we must leave the plane $\mathscr{F}$. Does r also have a meaning inside the plane $\mathscr{F}$? Here it does not in general describe a distance from the center: $\mathscr{F}$ does not even have to possess a center. But the length $L(p)$ of the orbit $\mathscr{O}(p)$ is $2\pi r(p)$. $L(p)$ can in principle be measured without leaving the plane $\mathscr{F}$. This length provides then the geometric meaning of r.

2.7.2 Space–Times

A simple example of a rotationally symmetric space–time is the Minkowski space–time. If we introduce the spherical coordinates t, r, ϑ, φ, we obtain the Minkowski metric written in the following form:

$$ds^2 = dt^2 - dr^2 - r^2 \left(d\vartheta^2 + \sin^2 \vartheta\, d\varphi^2\right).$$

The orbits are now 2-surfaces $r = \text{const}, t = \text{const}$, the functions r and t are invariants of the rotation group. The term $r^2(d\vartheta^2 + \sin^2 \vartheta\, dr^2)$ in the metric is the metric within the orbit and $dt^2 - dr^2$ measures the distances between the orbits (t, r) and $(t + dt, r + dr)$. In addition we were able to choose the coordinates ϑ and φ along the different orbits such that the surfaces $\vartheta = \text{const}, \varphi = \text{const}$ are orthogonal to the orbits. This also implies that the components $g_{t\vartheta}, g_{t\varphi}, g_{r\vartheta}$, and $g_{r\varphi}$ vanish. The function r is again related to the size of the orbits: the area of an orbit at point p is $4\pi r^2$.

Now we are intuitively prepared to make the following definition:

Definition 13 The space–time $(\mathscr{M}, g)$ is rotationally symmetric if there exist coordinates t, r, ϑ, and φ, in which the metric has the following form:

$$ds^2 = g_{00}dt^2 + 2g_{01}dt\, dr + g_{11}dr^2 - r^2 \left(d\vartheta^2 + \sin^2 \vartheta\, d\varphi^2\right), \tag{2.83}$$

where the components g_{00}, g_{01}, and g_{11} are independent of ϑ and φ.

The 2-surface $t = \text{const}, r = \text{const}$ are orbits of the rotation group. The metric decomposes in two parts: one measures the distances along the orbits and the other the distances between the orbits, and the orbital coordinates can be chosen in such a way that there are no mixing terms. The function r is related to the area $F(r)$ of the orbit at the point r: $F(r) = 4\pi r^2(p)$. Moreover, a space–time does not necessarily possess a center (e.g., black holes).

The metric (2.83) does not change its form if the coordinates ϑ and φ are "rotated". Such a rotation is given by the following transformation. Let us consider an imaginary unit sphere around the origin of the $\mathbb{E}^3$ space. The Cartesian axes are x, y, and z, and the corresponding spherical coordinates are denoted by ϑ and φ (i.e., ϑ is the angle between the position vector and the z-axes, etc.). The metric on the sphere's surface is given by

$$ds^2 = d\vartheta^2 + \sin^2 \vartheta \, d\varphi^2. \tag{2.84}$$

Let x', y', and z' be new, arbitrary Cartesian axes; these can be obtained from the old ones by a rotation $\mathbf{O}$ ($\mathbf{O}$ is an orthogonal matrix). Let, furthermore, ϑ' and φ' be the corresponding new spherical coordinates. The rotation leads to a coordinate transformation

$$\vartheta' = \vartheta'_{\mathbf{O}}(\vartheta, \varphi), \quad \varphi' = \varphi'_{\mathbf{O}}(\vartheta, \varphi), \tag{2.85}$$

which is complicated, but well defined by the rotation. We can now use the transformation (2.85) in the curved space–time (2.83) together with $t' = t$ and $r' = r$! It is obvious that the metric has again, in the new spherical coordinates, the form (2.84), and that this is also the case if we insert the transformation rules (2.85) in (2.84). The infinitesimal generators of the rotation around the three coordinate axes are the three vector fields ξ_x^μ, ξ_y^μ, and ξ_z^μ on $\mathbb{E}^3$, which satisfy, for instance, $r \mapsto r$, $\vartheta \mapsto \vartheta + \xi_x^\vartheta d\lambda$, and $\varphi \mapsto \varphi + \xi_x^\varphi d\lambda$.

By inserting in the Killing equation, we easily verify that the three vectors

$$\xi_x = (0, 0, -\sin\varphi, -\cot\vartheta \cos\varphi), \tag{2.86}$$
$$\xi_y = (0, 0, \cos\varphi, -\cot\vartheta \sin\varphi), \tag{2.87}$$
$$\xi_z = (0, 0, 0, 1) \tag{2.88}$$

are Killing vectors of the metric (2.83) (exercise). Note in particular that the Killing fields are independent of the choice of the components g_{00}, g_{01}, and g_{11} in the space–times (2.82); they have the same form for flat and curved space–times. It is possible to show that there do not exist coordinates that are adapted to all above three Killing fields simultaneously.

The previous consideration helps us to show an important property common to all geodesics (time-, light- and space-like) in a space–time with the metric (2.82):

Theorem 7 *For every geodesic $t(\lambda)$, $r(\lambda)$, $\vartheta(\lambda)$, and $\varphi(\lambda)$ in a rotationally symmetric space–time, there exists a rotation (2.85), such that the geodesic satisfies*

$$\vartheta'(\lambda) = \pi/2 \quad \forall \lambda. \tag{2.89}$$

Proof The three Killing fields (2.86), (2.87), and (2.88) lead to the first integrals

$$j_x = -g_{\mu\nu}\xi_x^\mu \dot{x}^\nu, \quad j_y = -g_{\mu\nu}\xi_y^\mu \dot{x}^\nu, \quad j_z = -g_{\mu\nu}\xi_z^\mu \dot{x}^\nu.$$

Substituting for the metric yields

$$j_x = -r^2(\lambda)\sin\varphi(\lambda)\,\dot{\vartheta}(\lambda) - r^2(\lambda)\sin\vartheta(\lambda)\cos\vartheta(\lambda)\cos\varphi(\lambda)\,\dot{\varphi}(\lambda),$$

$$j_y = r^2(\lambda)\cos\varphi(\lambda)\,\dot{\vartheta}(\lambda) - r^2(\lambda)\sin\vartheta(\lambda)\cos\vartheta(\lambda)\sin\varphi(\lambda)\,\dot{\varphi}(\lambda),$$

$$j_z = r^2(\lambda)\sin^2\vartheta(\lambda)\,\dot{\varphi}(\lambda),$$

where j_x, j_y, and j_z is a constant three-number. We have

$$j_x \sin\vartheta(\lambda)\cos\varphi(\lambda) + j_y \sin\vartheta(\lambda)\sin\varphi(\lambda) + j_z \cos\vartheta(\lambda) = 0,$$

as substitution shows. This can be interpreted as follows: the functions $\vartheta(\lambda)$ and $\varphi(\lambda)$ describe a curve on an auxiliary surface of a sphere of radius 1 and center $(0,0,0)$ in $\mathbb{E}^3$. The position vector

$$(\sin\vartheta(\lambda)\cos\varphi(\lambda), \sin\vartheta(\lambda)\sin\varphi(\lambda), \cos\vartheta(\lambda))$$

of each point on the curve must, because of the above equation, lie in the surface, which stands orthogonal to the vector (j_x, j_y, j_z) and passes through the origin. The motion takes place on a circle. We can now rotate the Cartesian axes in such a way that the z'-axis is parallel to the vector (j_x, j_y, j_z); then the new coordinate ϑ' along the circle has the value (2.89), qed.

Let, hence, ϑ and φ be the new coordinates, which are found in this way for an autoparallel. Then we obtain for the integrals $j_x = j_y = 0$ and $j_z = r^2(\lambda)\dot{\varphi}(\lambda)$. Our first three integrals are thus transformed to

$$\vartheta(\lambda) = \pi/2, \quad r^2(\lambda)\dot{\varphi}(\lambda) = j; \qquad (2.90)$$

there is now a single constant parameter left, j.

2.7.3 Geodesic Equation in the Static Case

We would now like to show how the integration of the geodesic equation in the case of a *static* and *rotationally symmetric* space–time reduces, because of the existence of Killing fields, to quadratures.

The metric of a static as well as rotationally symmetric space–time can be transformed, according to what was already said about these two cases, to the following form:

$$ds^2 = B(r)dt^2 - A(r)dr^2 - r^2\left(d\vartheta^2 + \sin^2\vartheta\,d\varphi^2\right). \qquad (2.91)$$

The metric admits four Killing fields: the three given in (2.86) (2.87), and (2.88), and

$$\xi_t = (1,0,0,0).$$

We call the fourth integral of motion e:

$$e = g_{\mu\nu}\xi_t^\mu \dot{x}^\nu.$$

Substitution for the metric yields

$$e = B\dot{t}. \tag{2.92}$$

There is another conserved quantity, (2.21), or

$$B\dot{t}^2 - A\dot{r}^2 - r^2(\dot{\vartheta}^2 + \sin^2\vartheta\,\dot{\varphi}^2) = \mu^2. \tag{2.93}$$

Let us now consider a fixed geodesic and let us choose spherical coordinates such that (2.90) holds. Then we get from (2.93)

$$B\dot{t}^2 - A\dot{r}^2 - r^2\dot{\varphi}^2 = \mu^2. \tag{2.94}$$

The three equations (2.90), (2.92), and (2.94) can be solved for $\dot{t}$, $\dot{r}$, and $\dot{\varphi}$ with the result

$$\dot{r} = \pm\sqrt{\frac{e^2}{A(r)B(r)} - \frac{j^2}{r^2 A(r)} - \frac{\mu^2}{A(r)}}, \tag{2.95}$$

$$\dot{t} = \frac{e}{B(r)}, \tag{2.96}$$

$$\dot{\varphi} = \frac{j}{r^2}. \tag{2.97}$$

These are the promised quadratures.

2.8 Asymptotically Flat Space–Times

In this section we study the gravitational field of a static, rotationally symmetric, *isolated* object. Isolated means that the object is alone in the universe and that its field decreases if we move away from it. In the Cartan–Friedrich theory, we have

$$R^\mu_{\nu\rho\sigma} \sim \frac{\partial^2\Phi}{\partial r^2} \sim \frac{1}{r^3} \to 0.$$

In general relativity, we want to demand, similarly, that the space–time at large distances from the center of the source be practically flat. Let us now formulate this condition more precisely. We will see that the metric on surfaces of normal bodies is already very similar to a flat metric. The reason is that such bodies are in a certain sense strongly diluted objects.

2.8.1 Eddington–Robertson Expansion

First we should determine what is meant by "large distances". Is this a kilometer, a light year, or a million light years? This will depend on the size of the source. Each source should be associated with a length, which takes into account its strength. Such a length we know already: it is the gravitational radius R_G. Let us therefore try to define the distance of a point with radial coordinate r from the source as large if $r \gg R_G$. The region where $r \gg R_G$ is called the *asymptotic region*. In the asymptotic region, the relation (R_G/r) is consequently a small parameter.

Where is the asymptotic region for concrete systems? The Newton constant has in our units ($c = 1$) the dimension meter/kilogram and the value $\approx 10^{-27}$. We then have for instance:

$$\text{Sun} \quad R_G = 1.477 \times 10^3 \, \text{m} \quad R = 6.960 \times 10^8 \, \text{m}$$
$$\text{Earth} \quad R_G = 4.437 \times 10^{-3} \, \text{m} \quad R = 6.371 \times 10^6 \, \text{m}.$$

All points above the surface of these two celestial bodies lie already deep inside the asymptotic region.

We assume that the functions $A(r)$ and $B(r)$ in the metric (2.91) are analytical functions of (R_G/r) in the asymptotic region. Hence, we can expand them in a series:

$$A(r) = a_0 + a_1(R_G/r) + a_2(R_G/r)^2 + \ldots$$
$$B(r) = b_0 + b_1(R_G/r) + b_2(R_G/r)^2 + \ldots.$$

We demand that the metric in the limit $r \to \infty$ becomes the Minkowski metric. This condition determines a_0 and b_0:

$$a_0 = b_0 = 1.$$

We demand furthermore that the metric correctly reproduces Newtonian physics, at least up to first order in the small parameter R_G/r. The comparison with (2.54) shows that for this to hold, it is sufficient to have $A(r) = 1$ and $B(r) = \exp 2\Phi$, where Φ is the Newton potential. For a static rotationally symmetric source of mass M, we have however (if we reintroduced c, we would then have to write $B(r) = c^2 \exp(2\Phi/c^2)$):

$$\Phi = -\frac{GM}{r} = -(R_G/r).$$

Up to first order, this means that

$$B(r) = 1 - 2(R_G/r),$$

hence $b_1 = -2$. We will see that the terms with a_1 and b_2 are small corrections to Newton's theory and that the terms of higher order are not measurable in the solar system with current instruments. We obtain the following expansion:

$$A(r) = 1 + 2\gamma(R_G/r) + \dots \tag{2.98}$$

$$B(r) = 1 - 2\alpha(R_G/r) + 2(\beta - \alpha\gamma)(R_G/r)^2 + \dots \tag{2.99}$$

where α, β, and γ are constants. Equations (2.98) and (2.99) are called *Eddington–Robertson expansion* [7, 8]. If it should reproduce in first-order Newton's theory, then α must be equal to 1. The form, in which the constants β and γ appear, has mainly historical reasons. Their physical and geometric meaning can be read off from their position in the metric. The constant γ is related to the curvature of the space; the spatial part of the metric,

$$ds^2 = (1 + 2\gamma R_G r^{-1})dr^2 + r^2(d\vartheta^2 + \sin^2\vartheta \, d\varphi^2),$$

is flat if $\gamma = 0$; if $\gamma < 0$, then the distance between the orbits r and $r + dr$ is smaller than dr; if $\gamma > 0$, this distance is larger than dr. The constant β represents the corrections of higher order to the Newton potential. Later we will describe methods of how these constants can be measured.

2.8.2 Energy and Momentum Balance

For our purposes, we can regard the expansion (2.98) and (2.99) as a definition for asymptotically flat space–times (a much more general definition is possible, [9]). Asymptotically flat space–times form an important class of models, for which many of the usual physical notions, such as for instance the gravitational waves and their energy, energy and momentum balance, interpretation of conservation laws, and many more, make good sense.

In general the issue of energy and momentum balance in general relativity is very different from what we are used to in Newton's or special relativity physics. A difficulty arises from the fact that energy and momentum are components of a 4-vector, i.e., a tensor, and that in general tensors cannot be compared at different points. But making the balance, we have to compare them at the beginning and at the end of the investigated process, and the beginning differs from the end at least in the time; they are given by different points in space–time.

Consider a particle that starts at a point, let's say p, is influenced by a field, and arrives at another point q of the space–time. Did it gain or loose energy? In flat space–time, this question always has a meaningful answer. It is possible to compare the 0-components of the particle's 4-momenta with respect to the same inertial frame at the two points. Indeed, the difference of these two 4-vectors is again a 4-vector and has an invariant meaning. We can make the comparison also without using an inertial frame: we place a vierbein in p and transport this vierbein parallelly to q. Then we decompose the two 4-momenta with respect to these two vierbeins and compare the components. In a curved space–time, we have the problem that no unique parallel transport exists—it depends on the connecting curve between p and q. We might also argue that by the experiment itself, exactly one connecting curve between p and q is chosen and uniquely determined, namely the orbit of the particle itself. Why shouldn't we use this to transport the vierbein from p to q? In this case, we would find out that in this way, we will never achieve an increase or decrease of energy or momentum in a gravitational field—the particle's 4-momentum as well as the vierbein are transported parallelly along the orbit, so that the components of the 4-momentum remain constant with respect to the vierbein. The definition would be formally correct but trivial.

Nonetheless, there are sometimes conditions that make the energy balance meaningful and non-trivial. Assume that the curvature is concentrated in a finite (compact) region $\mathscr{G}$ of space in a static space–time and that this space–time is flat outside this region. Let us consider a particle that starts outside of $\mathscr{G}$, flies through $\mathscr{G}$, and is recaptured again outside of $\mathscr{G}$. A vierbein can now be transported parallelly from p to q along a curve that runs outside of $\mathscr{G}$. Then the vierbein in q is determined uniquely by one given in p, and the balance is therefore well defined. It is this kind of balance, which is of physical interest.

Asymptotically flat space–times can be considered as a limit of the above-mentioned special conditions. Indeed, we can for instance make the following coordinate transformations in (2.91):

$$x = r\sin\vartheta\cos\varphi, \quad y = r\sin\vartheta\sin\varphi, \quad z = r\cos\vartheta. \tag{2.100}$$

(In spite of the fact that these equations are very similar to transformations from curved spherical coordinates to "rectilinear" Cartesian coordinates, the new coordinates t, x, y, and z are curvilinear again!) We then have

$$r = \sqrt{x^2 + y^2 + z^2}, \tag{2.101}$$

such that

$$dr = \frac{x\,dx + y\,dy + z\,dz}{r},$$

and we obtain that

$$ds^2 = dt^2 - dx^2 - dy^2 - dz^2$$
$$+ (B-1)dt^2 - (A-1)\left(\frac{x\,dx + y\,dy + z\,dz}{r}\right)^2. \tag{2.102}$$

If (2.98) and (2.99) hold, all terms in the second line of (2.102) vanish in the limit $r \to \infty$, and the metric will have the form of the Minkowski metric in an inertial frame. If the particle then starts at $r = \infty$ and ends at $r = \infty$, we can draw the balance for energy and momentum by comparing the components of the 4-momentum at the beginning and at the end with respect to the coordinates t, x, y, and z.

Are the coordinates t, x, y, and z in asymptotic regions uniquely determined? First, the components of the metric do not depend on t, i.e., the source does not move relative to this coordinate, and we have an analogy to the *rest frame* of a central body in general relativity, such as we know it from special relativity. This way the boosts are excluded, and the coordinate t is determined up to translations

$t \to t + \Delta t$. Second, the function r, defined by (2.101), is constant along the orbits of the group of rotations. This excludes spatial translations. Since everything is rotationally symmetric, the source's center of mass in an analogue situation in special relativity would lie at $r = 0$, and this way we are dealing with a general-relativity analogy of a *mass-centered* reference frame. Only the spatial rotations are left. A coordinate system of this kind is called *asymptotic reference frame*.

Let us consider an autoparallel, which is given by the functions $t(\lambda), r(\lambda), \vartheta(\lambda)$, and $\varphi(\lambda)$ and satisfies

$$\lim_{\lambda \to \infty} r(\lambda) = \infty \qquad \lim_{\lambda \to -\infty} r(\lambda) = \infty. \qquad (2.103)$$

If λ is a physical parameter, then $\dot{t}(\lambda), \dot{r}(\lambda), \dot{\vartheta}(\lambda)$, and $\dot{\varphi}(\lambda)$ are the components of the particle's 4-momentum. We can define the following:

Definition 14 For a test particle that satisfies condition (2.103), the asymptotic 4-momentum p^μ_- at the beginning and p^μ_+ at the end is defined by

$$p^\mu_\pm = \lim_{\lambda \to \pm\infty} (\dot{t}(\lambda), \dot{x}(\lambda), \dot{y}(\lambda), \dot{z}(\lambda)), \qquad (2.104)$$

and the asymptotic angular momentum by

$$j^k_\pm = \lim_{\lambda \to \pm\infty} [y(\lambda)\dot{z}(\lambda) - \dot{y}(\lambda)z(\lambda),$$
$$z(\lambda)\dot{x}(\lambda) - \dot{z}(\lambda)x(\lambda), x(\lambda)\dot{y}(\lambda) - \dot{x}(\lambda)y(\lambda)], \qquad (2.105)$$

where the coordinates are understood as those of the asymptotic reference frame.

In order to simplify things, we will rotate the asymptotic reference frame so that

$$\vartheta(\lambda) = \frac{1}{2}\pi$$

and the transformation rules (2.100) become

$$x(\lambda) = r(\lambda)\cos\varphi(\lambda), \quad y(\lambda) = r(\lambda)\sin\varphi(\lambda), \quad z(\lambda) = 0. \qquad (2.106)$$

The components of the tangent vector $\dot{x}^\mu$ to the autoparallel are given with respect to the asymptotic reference frame by

$$\dot{t}(\lambda),$$
$$\dot{x}(\lambda) = \dot{r}(\lambda)\cos\varphi(\lambda) - r(\lambda)\sin\varphi(\lambda)\dot{\varphi}(\lambda), \qquad (2.107)$$
$$\dot{y}(\lambda) = \dot{r}(\lambda)\sin\varphi(\lambda) + r(\lambda)\cos\varphi(\lambda)\dot{\varphi}(\lambda), \qquad (2.108)$$
$$\dot{z}(\lambda) = 0.$$

According to Definition 14, we know that $p^0_- := \dot{t}(-\infty)$ is the particle's energy measured at the beginning in the asymptotic reference frame. Similarly, $j^z_- := x(-\infty)\dot{y}(-\infty) - y(-\infty)\dot{x}(-\infty)$ is the z-component of the particle's angular momentum with respect to the rotated asymptotic reference frame, and therefore the whole angular momentum.

We can draw the energy and momentum balance in our static and rotationally symmetric space–time without explicitly solving the autoparallel equation. From the conservation law (2.96), we obtain that

$$e = B(\infty)\dot{t}(-\infty) = \dot{t}(-\infty)$$

and

$$e = B(\infty)\dot{t}(\infty) = \dot{t}(\infty).$$

This first implies that $e = p_+^0 = p_-^0$, hence the asymptotic energy is conserved. Second, we see that the first integral e has the meaning of the energy with respect to the asymptotic reference frame (asymptotic energy). Energy conservation follows here from the symmetry of the field with respect to time translations (static space–time).

The situation is similar for the angular momentum: (2.106), (2.107), and (2.108) yield

$$x(\lambda)\dot{y}(\lambda) - y(\lambda)\dot{x}(\lambda) = r^2(\lambda)\dot{\varphi}(\lambda) = j.$$

The angular momentum is therefore conserved: $j = j_+^z = j_-^z$ (which follows from the rotation symmetry), and j is nothing but the angular momentum with respect to the asymptotic reference frame.

This way we can draw the energy and momentum balance in the asymptotically flat space–times. The Eddington–Robertson expansion enables us moreover to insert concrete functions of r for $A(r)$ and $B(r)$.

2.9 Motion of Planets

Our next task is to determine the metric in a region around the Sun by measurements and observations. The Sun can be considered in a very good approximation as static and rotationally symmetric. The field around the Sun is described by a line element (2.91). Moreover, the space–time outside the Sun is an asymptotic region and we can use (2.98) and (2.99). The field of the planets does not disturb the static and rotationally symmetric picture much. For most purposes, this disturbance can be neglected. The planets themselves can be considered as mass points. Then their motion is described by (2.95), (2.96), and (2.97).

It is useful to eliminate the parameter λ from these equations. We divide (2.96) and (2.97) by (2.95) and obtain

$$dt = \pm \frac{e}{\mu} \frac{\sqrt{A}}{B} \frac{dr}{\sqrt{e^2 \mu^{-2} B^{-1} - j^2 \mu^{-2} r^{-2} - 1}}, \tag{2.109}$$

$$d\varphi = \pm \frac{j}{\mu} \frac{\sqrt{A}}{r^2} \frac{dr}{\sqrt{e^2 \mu^{-2} B^{-1} - j^2 \mu^{-2} r^{-2} - 1}}. \tag{2.110}$$

2.9.1 Comparison with Newton's Theory

The fully relativistic equations (2.109) and (2.110) should now be applied to the special case of small velocities $v \ll 1$ and weak gravitational fields $(R_G/r) \ll 1$. In this case, Newton's theory will be a good approximation. The gravitational field must also be weak in order to assure that the gravitational acceleration does not produce velocities that are too high. From observations, it follows that v and (R_G/r) are not independent: the kinetic and potential energies of the planets are comparable:

$$\mu v^2 \approx \mu(R_G/r).$$

(This also follows from Newtonian mechanics—being known as the *virial theorem* [2], but we don't want to motivate the above relation with Newtonian mechanics since our aim is to first derive it in the following.) If we denote the small parameter by δ and if we set

$$v \approx \delta,$$

then we have

$$R_G/r \approx \delta^2.$$

Let us expand the right-hand side of (2.109) and (2.110) in powers of δ by using relations (2.98) and (2.99). First we note that e is the total relativistic energy of the planet with respect to infinity, and, in the non-relativistic approximation, the following holds

$$e \approx \mu + \varepsilon,$$

where

$$\varepsilon = \frac{1}{2}\mu v^2$$

is the kinetic energy. Furthermore we have

$$j \approx \mu r^2 \dot{\varphi} \approx \mu r v \approx \mu r \delta,$$

where the dot denotes the derivative with respect to time. The expression under the square root in (2.109) and (2.110) can be transformed as follows:

$$e^2 \mu^{-2} B^{-1} - j^2 \mu^{-2} r^{-2} - 1 \approx \left(1 + \frac{2\varepsilon}{\mu}\right)(1 + 2\alpha(R_G/r)) - j^2 \mu^{-2} r^{-2} - 1$$

$$\approx 2\varepsilon\mu^{-1} + 2\alpha(R_G/r) - j^2 \mu^{-2} r^{-2}.$$

Here all terms of order δ^2 and higher corrections were neglected. The leading terms in (2.109) and (2.110) are

$$dt \approx \frac{dr}{\sqrt{2\varepsilon\mu^{-1} + 2\alpha(R_G/r) - j^2 \mu^{-2} r^{-2}}}, \qquad (2.111)$$

$$d\varphi \approx \frac{j}{\mu\, r^2 \sqrt{2\varepsilon\mu^{-1} + 2\alpha(R_G/r) - j^2 \mu^{-2} r^{-2}}}. \qquad (2.112)$$

An important property of this relation is that the constants β and γ do not appear in it: they first appear in terms of higher order, which we neglected. In the case of β, this is easy to understand since β appeared already as a coefficient in a term of higher order in (2.99). The constant γ appears only in corrections of higher order since the velocity was considered to be small with respect to the velocity of light. The test particle then propagates in space–time much more strongly in time than in spatial directions. Since γ describes the curvature of the space, it must be suppressed relative to α. As can be easily shown (exercise), (2.111) and (2.112) (if $\alpha = 1$) coincide with the relations provided by Newton's theory.

2.9.2 Perihelion Shift

We will now treat a relativistic correction to Newton's theory. Consider a planet with orbit that is given by the functions $t(\lambda)$, $r(\lambda)$, $\vartheta(\lambda)$, and $\varphi(\lambda)$. It is worthwhile to study the radial equation (2.95). We can write it in the following form:

$$\dot{r}^2 + V_{\text{eff}} = E,$$

where

$$V_{\text{eff}}(r) = e^2 - \frac{e^2}{AB} + \frac{j^2}{Ar^2} + \frac{\mu^2}{A} - \mu^2,$$

and

$$E = e^2 - \mu^2.$$

(We choose the constant E such that $V_{\text{eff}}(\infty) = 0$.)

This equation has the form of the radial equation for a particle with mass $1/2$ and potential V_{eff} in Newton's theory. It is known that such a particle will move in one direction of height E until it hits upon the potential curve in r–V graph, then it turns around (turning point), moves backwards until it hits again upon a potential, etc. For the planetary motion, we need exactly such a segment between two turning points. Let us hence assume that there exists in the orbit a segment $[\lambda_1, \lambda_2]$ of λ such that the following conditions are satisfied:

(i) $\dot{r}(\lambda) > 0 \quad \lambda \in (\lambda_1, \lambda_2)$,
(ii) $\dot{r}(\lambda_1) = \dot{r}(\lambda_2) = 0$.

We denote by R_1 the smaller of the values $r(\lambda_1)$, $r(\lambda_2)$, and by R_2 the larger one. R_1 is called "perihelion" and R_2 "aphelion". The whole orbit of the planet then consists of copies of these segments and their time reversal. In order to see this, we would like to plot the graphs of the three functions $t(\lambda)$, $r(\lambda)$, and $\vartheta(\lambda)$ around the point λ_2. The initial data for the autoparallel at this point are given by

$$t = t_2, \, r = R_2, \, \vartheta = \tfrac{1}{2}\pi, \, \varphi = \varphi_2,$$
$$\dot{t} = \dot{t}_2, \, \dot{r} = 0, \quad \dot{\vartheta} = 0, \quad \dot{\varphi} = \dot{\varphi}_2.$$

The Lagrangian

$$L = \frac{1}{2}B(r)\dot{t}^2 - \frac{1}{2}A(r)\dot{r}^2 - \frac{1}{2}r^2\dot{\vartheta}^2 - \frac{1}{2}r^2\sin^2\vartheta\,\dot{\varphi}^2$$

is invariant with respect to any of the following three reflections:

$$\varphi - \varphi_0 \rightarrow -(\varphi - \varphi_0), \quad t - t_0 \rightarrow -(t - t_0), \quad (\lambda - \lambda_2) \rightarrow -(\lambda - \lambda_2).$$

The reflected curves represent again a solution to the autoparallel equation. Together the original and the reflected segments make up a smooth curve. The curve composed in this way therefore constitutes an extension of the original one (since it corresponds to the same initial data at point λ_2).

Next we would like to calculate the angle $\delta\varphi$ between the two successive perihelions. If $\Delta\varphi$ is the difference of the angle coordinate φ between the aphelion and the perihelion, then we have

$$\delta\varphi = 2\Delta\varphi - 2\pi.$$

The difference $\delta\varphi$ becomes positive if the perihelion rotates in the positive direction, zero if it does not rotate at all, and negative if the perihelion rotates backward. We can determine $\Delta\varphi$ from (2.110):

$$\Delta\varphi = \int_{R_1}^{R_2} \frac{j}{\mu}\sqrt{A}\,\frac{dr}{r^2\sqrt{e^2\mu^{-2}B^{-1} - j^2\mu^{-2}r^{-2} - 1}}.$$

The unknown constants e/μ and j/μ can be expressed by the values $R_{1,2}$. Equation (2.95) implies that

$$e^2\mu^{-2}B^{-1} - j^2\mu^{-2}r^{-2} - 1 = 0$$

at the points $r = R_{1,2}$. Therefore we have two linear equations for $(e/\mu)^2$ and $(j/\mu)^2$. A longer calculation, which consists of inserting for e/μ and j/μ in the integral, the expansion with respect to powers of the smaller parameter R_G/r and same quadratures, yields

$$\delta\varphi = 6\pi R_G \frac{1}{2}\left(\frac{1}{R_1} + \frac{1}{R_2}\right)\frac{1}{3}(2\alpha + 2\gamma - \beta\alpha^{-1}). \tag{2.113}$$

If the perihelion rotates in the positive direction we must have $\beta < 2\alpha^2 + 2\alpha\gamma$, if it rotates in the negative direction then $\beta > 2\alpha^2 + 2\alpha\gamma$, and if the orbit closes after one rotation then $\beta = 2\alpha^2 + 2\alpha\gamma$.

Is the perihelion shift measurable at all? The big advantage of the perihelion shift is that it is cumulative. Even though the angle difference between two successive perihelions is very small, it becomes a large angle after a large number of rotations. For Mercury we have for instance

$$\frac{1}{2}\left(\frac{R_G}{R_1}+\frac{R_G}{R_2}\right)=1.8\times 10^{-11},$$

and it makes 415 rotations per year. This leads to

$$415\delta\varphi=42.98''\,\frac{1}{3}\left(2\alpha+2\gamma-\beta\alpha^{-1}\right). \tag{2.114}$$

This is the theoretical value for Mercury. The relevant results from optical observations of Mercury were obtained since 1667. After subtraction of the disturbing influences of other planets, the residual shift is determined today from these data with the following precision:

$$\frac{415\delta\varphi}{42.98''}=1.0034\pm 0.0033.$$

In addition, there exist since 1966 radar measurements, which lead to the following values

$$\frac{415\delta\varphi}{42.98''}=1.000\pm 0.002. \tag{2.115}$$

This means that our theoretical models of the Sun's field (which are distinguished from each other by the parameters α, β, and γ) are only in accordance with observation if $\alpha\approx\gamma\approx 1$ leads to $\beta\approx 1$. For more discussions, see [6, 10, 11].

The system Sun–Mercury is not the only one where relativistic corrections to Newton's theory can be observed. Since 1974 the so-called *double pulsar* PSR 1913+16 has been studied intensively. It is a double-star system. Each of the two neutron stars has approximately the mass of the Sun and a radius of 10 km. They turn around each other with a period of 8 h. The fields are hence a lot stronger (but still "asymptotic"), and the velocities are a lot larger (but still non-relativistic). Other systems of this kind were discovered in the year 1990 (PSR 2127+11C and PSR 1534+12). Several effects can be observed, not only "$\alpha-\beta-\gamma$", for instance gravitational waves (indirect, by decreasing the period). For details see [9].

2.10 Light Signals in the Solar System

Let us consider a light signal that passes the Sun. Its orbit is assumed to be described by the functions $t(\lambda),r(\lambda),\vartheta(\lambda)$, and $\varphi(\lambda)$. These functions then should satisfy (2.95), (2.96), and (2.97) with $\mu=0$. For the autoparallel, the relations (2.103) should again hold. We further assume that the photon approaches the source for $\lambda\in(-\infty,\lambda_0)$,

$$\dot{r}<0,$$

then reaches a minimal value R of the coordinate r at $\lambda=\lambda_0$, and then moves away to $r=\infty$ for $\lambda\in(\lambda_0,\infty)$,

$$\dot{r}>0.$$

Again the orbit is symmetric with respect to reflections in the auxiliary space (t, r, φ) around the point where $r = R$. Hence, in the relation (2.95), first the lower sign is the valid one and later the upper one. For R we obtain from (2.95) the following equation:

$$\frac{e^2}{j^2} = \frac{B(R)}{R^2}, \tag{2.116}$$

since $\dot{r}(\lambda_0) = 0$. Equations (2.95), (2.96) (2.97), and (2.116) yield

$$dt = \pm \sqrt{\frac{A(r)B(R)}{B(r)}} \frac{r\, dr}{\sqrt{B(R)r^2 - R^2 B(r)}}, \tag{2.117}$$

$$d\varphi = \pm \frac{R\sqrt{A(r)B(r)}}{r} \frac{dr}{\sqrt{B(R)r^2 - R^2 B(r)}}. \tag{2.118}$$

In this way, we were able to express the unknown constant e/j by the known *impact parameter* R, and the parameter λ could be excluded.

2.10.1 Deflection of Light

First we want to calculate how the gravitational field of the source deflects the photons, i.e., how the direction of the signal at the beginning differs from that at the end. This is in principle the balance of the linear momentum, since the direction of the 3-momentum coincides with the direction of the signal.

Let us study the pair of functions $(\dot{x}(\lambda), \dot{y}(\lambda))$. Equations (2.107) and (2.108) yield

$$\begin{pmatrix} \dot{x}(\lambda) \\ \dot{y}(\lambda) \end{pmatrix} = \dot{r}(\lambda) \begin{pmatrix} \cos\varphi(\lambda) - r(\lambda)\frac{d\varphi}{dr}\sin\varphi(\lambda) \\ \sin\varphi(\lambda) + r(\lambda)\frac{d\varphi}{dr}\cos\varphi(\lambda) \end{pmatrix}.$$

The deflection is obtained as the angle between the vectors $(\dot{x}(-\infty), \dot{y}(-\infty))$ and $(\dot{x}(\infty), \dot{y}(\infty))$. Because of (2.118), we have

$$r(\pm\infty)\frac{d\varphi}{dr}(\pm\infty) = 0.$$

Hence

$$(\dot{x}(\pm\infty), \dot{y}(\pm\infty)) = \dot{r}(\pm\infty)(\cos\varphi(\pm\infty), \sin\varphi(\pm\infty)).$$

Since $\dot{r}(-\infty) < 0$ and $\dot{r}(+\infty) > 0$ and $|\dot{r}(-\infty)| = |\dot{r}(\infty)|$ because of symmetry under reflections, we can write

$$(\dot{x}(-\infty), \dot{y}(-\infty)) = |\dot{r}(-\infty)| [\cos(\varphi(-\infty) + \pi), \sin(\varphi(-\infty) + \pi)]$$

$$(\dot{x}(\infty), \dot{y}(\infty)) = |\dot{r}(\infty)|(\cos\varphi(\infty), \sin\varphi(\infty)).$$

The directions of these two vectors at very distant points can be compared despite the curvature of space–time, since their components are given with respect to asymptotic reference frame. The light deviation $\delta\varphi$ is hence given by

$$\delta\varphi = \varphi(\infty) - \varphi(-\infty) - \pi. \tag{2.119}$$

The difference $\varphi(\infty) - \varphi(-\infty)$ is calculated as follows:

$$\varphi(\infty) - \varphi(-\infty) = \int_{-\infty}^{\infty} d\lambda\, \dot{\varphi} = \int_{-\infty}^{\infty} d\lambda\, \dot{r}\frac{d\varphi}{dr} = \int_{\infty}^{R} dr\,\frac{d\varphi}{dr} + \int_{R}^{\infty} dr\,\frac{d\varphi}{dr}$$

$$= 2\int_{R}^{\infty} dr\, \frac{R\sqrt{A(r)B(r)}}{r\sqrt{B(R)r^2 - R^2 B(r)}}$$

(the last equality holds because of symmetry under reflections).

The calculation of the integral up to first-order corrections in (R_G/R) is a long but relatively simple exercise. As a result, we obtain (2.120), after inserting the result into (2.119):

$$\delta\varphi = 4\,\frac{R_G}{R}\frac{\alpha + \gamma}{2}. \tag{2.120}$$

The whole effect consists of two summands: the α-part and the γ-part. The α-part for $\alpha = 1$ can be derived from the Newton theory and the equivalence principle by calculating the deflection of a massive particle, which moves with the speed of light. The orbit will be independent of the particle's mass. This way we can pass to the limit $\mu \to 0$. For the Sun, whose surface is just touched by the signal, if R equals the Sun's radius, we have (if the radians are transformed into degrees):

$$\delta\varphi = 1.751'' \times \frac{\alpha + \gamma}{2}. \tag{2.121}$$

Examples of measurements are (for more discussion, see [6, 10, 11])

1. the solar eclipse observed in 1973 with the result

$$\frac{\alpha + \gamma}{2} = 0.95 \pm 0.11,$$

2. the VLBI method (1991) with

$$\frac{\alpha + \gamma}{2} = 1.0001 \pm 0.001.$$

The measurements are on the one hand a test that the space has to be *curved*: α alone cannot explain the measurements, it is necessary to set in addition $\gamma \approx 1$.

The light deflection is on the other hand of high importance for theoretical physics. According to special relativity theory, no causal signal can run faster than light. Therefore, the boundary of the light cone for future events corresponding to a point p is the absolute boundary of the region in space–time, which can be influenced from the point p. Owing to the effects of light deflections, gravitation deforms this boundary. This means that the gravitational field determines the causal structure. Black holes yield examples of causal structures that are very different from that of Minkowski space–time.

2.10.2 Radar Echo Delay

Today it is possible to send radar signals to different objects in the solar system, to receive the reflected signals, and to measure very precisely the difference between the sending and receiving times. Let us consider the following experiment: we send a signal from the Earth at radius R_1, which passes very close to the Sun (impact parameter R). This signal is reflected by a body at radius R_2 and returns along the same path to the Earth. Since the planets and satellites move slowly with respect to the velocity of light, they can be considered as static.

The coordinate time which the radar signal starting from R_1 needs to reach R is given by (2.117):

$$t(R, R_1) = \int_R^{R_1} r \sqrt{\frac{A(r)B(R)}{B(r)}} \frac{dr}{\sqrt{B(R)r^2 - R^2 B(r)}}. \tag{2.122}$$

The total time Δs passed until the signal returned back to the Earth is then given by

$$\Delta s = 2\left[t(R, R_1) + t(R, R_2)\right] \sqrt{B(R_1)}, \tag{2.123}$$

where the factor $\sqrt{B(R_1)}$ transforms the coordinate time into the observer's proper time. A longer calculation yields

$$\Delta s = 2D + 2\alpha R_G \left(1 - \frac{R_2}{R_1}\right) + 2(\alpha + \gamma) R_G \log \frac{4R_1 R_2}{R^2}, \tag{2.124}$$

where

$$D = \sqrt{R_1^2 - R^2} + \sqrt{R_2^2 - R^2}$$

is the "Euclidean distance" between the Earth and the reflecting object. If we set $\alpha = 1$, this formula is well suited to measure γ. We see that the curvature of space enlarges the path of the light.

The delay of the radar signals was not—in contrast to the other effects that we discussed so far—predicted by Einstein. The existence of such an effect was first theoretically derived by I. I. Shapiro [12, 13]. In order to measure it, three different reflecting objects were used:

(a) the planets Mercury and Venus (passive mirrors),
(b) Sun satellites as active retransmitter (Mariner 6, 7, 1975),

$$\frac{\alpha + \gamma}{2} = 1.00 \pm 0.02,$$

(c) Satellites of planets (Viking of Mars, Mariner 9 of Venus).
   The Viking result [12, 13] is

$$\frac{\alpha + \gamma}{2} = 1.00 \pm .01.$$

This approximately corresponds to the result of measurements of γ by light deflection.

2.11 Exercises

1. Let $\mathcal{M}$ be an n-manifold with metric $g_{\mu\nu}$. Prove that the signature of $g_{\mu\nu}(x)$ is independent of $x \in \mathcal{M}$.

   Hint: use the following theorem from linear algebra— the eigenvalues $\lambda_1, \ldots, \lambda_n$ and the eigenvectors $V_1^\mu, \ldots, V_n^\mu$ of a symmetric matrix $g_{\mu\nu}$,

   $$g_{\mu\nu} V_i^\nu = \lambda_i \delta_{\mu\nu} V_i^\nu,$$

   depend continuously on $g_{\mu\nu}$; λ_i are all real, and V_i^μ satisfy

   $$\delta_{\mu\nu} V_i^\mu V_j^\nu = \delta_{ij}.$$

2. Let $\mathcal{M}$ be an n-manifold with metric $g_{\mu\nu}$, $p \in \mathcal{M}$, and $\{x^\mu\}$ a coordinate system around p.

   Prove that the following two equations are equivalent:

   $$\partial_\rho g_{\mu\nu}(p) = 0, \quad \forall \rho, \mu, \nu;$$
   $$\{^\rho_{\mu\nu}\}(p) = 0, \quad \forall \rho, \mu, \nu.$$

3. Let $\mathcal{M}$ be an n-manifold with affine connection $\Gamma^\mu_{\rho\sigma}$ and metric $g_{\mu\nu}$.
   Prove that the following two statements are equivalent:

   (a) The affine connection is metric

   $$\Gamma^\mu_{\rho\sigma} = \{^\mu_{\rho\sigma}\};$$

   (b) $g_{\mu\nu} u^\mu v^\nu$ is constant along any curve C in $\mathcal{M}$ if u^μ and v^μ are vectors that are parallel transported along C.

4. Let $dl^\mu = (\Delta x^0 - \frac{1}{2} dx^0, dx^k)$ be the vector, which is orthogonal to the 4-velocity of the observer x^k. Prove that for the distance dl between the two observers x^k and $x^k + dx^k$ holds

   $$dl^2 = -\hat{g}_{kl} dx^k dx^l = -g_{\mu\nu} dl^\mu dl^\nu.$$

5. A space–time is assumed to have the metric

$$ds^2 = dt^2 - R^2 \left(d\chi^2 + \sin^2\chi \, d\vartheta^2 + \sin^2\chi \sin^2\vartheta \, d\varphi^2 \right),$$

where R is a constant. This is the metric of the so-called Einstein Universe whose manifold is given by $\mathcal{M} = \mathbb{R} \times S_R^3$, where S_R^3 denotes the three- dimensional surface of a sphere, i.e., imbedded in $\mathbb{E}^4$ we have

$$S_R^3 = \left\{ x \in \mathbb{E}^4 \,|\, \|x\|^2 = R^2 \right\},$$

and the metric on S_R^3 is obtained if we parameterize with the four-dimensional spherical coordinates $0 \le \chi \le \pi, 0 \le \vartheta \le \pi$, and $0 \le \varphi \le 2\pi$.
Calculate

(a) autoparallel equation,
(b) the components of the affine connection,
(c) the metric $\hat{g}_{kl}$ for the observer along the t-curve,
(d) all light rays that satisfy the condition $\dot{\vartheta} = \dot{\varphi} = 0$; how long does it take for the light to pass around the entire universe?

6. Write the static metric as follows

$$ds^2 = V^2 dt^2 - \gamma_{kl} dx^k dx^l,$$

$k,l = 1,2,3$, where $V = V(x^1,x^2,x^3)$ and $\gamma_{kl} = \gamma_{kl}(x^1,x^2,x^3)$.
Calculate all Christoffel symbols $\Gamma^\mu_{\rho\sigma}$ of the 4-metric dependent on V and γ_{kl} and show that

$$\Gamma^k_{ij} = \gamma^k_{ij},$$

where γ^k_{ij} are the Christoffel symbols of the 3-metric γ_{kl}.
7. Rewrite the equation of motion of a test particle in the static space–time of Exercise 6 as a system of three equations for the three functions $x^1(\lambda)$, $x^2(\lambda)$, and $x^3(\lambda)$ ($x^0(\lambda)$ is no longer present; the appearance of the integration constant in the equation should not be confusing).
Find a conserved quantity for this system that only depends on $x^k(\lambda)$ and $\dot{x}^k(\lambda)$, $k = 1,2,3$.
8. Use the results of Exercise 7 to determine the 4-curvature tensor.

(a) Show that

$$R^k_{lmn} = r^k_{lmn},$$

where r^k_{lmn} is the curvature tensor of the 3-affine connection γ^k_{mn}.
(b) Show that

$$R^0_{lmn} = 0.$$

(c) Show that

$$R^0_{k0l} = \frac{-V_{,kl} + \gamma^m_{lk} V_{,m}}{V}.$$

9. Calculate the vector fields ξ_x^k, ξ_y^k, and ξ_z^k that generate the rotations around the three axes x, y, and z of $\vec{\mathbb{E}}^3$. Calculate in Cartesian coordinates x, y, and z and transform the result into spherical coordinates r, ϑ, and φ.
Prove that

(a) the three vector fields ξ_x^k, ξ_y^k, and ξ_z^k are Killing vectors of the metric

$$ds^2 = dx^2 + dy^2 + dz^2;$$

(b) the three vector fields η_x^μ, η_y^μ, and η_z^μ defined by

$$\eta_x^0 = \eta_y^0 = \eta_z^0 = 0$$

and

$$\eta_x^k = \xi_x^k, \quad \eta_y^k = \xi_y^k, \quad \eta_z^k = \xi_z^k$$

(in spherical coordinates!) are Killing vectors of the metric

$$ds^2 = g_{00}(t,r)dt^2 + 2g_{01}(t,r)dt\,dr - g_{11}(t,r)dr^2 - r^2(d\vartheta^2 + \sin^2\vartheta\,d\varphi^2).$$

Exercises 10–13 deal with an arbitrary static, rotationally symmetric, and asymptotically flat space–time.

10. Consider a free particle of mass m, which decays at a given point in space–time into two free particles of masses m_1 and m_2. Assume that the trajectories of the three particles are parameterized by the physical parameter; let E, E_1 and E_2, L_x, L_y and L_z, L_{1x}, L_{1y} and L_{1z}, L_{2x}, L_{2y} and L_{2z} be the corresponding integrals of motion.
Prove the equations:

$$E = E_1 + E_2,$$
$$L_x = L_{1x} + L_{2x},$$
$$L_y = L_{1y} + L_{2y},$$
$$L_z = L_{1z} + L_{2z}.$$

11. A free particle of mass μ starts from $r = \infty$ with energy E, then reaches a point with coordinates $r = R$, and decays in a bound (i.e., it can never reach $r = \infty$) free particle of mass μ_1 and a second free particle of mass μ_2, which does reach $r = \infty$ with energy E_2.

(a) Can the asymptotic observers gain energy through this process?
(b) The same question for angular momentum L_x, L_y, L_z.

12. Calculate the 4-acceleration of a circular motion of radius $r = R$ for a free particle. (Is it necessary to calculate a lot?)

13. Calculate the energy E and the angular momentum L of a purely circular motion of radius $r = R$ for a free particle of mass μ in the asymptotic region. Use the metric where all quadratic and higher terms in R_G/r are negligible.
Hint: analogy to the "Kepler problem and effective potential" of mechanics.

14. Write the Newtonian Lagrangian for the motion of a mass point of mass μ in the gravitational field of a central mass $\mathcal{M}$. Use the conservation laws for the energy E and the angular momentum L in order to transform the dynamic equations into the form

$$dt = F_t(\mu, M, E, L; r)dr,$$
$$d\varphi = F_\varphi(\mu, M, E, L; r)dr.$$

Find the functions F_t and F_φ and compare them with the relativistic formulas.

15. The double pulsar PSR1913+16 [10] is a system of two neutron stars with masses of about 1.4 M_{Sun} each and with radii of about 10 km, which circle around each other. The periastron shift was observed: $\delta\varphi \approx 4.2^o$ per year! Use our formula for the periastron shift and the Newtonian Kepler's law in order to estimate (a) the distance between the two stars and (b) the period of the motion.

16. Illustration of the curvature of space around a star: find the rotation surface in $\mathbb{E}^3$, whose metric coincides up to terms of order R_G/r with the one that is carried by the equatorial surface $\vartheta = \pi/2$, t =const (Robertson–Eddington expansion).

References

1. A. P. French, *Special Relativity*, Massachusetts Institute of Technology, Cambridge, MA, 1968.
2. H. Goldstein, *Classical Mechanics*, Addison-Wesley, Reading, MA, 1980.
3. S. W. Hawking and G. F. R. Ellis, *The Large Scale Structure of Space-Time*, Cambridge University Press, Cambridge, 1973.
4. I. M. Gel'fand, *Lectures on Linear Algebra*. Interscience Publishers, New York, 1961.
5. J. L. Synge, *Relativity: The General Theory*. North-Holland, Amsterdam, 1960.
6. I. Ciufolini and J. A. Wheeler, *Gravitation and Inertia*. Princeton University Press, Princeton, NJ, 1995.
7. A. S. Eddington, *The Mathematical Theory of Relativity*, Cambridge University Press, Cambridge, UK, 1922.
8. H. P. Robertson, *Relativity and Cosmology*, Deutsch and Klemperer, 1962.
9. R. M. Wald, *General Relativity*, University of Chicago Press, Chicago, IL, 1984.
10. C. M. Will, *Theory and Experiment in Gravitational Physics*. Cambridge University Press, Cambridge, 1993.
11. B. Schutz, *Gravity from the Ground up*, Cambridge University Press, Cambridge, UK, 2003.
12. I. I. Shapiro, Phys. Rev. Lett. **13**, (1964) 789.
13. I. I. Shapiro, et al.: J: Geophys. Res. **82**, (1977) 4329–4334.

Chapter 3
Field Dynamics

In the previous chapter we studied the movement of mass points in curved space–times. We gained insight into how the curvature of thespace–time affects the dynamics of mass points and light rays. In modern theoretical physics however, not particles but *fields* are the elements of nature. In this chapter we shall turn our attention to fields. We shall not build a field theory systematically, but introduce some basic notions that are essential for the further development of general relativity. Among these are the covariant derivative for arbitrary tensor fields, the stress-energy tensor, and some material on variation principles. We will assume knowledge of the most important facts about relativistic field theories, such as electromagnetism [1]. The focus will lie on the field theories, which are important for astrophysics. These are mainly electrodynamics (magnetic fields become quite strong and important in space—look, e.g., at the X-ray plates of the Sun [2] to see some magnetic field that is not tiny!) and hydrodynamics.

3.1 Electrodynamics

In this section we will study the dynamics of electromagnetic fields under the influence of gravity. We will exclusively use the formalism of 4-vectors and tensors ([1], p. 377). For example, the potential ϕ and the vector potential A_k form the 4-potential A_μ with four components $A_0 = -\phi, A_k$. The electromagnetic field in special relativity is described by this 4-potential $A_\mu(x)$. It is a covector field, that is an assignment of a covector, which is given by its components $A_\mu(x)$ relative to an inertial frame, to every point x of Minkowski space–time. The *tensor of the electromagnetic field* is then defined by

$$F_{\rho\sigma} = \partial_\rho A_\sigma - \partial_\sigma A_\rho \ . \tag{3.1}$$

The meaning of the components of $F_{\mu\nu}$ in an inertial frame follows from the so-called $3 + 1$ splitting

$$E_k = F_{k0} \ , \quad B_k = (1/2)\sum_{r,s}\varepsilon_{krs}F_{rs} \ ,$$

Hájíček, P.: *Field Dynamics*. Lect. Notes Phys. **750**, 93–124 (2008)
DOI 10.1007/978-3-540-78659-7_3

where E_k and B_k denote the electric and magnetic field respectively. ε_{krs} is a very useful expression, which is totally antisymmetric in all three indices $k = 1,2,3$, $r = 1,2,3$, and $s = 1,2,3$, with $\varepsilon_{123} = 1$. This determines ε_{krs} uniquely. One can show the important identity

$$\sum_k \varepsilon_{kmn}\varepsilon_{krs} = \delta_{mr}\delta_{ns} - \delta_{ms}\delta_{nr} . \tag{3.2}$$

(Proof: exercise). The physical content of the potential is not affected by a *gauge transformation*:

$$A_\mu \mapsto A_\mu + \partial_\mu \Lambda , \tag{3.3}$$

where Λ is an arbitrary function, $F_{\mu\nu}$ is gauge-invariant.

The dynamics of the (free) field are determined by the action principle $\delta S = 0$, where

$$S = -\frac{1}{16\pi} \int d^4x\, F^{\rho\sigma} F_{\rho\sigma} , \tag{3.4}$$

and

$$F^{\rho\sigma} = g^{\rho\kappa} g^{\sigma\lambda} F_{\kappa\lambda} . \tag{3.5}$$

Since we are working in an inertial frame, the metric $g_{\mu\nu}$ of flat space–time has the canonical form $g_{\mu\nu} = \eta_{\mu\nu}$.

3.1.1 Equivalence Principle

We want to generalize the Lagrangian (3.4) in such a way that it is still valid in the presence of gravity. To this end, we use the equivalence principle in form of the principle of general covariance: we transform the Lagrangian (3.4) to an arbitrary curvilinear coordinate system in flat space–time and replace the transformation coefficients X_ν^μ by expressions in the metric components. Then we postulate that the resulting Lagrangian induces the correct dynamics in curved space–times (minimal coupling).

Thus, let $\{x^\mu\}$ be the coordinates of an inertial frame and $\{x'^\mu\}$ coordinates of an arbitrary reference frame. $A_\mu(x)$ transforms like a covector, that is

$$A_\mu(x) = X_\mu^{\rho'} A'_\rho(x') ,$$

where x and x' are the coordinates of the same point with respect to the two systems. We obtain

$$F_{\mu\nu} = X_\mu^{\rho'} X_\nu^{\sigma'} F'_{\rho\sigma} , \tag{3.6}$$

where

$$F'_{\rho\sigma} = \partial'_\rho A'_\sigma - \partial'_\sigma A'_\rho \tag{3.7}$$

is defined in curvilinear coordinates exactly as in an inertial frame (we denoted $\partial/\partial x'^\rho$ as ∂'_ρ). Equations (3.6) and (3.7) hold in general: anti-symmetrized derivatives of a covector field transform as a tensor. Hence, we can transform the scalar $F^{\rho\sigma}F_{\rho\sigma}$ in the Lagrangian (3.4):

$$F^{\rho\sigma}F_{\rho\sigma} = g^{\rho\kappa}g^{\sigma\lambda}F_{\kappa\lambda}F_{\rho\sigma} = g'^{\rho\kappa}g'^{\sigma\lambda}F'_{\kappa\lambda}F'_{\rho\sigma} \,,$$

since g and F are tensors. This expression no longer contains the transformation coefficients $X_\mu^{\rho'}$ or derivatives thereof. We are left with the task to transform the volume element d^4x to curvilinear coordinates. We have the following theorem.

Theorem 8 *The value of the following expression is independent of coordinates:*

$$\mathrm{d}^4x\sqrt{-g}$$

(Proof: exercise 2). Here we introduced the useful shorthand

$$g = \det(g_{\mu\nu}) \,. \tag{3.8}$$

Hence, if x are inertial and x' are curvilinear coordinates then we have

$$\mathrm{d}^4x = \mathrm{d}^4x'\sqrt{-g'} \,. \tag{3.9}$$

This important equation expresses the invariant volume element with respect to arbitrary coordinates in terms of the metric components in these coordinates. We remark that (3.9) can be derived in the same way for arbitrary dimension n of the manifold in question. We just have to replace the number 4 by n. Then the right-hand side can be used to compute the volume element of a surface for example.

With (3.9) we can write the action (3.4) with respect to arbitrary coordinates (where we omit the primes):

$$S = -\frac{1}{16\pi}\int \mathrm{d}^4x\sqrt{-g}g^{\rho\kappa}g^{\sigma\lambda}F_{\kappa\lambda}F_{\rho\sigma} \,. \tag{3.10}$$

Postulate 3.1 *The dynamics of the electromagnetic field in curved space–times is determined by the variation principle with action (3.10). The action is a function of the potential $A_\mu(x)$, where*

$$F_{\mu\nu} = \partial_\mu A_\nu - \partial_\nu A_\mu \,. \tag{3.11}$$

The action (3.10) is invariant with respect to gauge transformations (3.3). The measurable content of the field tensor at a point p with respect to a local inertial frame at p is

$$F_{k0} = E_k, \quad F_{kl} = \sum_m \varepsilon_{klm}B_m, \tag{3.12}$$

where E_k and B_k represent the components of the electric and magnetic field for the corresponding observer.

3.1.2 The Maxwell Equations

We obtain the second set of the generalized Maxwell equations by variation of the action (3.10) with respect to A_μ. To simplify the computation, we assume that the variation $\delta A_\mu(x)$ has compact support in our curved space–time. Then it vanishes along the boundary, and thus all boundary integrals resulting from partial integration must vanish. We compute the variation for a constant metric $\delta g_{\mu\nu}(x) = 0$, and obtain step by step:

$$
\begin{aligned}
\delta S &= -\frac{1}{16\pi}\int d^4x\sqrt{-g}\,g^{\rho\kappa}g^{\sigma\lambda}\,\delta\left(F_{\kappa\lambda}F_{\rho\sigma}\right)\\
&= -\frac{1}{8\pi}\int d^4x\sqrt{-g}\,g^{\rho\kappa}g^{\sigma\lambda}F_{\kappa\lambda}\,\delta F_{\rho\sigma}\\
&= -\frac{1}{8\pi}\int d^4x\sqrt{-g}\,g^{\rho\kappa}g^{\sigma\lambda}F_{\kappa\lambda}\left(\partial_\rho\delta A_\sigma - \partial_\sigma\delta A_\rho\right)\\
&= -\frac{1}{4\pi}\int d^4x\sqrt{-g}\,g^{\rho\kappa}g^{\sigma\lambda}F_{\kappa\lambda}\,\partial_\rho\delta A_\sigma\\
&= \frac{1}{4\pi}\int d^4x\,\partial_\rho\left(\sqrt{-g}\,g^{\rho\kappa}g^{\sigma\lambda}F_{\kappa\lambda}\right)\delta A_\sigma\;.
\end{aligned}
$$

Hence, the equation

$$
\partial_\rho\left(\sqrt{-g}\,g^{\rho\kappa}g^{\sigma\lambda}F_{\kappa\lambda}\right) = 0 \tag{3.13}
$$

follows. This is the desired generalization of the second set of Maxwell's equations in curved space–times. For the flat space–time and with respect to an inertial frame or for curved space–times in a local inertial frame, the correct equations $\partial_\rho F^\rho_\sigma = 0$ for the free field result. The first set of Maxwell's equations follows directly from the definition of $F_{\rho\sigma}$, since (3.11) implies

$$
\partial_\tau F_{\rho\sigma} + \partial_\rho F_{\sigma\tau} + \partial_\sigma F_{\tau\rho} = 0. \tag{3.14}
$$

Hence, the metric only enters the second set of equations.

Equations (3.13) and (3.14) are supposed to describe the dynamics of the electromagnetic field in the gravitational field. In particular, they should describe the deviation of light in the gravitational field of the Sun and the redshift. We remark without proof that they actually do so.

Equations (3.13) and (3.14) apparently hold in arbitrary coordinates, since we did not need any restrictions for their deduction. However, the form of the equations does not reflect this. We already know that the coordinate derivatives of the tensor fields involved do not transform like tensors. Thus, it is not a priori clear, whether the left-hand sides vanish in every coordinate system if they vanish in a particular one (exercise). We will return to this problem later.

3.1.3 The Stress-Energy Tensor

The stress-energy tensor in general relativity plays a similar role as the current 4-vector in electrodynamics: it is the source of the field. In non-relativistic theory, the mass density is such a source, it is the right-hand side of the Poisson equation.

In a non-relativistic theory, it is completely satisfactory that any given (extensive) quantity has a corresponding density. For example, the charge Q, which is a scalar, defines the charge density ρ, a scalar field, by the following relation. The total charge dQ in a volume element dV at a point p is $dQ = \rho(p)dV$. However, in special relativity the 3-volume is not absolute, it depends on the observer. For an observer with 4-velocity u^μ, it is orthogonal to u^μ. Consequently, the 3-volume element is defined not only by its magnitude dV but also by its normal u^μ [1]. It is well known that the total charge dQ in a 3-volume dV of an observer with 4-velocity u^μ is given by $dQ = j^\mu(p)u_\mu dV$, where $j^\mu(x)$ is a vector field called the *current of electric charge*. In special relativity, this 4-component current plays the role of the charge density $\rho(x)$.

Similar rules hold for tensor quantities: they determine currents having one extra index. The energy, for example, is the 0-component of the momentum 4-vector, P^μ. Hence, the corresponding density must be a 4-tensor $T^{\mu\nu}(x)$ of second order, determined by the equation

$$dP^\mu = T^{\mu\nu}(p)u_\nu dV , \qquad (3.15)$$

where dP^μ is the total momentum in direction μ in the 3-volume element dV of the observer at a point p with 4-velocity u^μ. We adopt this equation in general relativity.

In general, we can obtain the current, that is the source of a field φ, by variation of the actions of the other fields with respect to φ. This originates from the fact that the total action of a field system can be represented as the sum of the actions of the individual fields,

$$S_{\text{tot}} = S[\varphi] + S_1[\psi_1, \ldots, \psi_n, \varphi] ,$$

where $\psi_1, \ldots, \psi_n$ are the fields in the system which are non-trivially coupled to φ. The field equation for φ then becomes

$$\frac{\delta S}{\delta \varphi} = -\frac{\delta S_1}{\delta \varphi} .$$

The left-hand side of this equation agrees with the left-hand side of the free dynamical equation,

$$\frac{\delta S}{\delta \varphi} = 0$$

for the field φ. Then the right-hand side is the sum of the sources formed by the other fields.

The electromagnetic field carries energy and thus has a non-trivial coupling to gravity—it generates a gravitational field. We obtain the corresponding source term by variation of the action (3.10) with respect to $g_{\mu\nu}$ (or with respect to $g^{\mu\nu}$). That is, we have to compute the variation of S with $\delta g^{\mu\nu} \neq 0$ and $\delta A_\mu(x) = 0$. To begin with, we infer

$$\delta S = -\frac{1}{16\pi} \int d^4 x \delta \left(\sqrt{-g} g^{\rho\kappa} g^{\sigma\lambda}\right) F_{\kappa\lambda} F_{\rho\sigma} .$$

Before we continue, we need the variation of the determinant. We have the important formula (exercise):

$$\delta\sqrt{-g} = -(1/2)\sqrt{-g} g_{\mu\nu} \delta g^{\mu\nu} . \tag{3.16}$$

Then we can compute the variation as follows:

$$\delta S = \frac{1}{16\pi} \int d^4 x \delta \left(\sqrt{-g} g^{\rho\kappa} g^{\sigma\lambda}\right) F_{\kappa\lambda} F_{\rho\sigma}$$

$$= -\frac{1}{16\pi} \int d^4 x \left(-(1/2)\sqrt{-g} g_{\mu\nu} g^{\rho\kappa} g^{\sigma\lambda} + \sqrt{-g}\, \delta^\rho_\mu \delta^\kappa_\nu g^{\sigma\lambda}\right.$$

$$\left. + \sqrt{-g}\, \delta^\sigma_\mu \delta^\lambda_\nu g^{\rho\kappa}\right) F_{\kappa\lambda} F_{\rho\sigma} \delta g^{\mu\nu}$$

$$= -\frac{1}{16\pi} \int d^4 x \sqrt{-g} \left(F_\nu{}^\sigma F_{\mu\sigma} + F_\nu^\rho F_{\rho\mu} - (1/2) g_{\mu\nu} F^{\rho\sigma} F_{\rho\sigma}\right) \delta g^{\mu\nu}$$

$$= +\frac{1}{2} \int d^4 x \sqrt{-g} \left[-\frac{1}{4\pi} \left(F_{\mu\sigma} F_\nu{}^\sigma - \frac{1}{4} g_{\mu\nu} F^{\rho\sigma} F_{\rho\sigma}\right)\right] \delta g^{\mu\nu} .$$

The expression in square brackets is called the *electromagnetic stress-energy tensor*, $T^{EM}_{\mu\nu}$:

$$T^{EM}_{\mu\nu} = -\frac{1}{4\pi} \left(F_{\mu\sigma} F_\nu{}^\sigma - \frac{1}{4} g_{\mu\nu} F^{\rho\sigma} F_{\rho\sigma}\right) . \tag{3.17}$$

Clearly, it is a tensor that is symmetric in both indices.

We can write the resulting relation as follows:

$$\frac{\delta S_M}{\delta g^{\mu\nu}(x)} = (1/2)\sqrt{-g(x)}\, T_{\mu\nu} . \tag{3.18}$$

The term on the left-hand side is the variational derivative of the functional S with respect to the variables $g^{\mu\nu}(x)$ (which are distinguished from each other not only by the indices but also by their arguments x^μ). This variational derivative is defined as the coefficient of δS in the variation of the variable $g^{\mu\nu}(x)$. This is analogous to the partial derivative of a function of multiple variables: the partial derivative with respect to a variable is the coefficient of the differential of this variable in the total differential of the function. Equation (3.18) can be regarded as the general definition of the stress-energy tensor. It determines the numerical factors in front of the action integrals, that is the normalization of the actions for the individual fields, since the form of the stress-energy tensor for every field is known.

By the usual dimension estimate (in a local inertial frame with $c = 1$), we find that in general the variational derivative (3.18) has the dimensions of energy density. The dimension of the action is $[E \times T]$, the dimension of d^4x is $L^3 \times T$, whereas $\delta g^{\mu\nu}(x)$ and $\sqrt{-g}$ are dimensionless. Therefore the dimension of $\delta S / (\sqrt{-g} \delta g^{\mu\nu}(x))$ is $[E \times T]/[L^3 \times T] = [E/L^3]$.

The meaning of the components of the stress-energy tensor, given by (3.15), can be verified by computing these components with respect to a local inertial frame. In such a frame, we have the relations (3.12). First we compute

$$T_{00}^{\text{EM}} = -\frac{1}{4\pi} \left[F_{0k} F_0{}^k - (1/4) \left(F^{0k} F_{0k} + F^{k0} F_{k0} + F_{kl} F^{kl} \right) \right]$$

$$= \frac{1}{4\pi} \left[\sum_k F_{0k} F_{0k} + (1/4) \left(-2 \sum_k F_{0k} F_{0k} + \sum_{kl} F_{kl} F_{kl} \right) \right] .$$

From (3.2) we infer that

$$\sum_{kl} \varepsilon_{klm} \varepsilon_{kln} = 2\delta_{mn} .$$

Together, we obtain

$$T_{00}^{\text{EM}} = \frac{1}{8\pi} \left(E^2 + B^2 \right) .$$

This is the energy density (or mass density, $c = 1$) of the electromagnetic field. In a similar way, we obtain that

$$T_{0k}^{\text{EM}} = -\frac{1}{4\pi} \left[\vec{E} \times \vec{B} \right]_k .$$

This is the so-called Poynting vector, giving the 3-momentum density, or equivalently, the energy current density.

In special relativity, the stress-energy tensor satisfies the divergence equation. This follows from the Maxwell equations and the following identity:

$$\partial_\nu \left(F^\mu{}_\rho F^{\nu\rho} - \frac{1}{4} \eta^{\mu\nu} F^{\rho\sigma} F_{\rho\sigma} \right)$$

$$= -F^\mu{}_\rho \left(\partial_\nu F^{\rho\nu} \right) - \frac{1}{2} \eta^{\mu\nu} F^{\rho\sigma} \left(\partial_\nu F_{\rho\sigma} + \partial_\rho F_{\sigma\nu} + \partial_\sigma F_{\nu\rho} \right) .$$

This immediately implies the claim. Via the Gauss theorem, the divergence equation implies conservation of total energy:

$$E := \int_\Sigma d^3x \, T^{00} , \quad P^k := \int_\Sigma d^3x \, T^{0k} ,$$

where Σ is a time slice $x^0 = \text{const}$.

3.2 Variation Principle

We start by defining and examining the transformation properties of tensor fields. The transformation properties are their most important properties. It influences other, less formal properties of the fields like their dynamics. Indeed, the dynamics are determined by a Lagrange function (Lagrangian is, roughly, the integral of Lagrange function over the space), and the Lagrange function has to be invariant (as we will see later). How an invariant can be constructed from the components and the derivatives of a field depends on their respective transformation behavior.

Here, we focus our attention to tensor fields. Spinor fields are at least as important to theoretical physics [3], but they do not play much role in astrophysics.

3.2.1 Transformation Properties of Tensor Fields

Consider a fixed space–time M with metric $g_{\mu\nu}$ and choose a coordinate system $\{x^\mu\}$.

Definition 15 A tensor field of type (p,q) is a map which assigns to every point r in a set $G \subset M$ a tensor $T^{\mu\cdots}_{\nu\cdots}$ in r of type (p,q). G is called the domain of the tensor field.

In a coordinate system $\{x^\mu\}$, a tensor field is described by 4^{p+q} C^∞ functions $T^{\mu\cdots}_{\nu\cdots}(x^0,x^1,x^2,x^3)$, where $T^{\mu\cdots}_{\nu\cdots}(r)$ are the components of the tensor at the point r and (x^0,x^1,x^2,x^3) are the coordinates of r with respect to $\{x^\mu\}$. These functions are called the *component functions of the tensor field* with respect to $\{x^\mu\}$.

For example, $\Phi(r)$, a scalar field, is given by a function $\Phi(x^0,x^1,x^2,x^3)$ (a tensor of type $(0,0)$), and $A_\mu(r)$, a vector potential, is given by four functions $A_0(x^0,x^1, x^2,x^3)$, ..., $A_3(x^0,x^1,x^2,x^3)$.

Let $\{x'^\mu\}$ be a different coordinate system. With respect to $\{x'^\mu\}$, the same tensor field is represented by different functions. We obtain the transformation from one set of functions to the other in two steps. First, the arguments of the functions have to be transformed:

$$x^\mu = x^\mu\left(x'^0,x'^1,x'^2,x'^3\right) . \tag{3.19}$$

Second, we have to form linear combinations of the resulting functions according to the transformation rule for the respective tensor type. For example, the vector potential transforms like

$$A'_\mu\left(x'^0,x'^1,x'^2,x'^3\right) = X^\rho_{\mu'}\left(x'^0,x'^1,x'^2,x'^3\right) A_\rho\left(x^0(x'),x^1(x'),x^2(x'),x^3(x')\right) ,$$

where we have to insert (3.19) for x^μ on the right-hand side. The (single) component of the scalar field also transforms non-trivially:

$$\Phi'\left(x'^0,x'^1,x'^2,x'^3\right) = \Phi\left(x^0(x'),x^1(x'),x^2(x'),x^3(x')\right) .$$

3.2.2 The Action

The variation principle of the action S is used to derive the dynamics of a system. It serves as a common foundation for mechanics and field theory, in classical as well as in quantum theory.

Let (M,g) be a space–time and $\psi^a(x)$ a system of fields. The index a enumerates the components of a whole system of fields, including the metric. For example

a	1	2	3	4	...
$\psi^a(x)$	Φ	g_{00}	g_{01}	g_{02}	...,

is a system of a scalar field $\Phi(x)$ and the metric $g_{\mu\nu}(x)$.

At first, we recall the most important requirements for the action in theoretical physics.

The form of the action: In field theory, we will work with actions of the following form:

$$S = \int_V d^4x \sqrt{-g}\, L, \tag{3.20}$$

where V is an open set in M, x^μ are coordinates in a neighborhood of V, and L is the so-called *Lagrange function*, a function of the fields $\psi^a(x)$ and their first derivatives $\psi^a_{,\mu}(x)$, at the point x^μ:

$$L = L\left(\psi^a(x), \psi^a_{,\mu}(x)\right) . \tag{3.21}$$

In particular, the Lagrange function must not depend on higher derivatives.

The determinant $\sqrt{-g}$ has been separated from L. The reason is that $\sqrt{-g}\, d^4x$ is an invariant volume element, as we have seen in Sect. 3.1.1. However, we could let $\mathcal{L} = L\sqrt{-g}$ and work with $\mathcal{L}$ instead of L. $\mathcal{L}$ is the so-called *Lagrangian density*. The most important property of the Lagrange function (or Lagrangian density) is its functional character: It is independent of the coordinates we are working with.

General covariance: Let $L(x)$ be the composite function given by (3.21). Then, up to a divergence term, $L(x)$ is a scalar field on M, that is

$$L(x) = \sigma(x) + \frac{1}{\sqrt{-g}} \partial_\mu V^\mu(x) , \tag{3.22}$$

where $\sigma(x)$ is a scalar field of the form

$$\sigma(x) = \sigma\left(\psi^a(x), \psi^a_{,\mu}(x), \psi^a_{,\mu\nu}(x), \dots\right)$$

and the 4-component quantity $V^\mu(x)$ has to transform such that (3.22) holds in arbitrary coordinate systems. This assumption implies that with respect to coordinate transformations the action is invariant up to surface integrals along the boundary ∂V of V.

The special role of gravity: The Lagrange function L shall consist of two terms:

$$L = L_G + L_M , \tag{3.23}$$

where $L_G = L(g_{\mu\nu}, g_{\mu\nu,\rho})$ is the Lagrange function for the gravitational field and L_M describes "the rest of the world" (matter). Note that the Lagrange function of the gravitational field does not depend on any other field than the metric.

The field equations: Let $\psi^a(x)$ be a system of fields. Then the dynamically admissible fields $\psi^a(x)$ are those for which the action (3.20) is extremal with respect to all variations with compact support in V.

Now we can understand the significance of the invariance of the action with respect to coordinate transformations. Let $\psi^a(x)$ be an extremal field. Perform a coordinate change which is the identity outside a compact subset of the domain of integration V. Then the value of the action does not change, that is the transformed components $\psi'^a(x')$ are still extremal. The property of being a dynamically admissible field is invariant under such transformations.

Example: It is easy to compose scalars of the form (3.21) from a scalar field $\Phi(x)$ and its first derivatives $\Phi_{,\mu}(x)$. Every function $F(\Phi(x))$ is a scalar field. As $\Phi_{,\mu}(x)$ is a covector, the natural scalar to form with the metric is the square of the norm of the covector:

$$g^{\mu\nu}\Phi_{,\mu}(x)\Phi_{,\nu}(x) .$$

If we require that the scalar (3.21) be quadratic in $\Phi(x)$ and $\Phi_{,\nu}(x)$, we obtain a two-parameter family of Lagrange functions:

$$L = a\left(g^{\mu\nu}\Phi_{,\mu}(x)\Phi_{,\nu}(x) + b\Phi^2\right) .$$

The constant a normalizes the action such that the stress-energy tensor takes the right value. Coefficient b is a measurable parameter, $\sqrt{-b}$ is called the *mass* of the field.

3.2.3 Variation Formula

We need very general variations of the action. For example, the variation defined by (infinitesimal) coordinate transformations. Such variations do not commute with partial derivatives with respect to the coordinates $\delta\partial_\mu \neq \partial_\mu\delta$. Thus, we have to find a way to calculate δS that does not depend on this assumption.

We want to compute the change of

$$S = \int d^4x \mathscr{L}$$

generated by an infinitesimal variation of the independent, as well as the dependent variables:

$$x'^\mu = x^\mu + \delta x^\mu(x) , \tag{3.24}$$

$$\psi'^a(x') = \psi^a(x) + \delta\psi^a(x) , \tag{3.25}$$

$$\psi'^a_{,\mu}(x') = \psi^a_{,\mu}(x) + \delta\psi^a_{,\mu}(x) . \tag{3.26}$$

This kind of variation arises from comparing the new and old values of the fields (including the coordinates) in fixed points of space–time. This defines a variation of $\psi^a(x)$ which does not commute with coordinate derivatives. To see this, we differentiate both sides of (3.25) with respect to x^μ. On the left-hand side, we obtain, using (3.24) and (3.26),

$$\frac{\partial}{\partial x^\mu} \psi'^a(x') = \frac{\partial x'^\nu}{\partial x^\mu} \psi'^a_{,\nu}(x') = \psi'^a_{,\mu}(x') + \psi'^a_{,\nu}(x') \partial_\mu \delta x^\nu(x)$$

$$= \psi^a_{,\mu}(x) + \delta \psi^a_{,\mu}(x) + \psi^a_{,\nu}(x) \partial_\mu \delta x^\nu(x)$$

(where we neglected terms of second order in the variations). On the right-hand side we simply have

$$\psi^a_{,\mu}(x) + \partial_\mu \delta \psi^a(x) .$$

Hence the commutator satisfies

$$\partial_\mu \delta \psi^a(x) - \delta \psi^a_{,\mu}(x) = \psi^a_{,\nu}(x) \partial_\mu \delta x^\nu(x) . \tag{3.27}$$

This motivates the introduction of another type of variation δ_*, which is sometimes called *form-variation*. In our case, it is defined as follows:

$$\delta_* \psi^a(x) = \psi'^a(x) - \psi^a(x) , \tag{3.28}$$

$$\delta_* \psi^a_{,\mu}(x) = \psi'^a_{,\mu}(x) - \psi^a_{,\mu}(x) . \tag{3.29}$$

These are variations of the form of the component functions of the tensor fields, where we compare the values of these functions for the same value of their arguments. Then δ_* commutes with partial coordinate derivatives, which follows immediately from its definition. Thus, we are left with the task to compute the relationship of the two types of variation:

$$\delta \psi^a(x) = \psi'^a(x') - \psi^a(x) = \psi'^a(x) + \partial_\mu \psi'^a(x) \delta x^\mu(x) - \psi^a(x)$$

$$= \delta_* \psi^a(x) + \psi^a_{,\mu}(x) \delta x^\mu(x) . \tag{3.30}$$

Similarly, we infer

$$\delta \psi^a_{,\mu}(x) = \delta_* \psi^a_{,\mu}(x) + \psi^a_{,\mu\nu}(x) \delta x^\nu(x) . \tag{3.31}$$

With these tools at hand, we are able to compute the variation of the action as follows:

$$\delta S = \int_{V'} d^4 x' \mathscr{L}(\psi'^a(x'), \psi'^a_{,\mu}(x')) - \int_V d^4 x \mathscr{L}(\psi^a(x), \psi^a_{,\mu}(x)) .$$

The first integral is over all values of x' which correspond to the set V for x. In the first integrand we substitute $x'^\mu = x^\mu + \delta x^\mu(x)$ and find

$$\mathscr{L}\left(\psi'^a(x'), \psi'^a_{,\mu}(x')\right)$$
$$= \mathscr{L}\left(\psi^a(x), \psi^a_{,\mu}(x)\right) + \frac{\partial \mathscr{L}}{\partial \psi^a(x)} \delta \psi^a(x) + \frac{\partial \mathscr{L}}{\partial \psi^a_{,\mu}(x)} \delta \psi^a_{,\mu}(x) \, .$$

Furthermore,

$$d^4x' = \det\left(X^{\mu'}_\nu\right) d^4x = \det\left(\delta^\mu_\nu + \partial_\nu \delta x^\mu\right) d^4x = \det(\delta^\mu_\nu) d^4x$$
$$+ \left.\frac{\partial \det(A^\mu_\nu)}{\partial A^\mu_\nu}\right|_{A^\mu_\nu = \delta^\mu_\nu} \partial_\nu \delta x^\mu \, d^4x = d^4x + \delta^\nu_\mu \partial_\nu \delta x^\mu d^4x$$
$$= \left(1 + \partial_\mu \delta x^\mu\right) d^4x \, .$$

Altogether we obtain (where we neglected terms of higher order in the variations)

$$\delta S = \int_V d^4x \left(\mathscr{L} \partial_\mu \delta x^\mu + \frac{\partial \mathscr{L}}{\partial \psi^a(x)} \delta \psi^a(x) + \frac{\partial \mathscr{L}}{\partial \psi^a_{,\mu}(x)} \delta \psi^a_{,\mu}(x)\right) \, .$$

Now we express the variation in terms of δ_*:

$$\delta S = \int_V d^4x \left[\mathscr{L} \partial_\mu \delta x^\mu + \left(\frac{\partial \mathscr{L}}{\partial \psi^a(x)} \frac{\partial \psi^a(x)}{\partial x^\mu} + \frac{\partial \mathscr{L}}{\partial \psi^a_{,\nu}} \frac{\partial \psi^a_{,\nu}(x)}{\partial x^\mu}\right) \delta x^\mu(x)\right.$$
$$\left. + \frac{\partial \mathscr{L}}{\partial \psi^a(x)} \delta_* \psi^a(x) + \frac{\partial \mathscr{L}}{\partial \psi^a_{,\nu}(x)} \delta_* \psi^a_{,\mu}(x)\right] \, .$$

The expression in parentheses is the composed derivative of $\mathscr{L}$ with respect to x^μ:

$$\frac{\partial^c}{\partial x^\mu} \mathscr{L} = \frac{\partial \mathscr{L}}{\partial \psi^a(x)} \frac{\partial \psi^a(x)}{\partial x^\mu} + \frac{\partial \mathscr{L}}{\partial \psi^a_{,\nu}(x)} \frac{\partial \psi^a_{,\nu}(x)}{\partial x^\mu} \, .$$

It is still a partial derivative with respect to other coordinates x^λ, $\lambda \neq \mu$. If we use the fact that ∂_μ and δ_* commute, we can write the result in the following, simple form:

$$\delta S = \int_V d^4x \left[\frac{\partial^c}{\partial x^\mu} \left(\mathscr{L} \delta x^\mu + \frac{\partial \mathscr{L}}{\partial \psi^a_{,\mu}} \delta_* \psi^a(x)\right)\right.$$
$$\left. + \left(\frac{\partial \mathscr{L}}{\partial \psi^a(x)} - \frac{\partial}{\partial x^\mu} \frac{\partial \mathscr{L}}{\partial \psi^a_{,\mu}(x)}\right) \delta_* \psi^a(x)\right] \, . \tag{3.32}$$

This is the variation formula [4].

3.2.4 Field Equations of Matter

We now use the variation principle and the variation formula to derive the equations of motion for the matter fields.

We begin by splitting the system in matter fields $\varphi^a(x)$ and the metric $g_{\mu\nu}(x)$. We compute the variation of the action in V with

$$\delta x^\mu \equiv 0, \quad \delta g_{\mu\nu} \equiv 0. \tag{3.33}$$

Hence, coordinates and metric stay fixed. Because of (3.33), equation (3.30) gives $\delta_* \varphi^a = \delta \varphi^a$, and all that remains in the variation formula (3.32) is

$$\delta S = \int_V d^4 x \left[\frac{\partial^c}{\partial x^\mu} \left(\frac{\partial \mathscr{L}}{\partial \varphi^a_{,\mu}(x)} \delta \varphi^a(x) \right) + \left(\frac{\partial \mathscr{L}}{\partial \varphi^a(x)} - \frac{\partial^c}{\partial x^\mu} \frac{\partial \mathscr{L}}{\partial \varphi^a_{,\mu}(x)} \right) \delta \varphi^a(x) \right].$$

$$\tag{3.34}$$

S is extremal, if δS vanishes for arbitrary variations $\delta \varphi^a(x)$ with compact support in V. This implies

$$\frac{\partial \mathscr{L}}{\partial \varphi^a(x)} - \frac{\partial^c}{\partial x^\mu} \frac{\partial \mathscr{L}}{\partial \varphi^a_{,\mu}(x)} = 0. \tag{3.35}$$

Equation (3.35) is the desired field equation. The divergence term equals a surface integral on the boundary of the support of the variation and thus vanishes.

Example: Consider a scalar field $\Phi(x)$ with Lagrange function

$$L = (1/2)g^{\mu\nu}\Phi_{,\mu}\Phi_{,\nu} - (1/2)m^2\Phi^2.$$

We have

$$\frac{\partial \mathscr{L}}{\partial \Phi} = -\sqrt{-g}\, m^2\Phi, \quad \frac{\partial \mathscr{L}}{\partial \Phi_{,\mu}} = \sqrt{-g}\, g^{\mu\nu}\partial_\nu\Phi.$$

Thus the field (3.35) becomes

$$\frac{1}{\sqrt{-g}}\partial_\mu \left(\sqrt{-g}\, g^{\mu\nu}\partial_\nu\Phi \right) + m^2\Phi = 0. \tag{3.36}$$

This generalizes the Klein–Gordon equation to curved space–times.

3.3 Covariant Derivative

The dynamical laws for the fields—the so-called field equations—take the form of differential equations. The evolution of the fields depends on the difference of the excitation in neighboring points. (A simplified model would be a coupled system of pendulums.)

However, in curved space–times, we face the following difficulty. Consider the familiar example of a covector field $A_\mu(x)$. Its partial derivatives transform as follows:

$$\frac{\partial A'_\mu(x)}{\partial x'^\mu}(p) = X^\rho_{\mu'}(p)X^\sigma_{\nu'}(p)\frac{\partial A_\rho(x)}{\partial x^\sigma}(p) + X^\rho_{\mu'\nu'}(p)A_\rho(p) . \qquad (3.37)$$

We see that we can make this derivative arbitrarily large at any point p by suitably choosing the derivatives

$$X^\rho_{\mu'\nu'}(p) := \frac{\partial^2 x^\rho}{\partial x'^\mu \partial x'^\nu}(p)$$

at p. This even works if the derivative of the field $\frac{\partial A_\rho(x)}{\partial x^\sigma}(p)$ vanishes at p. Thus, the partial derivative of the components of a covector is not a suitable differential operator for the field equations. Therefore, the differential operators in the previously derived field equations must have a more geometric meaning. We will see that we can express them in terms of the so-called *covariant derivative*.

3.3.1 Definition of the Covariant Derivative

In special relativity the above objection to the partial derivatives with respect to coordinates does not hold. If we are working exclusively in an inertial frame, then all transformations are linear, and we have

$$\frac{\partial^2 x^\rho}{\partial x'^\mu \partial x'^\nu}(p) = 0, \quad \forall p.$$

The difference of the components of the tensor in neighboring points, computed in this way, is meaningful and useful.

In general relativity there is an analogy to the inertial frame, namely the local inertial frame. Although they are defined only locally, they are sufficient for our purposes. To illustrate this, consider two local inertial frames $\bar{x}^\mu$ and $\bar{x}'^\mu$ at p. The corresponding coordinate transformation is

$$\bar{x}'^\mu = \bar{x}'^\mu\left(\bar{x}^0, \cdots, \bar{x}^3\right) ,$$

where always

$$\frac{\partial^2 \bar{x}^\rho}{\partial \bar{x}'^\mu \partial \bar{x}'^\nu}(p) = 0 .$$

This follows from the definition of a local inertial frame as $\bar{\Gamma}'^\mu_{\nu\rho}(p) = 0$ and $\bar{\Gamma}^\mu_{\nu\rho}(p) = 0$, and the transformation rule of the affine connections given by (1.9). We find

$$X^{\bar{\alpha}'}_\rho \frac{\partial^2 \bar{x}^\rho}{\partial \bar{x}'^\mu \partial \bar{x}'^\nu}(p) = 0$$

which is equivalent to the previous equation. Hence, we have

$$\frac{\partial \bar{A}'_\mu(\bar{x}')}{\partial \bar{x}'^\mu}(p) = X^{\bar{\rho}}_{\bar{\mu}'}(p) X^{\bar{\sigma}}_{\bar{v}'}(p) \frac{\partial \bar{A}_\rho(\bar{x})}{\partial \bar{x}^\sigma}(p) \,.$$

Therefore the (first) derivatives with respect to local inertial frame coordinates transform like tensors, as in special relativity. We just have to use the local inertial frames at a point p to compute the derivative of a field at p.

In this way we can define the derivative in a restricted set of coordinate systems only. How can this definition be extended to general coordinate systems? This is as simple as to *postulate* that the derivative be a tensor. The logical structure is as follows. First, one computes the components of an object with respect to a particular coordinate system. Second, the components with respect to another, arbitrary coordinate system are given by those with respect to the original system and the transformation law. This makes the object well defined as a differential geometric object, as all its components are known in any coordinate system.

Definition 16 Let $T^{\mu\dots}_{v\dots}(x)$ be the component functions of a tensor field of type (p,q) in the coordinates x^μ. Then $\nabla_\rho T^{\mu\dots}_{v\dots}(x)$ are the component functions of a tensor field of type $(p, q+1)$ in x^μ, which satisfy the following condition. With respect to an arbitrary local inertial frame $\bar{x}^\mu$ at an arbitrary point r we have

$$\bar{\nabla}_\rho \bar{T}^{\mu\dots}_{v\dots}(\bar{x}) := \frac{\partial \bar{T}^{\mu\dots}_{v\dots}}{\partial \bar{x}^\rho}(\bar{x}) \,.$$

The tensor field defined in this way is the covariant derivative of the tensor field $T^{\mu\dots}_{v\dots}(x)$.

3.3.2 Direct Expression for the Covariant Derivative

The previous definition has some useful aspects, but is not very handy to use. For example, to show that the tensor field defined in this way has smooth component functions, provided the original field is smooth, would be very cumbersome. To do this, one has to know the covariant derivative in *one* coordinate system in a whole neighborhood of a point.

The definition obviously depends on the affine connection of the manifold. If we know the Γs, we compute the local inertial frame and the covariant derivative. But there is also a direct expression, which holds in a single coordinate system and contains the Γs. We now derive such an expression.

Let p be an arbitrary point, $\bar{x}^\mu$ a local inertial frame at p, and x^μ an arbitrary coordinate system near p. Then the definition implies that the components of the covariant derivative $\nabla_\rho T^{\mu\dots}_{v\dots}(p)$ satisfy

$$\nabla_\rho T^{\mu\dots}_{v\dots}(p) = \left(X^\mu_{\bar{\alpha}} \dots X^{\bar{\beta}}_v \dots X^{\bar{\gamma}}_\rho \partial_{\bar{\gamma}} \bar{T}^{\alpha\dots}_{\bar{\beta}\dots} \right)(p) \,.$$

The expression on the right-hand side can be rearranged as follows:

$$\left(X^{\mu}_{\bar{\alpha}}\ldots X^{\bar{\beta}}_{\nu}\ldots\left(X^{\bar{\gamma}}_{\rho}\partial_{\bar{\gamma}}\bar{T}^{\alpha\cdots}_{\beta\cdots}\right)\right)(p)=\left(X^{\mu}_{\bar{\alpha}}\ldots X^{\bar{\beta}}_{\nu}\ldots\frac{\partial}{\partial x^{\rho}}\left(\bar{T}^{\alpha\cdots}_{\beta\cdots}\right)\right)(p)$$

$$=\frac{\partial}{\partial x^{\rho}}\left(X^{\mu}_{\bar{\alpha}}\ldots X^{\bar{\beta}}_{\nu}\ldots\bar{T}^{\alpha\cdots}_{\beta\cdots}\right)(p)$$

$$-\left(\frac{\partial X^{\mu}_{\bar{\alpha}}}{\partial x^{\rho}}\ldots X^{\bar{\beta}}_{\nu}\ldots\bar{T}^{\alpha\cdots}_{\beta\cdots}\right)(p)-\cdots-\left(X^{\mu}_{\bar{\alpha}}\ldots\frac{\partial X^{\bar{\beta}}_{\nu}}{\partial x^{\rho}}\ldots\bar{T}^{\alpha\cdots}_{\beta\cdots}\right)(p)-\cdots,$$

where the dots in the sum represent similar terms, which come from the corresponding derivatives of the remaining transformation coefficients $X^{\bar{\kappa}}_{\lambda}$ or $X^{\kappa}_{\bar{\lambda}}$. Furthermore, the derivatives of the transformation coefficients satisfy

$$\frac{\partial X^{\mu}_{\bar{\alpha}}}{\partial x^{\rho}}=X^{\mu}_{\bar{\alpha}\bar{\gamma}}X^{\bar{\gamma}}_{\rho}=\frac{\partial}{\partial \bar{x}^{\alpha}}\left(X^{\mu}_{\bar{\gamma}}X^{\bar{\gamma}}_{\rho}\right)-X^{\mu}_{\bar{\gamma}}X^{\bar{\gamma}}_{\rho\lambda}X^{\lambda}_{\bar{\alpha}}=\frac{\partial \delta^{\mu}_{\rho}}{\partial x^{\alpha}}-X^{\mu}_{\bar{\gamma}}X^{\bar{\gamma}}_{\rho\lambda}X^{\lambda}_{\bar{\alpha}}=-X^{\mu}_{\bar{\gamma}}X^{\bar{\gamma}}_{\rho\lambda}X^{\lambda}_{\bar{\alpha}}$$

and

$$\frac{\partial X^{\bar{\beta}}_{\nu}}{\partial x^{\rho}}=X^{\bar{\beta}}_{\nu\rho}=X^{\bar{\beta}}_{\sigma}X^{\sigma}_{\bar{\gamma}}X^{\bar{\gamma}}_{\rho\nu}\ .$$

These derivatives can now be expressed in terms of the Γs, as the transformation rule for the affine connection yields

$$\Gamma^{\mu}_{\nu\rho}=X^{\mu}_{\bar{\gamma}}X^{\bar{\gamma}}_{\nu\rho}\ .$$

Hence

$$\frac{\partial X^{\mu}_{\bar{\alpha}}}{\partial x^{\rho}}=-\Gamma^{\mu}_{\lambda\rho}X^{\lambda}_{\bar{\alpha}}\ ,\qquad \frac{\partial X^{\bar{\beta}}_{\nu}}{\partial x^{\rho}}=\Gamma^{\sigma}_{\nu\rho}X^{\bar{\beta}}_{\sigma}\ .$$

This results in the desired general expression:

$$\nabla_{\rho}T^{\mu\cdots}_{\nu\cdots}=\partial_{\rho}T^{\mu\cdots}_{\nu\cdots}+\Gamma^{\mu}_{\lambda\rho}T^{\lambda\cdots}_{\nu\cdots}+\cdots-\Gamma^{\sigma}_{\nu\rho}T^{\mu\cdots}_{\sigma\cdots}-\cdots\ . \tag{3.38}$$

Let us discuss this important formula. The structure of the right-hand side is simple: the first term is the corresponding coordinate derivative of the field. In addition, there are correction terms, one for each index of the field. The sign "+" is used for contravariant indices, and the sign "−" for covariant indices. The correction terms are formed from products of the tensor with the Γs such that Γ carries the corresponding index of the tensor, a summation index appearing also in the tensor, and the index of the derivative. We always write the index of the derivative (here ρ) as the last lower index at Γ.

Examples: A scalar field $\Phi(x)$:

$$\nabla_{\rho}\Phi(x)=\partial_{\rho}\Phi(x)\ ,$$

a vector field $V^\mu(x)$:

$$\nabla_\rho V^\mu(x) = \partial_\rho V^\mu(x) + \Gamma^\mu_{\sigma\rho} V^\sigma(x) \,,$$

a covector field $U_\mu(x)$:

$$\nabla_\rho U_\mu(x) = \partial_\rho U_\mu(x) - \Gamma^\sigma_{\mu\rho} V^\sigma(x) \,,$$

etc.

3.3.3 Algebraic Properties

Through operations of tensor algebra—linear combination, product, and contraction—we can form new tensor fields from other tensor fields by application of these operations at every point to the components of the tensors there. How does the covariant derivative interact with these constructions?

3.3.3.1 Linear Combination

Let $T^{\mu\dots\nu}_{\rho\dots\sigma}(x)$ and $S^{\mu\dots\nu}_{\rho\dots\sigma}(x)$ be two tensor fields of type (p,q), a and b two constants (independent of coordinates). Then the linear combination $aT^{\mu\dots\nu}_{\rho\dots\sigma}(x) + bS^{\mu\dots\nu}_{\rho\dots\sigma}(x)$ is again a tensor fields of type (p,q) and

$$\nabla_\lambda \left(aT^{\mu\dots\nu}_{\rho\dots\sigma}(x) + bS^{\mu\dots\nu}_{\rho\dots\sigma}(x) \right) = a\nabla_\lambda T^{\mu\dots\nu}_{\rho\dots\sigma}(x) + b\nabla_\lambda S^{\mu\dots\nu}_{\rho\dots\sigma}(x) \,.$$

Hence, the covariant derivative is a *linear* operation. The proof is simple if we use the usual trick to compute the components of all tensors in a special coordinate system, find that the relations between them behave like a tensor, and hence infer that it is valid in arbitrary coordinates. Here, the special system is a local inertial frame, and we can directly use the definition of the covariant derivative. In a local inertial frame, the covariant derivative reduces to the coordinate derivative, and the coordinate derivative is linear. Thus the claim follows immediately.

We can also use (3.38) for the proof. We carry this out for two vector fields. The general proof works analogously.

Thus, let $V^\mu(x)$ and $U^\mu(x)$ be two vector fields. We compute the covariant derivative of a linear combination of these from (3.38):

$$\begin{aligned}
\nabla_\lambda \left(aV^\mu(x) + bU^\mu(x) \right) &= \partial_\lambda \left(aV^\mu(x) + bU^\mu(x) \right) + \Gamma^\mu_{\rho\lambda} \left(aV^\rho(x) + bU^\rho(x) \right) \\
&= a\partial_\lambda V^\mu(x) + b\partial_\lambda U^\mu(x) + a\Gamma^\mu_{\rho\lambda} V^\rho(x) + b\Gamma^\mu_{\rho\lambda} U^\rho(x) \\
&= a\nabla_\lambda V^\mu(x) + b\nabla_\lambda U^\mu(x) \,.
\end{aligned}$$

qed.

3.3.3.2 Tensor Product

Let $T^{\mu...\nu}_{\rho...\sigma}(x)$ and $S^{\alpha...\beta}_{\gamma...\delta}(x)$ be two tensor fields of type (p_1,q_1) and (p_2,q_2). Their tensor product satisfies

$$\nabla_\lambda \left(T^{\mu...\nu}_{\rho...\sigma}(x)S^{\alpha...\beta}_{\gamma...\delta}(x)\right) = \left(\nabla_\lambda T^{\mu...\nu}_{\rho...\sigma}(x)\right) S^{\alpha...\beta}_{\gamma...\delta}(x) + T^{\mu...\nu}_{\rho...\sigma}(x) \left(\nabla_\lambda S^{\alpha...\beta}_{\gamma...\delta}(x)\right) .$$

Hence, the covariant derivative satisfies the Leibniz rule like the coordinate derivative. Again, the general proof follows from the definition and the fact that partial coordinate derivatives satisfy the Leibniz rule.

Alternatively, here is a proof for two vector fields $V^\mu(x)$ and $U^\mu(x)$ using (3.38):

$$\nabla_\lambda \left(V^\mu(x)U^\alpha(x)\right)$$
$$= \partial_\lambda \left(V^\mu(x)U^\alpha(x)\right) + \Gamma^\mu_{\rho\lambda} \left(V^\rho(x)U^\alpha(x)\right) + \Gamma^\alpha_{\rho\lambda} \left(V^\mu(x)U^\rho(x)\right)$$
$$= \left(\partial_\lambda V^\mu(x)\right) U^\alpha(x) + V^\mu(x)\left(\partial_\lambda U^\alpha(x)\right) + \Gamma^\mu_{\rho\lambda} V^\rho(x)U^\alpha(x) + \Gamma^\alpha_{\rho\lambda} V^\mu(x)U^\rho(x)$$
$$= \left(\nabla_\lambda V^\mu(x)\right) U^\alpha(x) + V^\mu(x)\left(\nabla_\lambda U^\alpha(x)\right) ,$$

qed.

3.3.3.3 Contraction

Let $W^\mu_{\rho\sigma}(x)$ be a tensor field of type $(1,2)$. We can form a contraction in every point $U_\sigma(x) = W^\rho_{\rho\sigma}(x)$. $U_\sigma(x)$ is a tensor of type $(0,1)$. What is the relationship between the covariant derivatives of these two tensor fields? We have

$$\nabla_\lambda W^\mu_{\rho\sigma}(x) = \partial_\lambda W^\mu_{\rho\sigma}(x) + \Gamma^\mu_{\alpha\lambda} W^\alpha_{\rho\sigma}(x) - \Gamma^\alpha_{\rho\lambda} W^\mu_{\alpha\sigma}(x) - \Gamma^\alpha_{\sigma\lambda} W^\mu_{\rho\alpha} .$$

This is a tensor field of type $(1,3)$, and we contract the two indices μ and ρ:

$$\nabla_\lambda W^\rho_{\rho\sigma}(x) = \partial_\lambda W^\rho_{\rho\sigma}(x) + \Gamma^\rho_{\alpha\lambda} W^\alpha_{\rho\sigma}(x) - \Gamma^\alpha_{\rho\lambda} W^\rho_{\alpha\sigma}(x) - \Gamma^\alpha_{\sigma\lambda} W^\rho_{\rho\alpha}$$
$$= \partial_\lambda U_\sigma(x) - \Gamma^\alpha_{\sigma\lambda} U_\alpha(x) = \nabla_\lambda U_\sigma(x) .$$

In this case, we were able to show that forming the covariant derivative first and contracting afterwards yields the same result as forming the covariant derivative of the contraction. This rule—that contraction and covariant derivative commute—holds in general. However, we must take care to contract the same indices.

3.3.3.4 Composition of Covariant Derivatives

Forming the covariant derivative of a tensor field $T(x)$ of type (p,q) yields a tensor field of type $(p,q+1)$. Hence, the covariant derivative can be applied again, and the result will be a tensor field of type $(p, q+2)$. It turns out that the covariant derivatives do not commute, in contrast to the coordinate derivatives. Instead, we have the following formula:

$$\nabla_\lambda \nabla_\kappa T^{\mu \ldots \nu}_{\rho \ldots \sigma}(x) - \nabla_\kappa \nabla_\lambda T^{\mu \ldots \nu}_{\rho \ldots \sigma}(x)$$

$$= R^\mu_{\alpha \lambda \kappa} T^{\alpha \ldots \nu}_{\rho \ldots \sigma} + \ldots + R^\nu_{\alpha \lambda \kappa} T^{\mu \ldots \alpha}_{\rho \ldots \sigma} - R^\alpha_{\rho \lambda \kappa} T^{\mu \ldots \nu}_{\alpha \ldots \sigma} - \ldots - R^\alpha_{\sigma \lambda \kappa} T^{\mu \ldots \nu}_{\rho \ldots \alpha} , \quad (3.39)$$

where $R^\mu_{\alpha \lambda \kappa}$ is the curvature tensor of the affine connection. In particular, two covariant derivatives do not commute for smooth tensor fields whenever space–time is curved.

Equation (3.39) *cannot* be derived from the direct definition of the covariant derivative in a local inertial frame. It is not true that $\bar{\nabla}_\mu \bar{\nabla}_\nu \bar{T}(p) = \bar{\partial}_\mu \bar{\partial}_\nu \bar{T}(p)$, since we have to know the first derivative in a whole neighborhood of p to compute the second derivative.

3.3.4 Metric Affine Connections

So far, we worked with a general affine connection, and did not refer to a metric. In this section we shall consider the important case of a metric affine connection. We thus let M be an n-manifold with metric $g_{\mu \nu}(x)$. The components of the affine connection are given by the Christoffel symbols (cf. (1.16)):

$$\Gamma^\mu_{\rho \sigma} = \left\{ {}^\mu_{\rho \sigma} \right\} .$$

3.3.4.1 Covariant Derivative of the Metric

Theorem 9 *The covariant derivative of the tensor field δ^ρ_σ, with components the Kronecker-delta in each point with respect to every coordinate system, is zero:*

$$\nabla_\mu \delta^\rho_\sigma = 0 \qquad (3.40)$$

for every affine connection on $\mathcal{M}$. For the covariant derivative of the metric affine connection $\Gamma^\mu_{\rho \sigma}$ of $g_{\mu \nu}(x)$, we have in addition that

$$\nabla_\rho g_{\mu \nu}(x) = 0, \quad \nabla_\rho g^{\mu \nu}(x) = 0 \qquad (3.41)$$

Proof All three equations directly follow from the definition of the covariant derivative, as the components of all three tensors in a local inertial frame have vanishing derivatives with respect to all coordinates.

A different proof works as follows. Equation (3.38) yields that

$$\nabla_\mu \delta^\rho_\sigma = \partial_\mu \delta^\rho_\sigma + \Gamma^\rho_{\lambda \mu} \delta^\lambda_\sigma - \Gamma^\lambda_{\sigma \mu} \delta^\rho_\lambda = 0 .$$

Similarly,

$$\nabla_\rho g_{\mu\nu}(x) = \partial_\rho g_{\mu\nu} - \Gamma^\lambda_{\mu\rho} g_{\lambda\nu} - \Gamma^\lambda_{\nu\rho} g_{\mu\lambda} . \tag{3.42}$$

Identity (2.12) implies that the right-hand side vanishes. Using this result, (3.42) yields the first equation of (3.41). By covariantly differentiating the identity

$$g_{\mu\nu}(x) g^{\nu\rho}(x) = \delta^\rho_\mu$$

we obtain that

$$\left(\nabla_\lambda g_{\mu\nu}(x)\right) g^{\nu\rho}(x) + g_{\mu\nu}(x) \left(\nabla_\lambda g^{\nu\rho}(x)\right) = \nabla_\lambda \delta^\rho_\mu .$$

Inserting the previously derived equations, we arrive at

$$g_{\mu\nu}(x) \left(\nabla_\lambda g^{\nu\rho}(x)\right) = 0 .$$

As $g_{\mu\nu}$ is a regular matrix, the second equation of (3.41) follows immediately, qed.

3.3.4.2 Covariant Divergence

Here we have a closer look at the differential operators appearing in the variation of the invariant Lagrange functions. Consider a vector field $V^\mu(x)$. Its covariant derivative $\nabla_\nu V^\mu(x)$ is a tensor field of type $(1,1)$ and hence can be contracted to yield the so-called *covariant divergence* $\nabla_\mu V^\mu(x)$ of $V^\mu(x)$. This operation maps each vector field into a scalar field. The meaning of the covariant divergence of a vector field is simply that of a source for this field. (The total flow of the vector through the surface of an infinitesimal n-cube is $(\nabla_\mu V^\mu)\sqrt{-g}\, d^n x$).

The covariant divergence can be expressed in terms of the metric by substituting into the equation

$$\nabla_\mu V^\mu(x) = \partial_\mu V^\mu(x) + \Gamma^\mu_{\alpha\mu} V^\alpha(x) , \tag{3.43}$$

from the frequently used formula:

$$\Gamma^\mu_{\alpha\mu} = \frac{1}{\sqrt{-g}} \frac{\partial \sqrt{-g}}{\partial x^\alpha} . \tag{3.44}$$

Hence, we obtain for the covariant divergence

$$\nabla_\mu V^\mu(x) = \partial_\mu V^\mu(x) + \frac{1}{\sqrt{-g}} \frac{\partial \sqrt{-g}}{\partial x^\alpha} V^\alpha(x)$$

$$= \frac{1}{\sqrt{-g}} \left(\sqrt{-g}\, \partial_\mu V^\mu(x) + (\partial_\mu \sqrt{-g}) V^\mu(x)\right) ,$$

or equivalently

$$\nabla_\mu V^\mu(x) = \frac{1}{\sqrt{-g}} \partial_\mu \left(\sqrt{-g}\, V^\mu(x)\right) . \tag{3.45}$$

This formula is better suited to compute the covariant divergence than (3.43).

An important application of the covariant divergence can be deduced from (3.45). Let u^μ be a *normalized* surface-orthogonal vector field and let x^μ be coordinates adapted to u^μ such that, first,

$$u^\mu = \delta_0^\mu$$

and second, the hypersurfaces $x^0 = $ const are orthogonal to u^μ, that is

$$ds^2 = (dx^0)^2 + g_{kl} dx^k dx^l .$$

These coordinates exist in a neighborhood of every point. Then,

$$\nabla_\mu u^\mu = \frac{1}{\sqrt{^3 g}} \partial_0 \sqrt{^3 g} ,$$

where $^3 g$ is the determinant of the 3-metric g_{kl} on the hypersurface $x^0 = $ const, induced by the ambient metric. Thus, the divergence equals the relative growth of the 3-volume $\det(g_{kl}) d^3 x$ in adapted coordinates.

If we have a tensor field instead of a vector, a similar formula can only be derived in special cases. For example, consider a field $T^{\mu\nu}(x)$ of type $(2,0)$:

$$\nabla_\mu T^{\mu\nu}(x) = \partial_\mu T^{\mu\nu}(x) + \Gamma_{\alpha\mu}^\mu T^{\alpha\nu}(x) + \Gamma_{\alpha\mu}^\nu T^{\mu\alpha}(x)$$
$$= \frac{1}{\sqrt{-g}} \partial_\mu \left(\sqrt{-g} T^{\mu\nu}(x) \right) + \Gamma_{\alpha\mu}^\nu T^{\mu\alpha}(x) .$$

The last term on the right-hand side vanishes if the tensor field $T^{\mu\nu}(x)$ is antisymmetric, as Γ is symmetric in its lower two indices.

In general, if $T^{\mu \dots \nu}(x)$ is a totally antisymmetric tensor field of type $(p,0)$, $p \leq n$, that is

$$T^{\mu \dots \rho \dots \sigma \dots \nu}(x) = -T^{\mu \dots \sigma \dots \rho \dots \nu}(x)$$

for each pair of indices ρ and σ, then the covariant divergence can be expressed as

$$\nabla_\mu T^{\mu \dots \nu}(x) = \frac{1}{\sqrt{-g}} \partial_\mu \left(\sqrt{-g} T^{\mu \dots \nu}(x) \right) \tag{3.46}$$

in terms of the metric.

The electromagnetic tensor $F_{\mu\nu}$, for example, is totally antisymmetric. Its covariant form $F^{\mu\nu} = g^{\mu\kappa} g^{\nu\lambda} F_{\kappa\lambda}$ is also antisymmetric and of type $(2,0)$. Hence,

$$\nabla_\mu \left(g^{\mu\kappa} g^{\nu\lambda} F_{\kappa\lambda} \right) = \frac{1}{\sqrt{-g}} \partial_\mu \left(\sqrt{-g} g^{\mu\kappa} g^{\nu\lambda} F_{\kappa\lambda} \right) .$$

Up to a constant factor, this is the differential operator in the generalized version of the Maxwell equations (3.13). We can rewrite this equation as follows:

$$\nabla_\mu F^{\mu\nu} = 0 . \tag{3.47}$$

Note that this is equivalent to the equation

$$\nabla_\mu F_\nu^\mu = 0 \ .$$

where $F_\nu^\mu = g_{\nu\kappa} F^{\mu\kappa}$. This is due to the fact that the metric is covariantly constant, (3.41). However, only (3.47) can be written in the form (3.46).

3.3.4.3 Covariant Laplacean

Let $T_{\rho\ldots\sigma}^{\mu\ldots\nu}(x)$ be an arbitrary tensor field of type (p,q). Its double covariant derivative is a tensor field of type $(p,q+2)$. Then we can form the following tensor field of type (p,q):

$$g^{\kappa\lambda} \nabla_\kappa \nabla_\lambda T_{\rho\ldots\sigma}^{\mu\ldots\nu}(x) = \Delta T_{\rho\ldots\sigma}^{\mu\ldots\nu}(x) \ .$$

The corresponding differential operator is called *covariant Laplacean*.

Example: A scalar field $\Phi(x)$. Due to (3.41), we have

$$\Delta\Phi = g^{\kappa\lambda} \nabla_\kappa \nabla_\lambda \Phi(x) = \nabla_\kappa \left(g^{\kappa\lambda} \nabla_\lambda \Phi(x) \right) \ .$$

The expression in parentheses is a vector field, and thus the right-hand side has the form of a covariant divergence. Therefore, (3.45) implies

$$\Delta\Phi(x) = \frac{1}{\sqrt{-g}} \partial_\kappa \left(\sqrt{-g} g^{\kappa\lambda} \partial_\lambda \Phi(x) \right) \ . \tag{3.48}$$

This is the differential operator which appears in the generalized Klein–Gordon equation (3.36). It is very useful and frequently used.

3.4 The Stress-Energy Tensor

3.4.1 Definition

In field theory on Minkowski space–time, the so-called canonical stress-energy tensor is defined as a quantity Θ_ν^μ comprised of four Noether currents. These currents are connected to the invariance of the action with respect to the four Poincaré translations but they are not uniquely defined [4]. Only integrals thereof on Cauchy hypersurfaces are well defined. These integrals play the role of total energy and momentum. The stress-energy tensor describes for example how energy and momentum of a field are distributed in space–time. Such a distribution can only make physical sense in general if there is a procedure to measure it. The theory of gravity can supply such a procedure, since energy density is a source of gravity. In principle, it is thus possible to measure the energy distribution in space–time by measuring the resulting gravitational field.

For a field model including gravity, the action has two parts:

$$S = S_G + S_M \,,$$

where S_G is the action of gravity and S_M is the action of the other fields in the model. By variation with respect to the metric, we obtain

$$\frac{\delta S_G}{\delta g_{\mu\nu}(x)} = -\frac{\delta S_M}{\delta g_{\mu\nu}(x)} \,.$$

The left-hand side yields the equations of motion for the metric, and therefore the right-hand side is the corresponding source term. The stress-energy tensor is somehow related to this right-hand side. From the example of electrodynamics, we are motivated to postulate

$$T^{\mu\nu}(x) = -\frac{2}{\sqrt{-g}} \frac{\delta S_M}{\delta g_{\mu\nu}(x)} \,. \tag{3.49}$$

The factor $(\sqrt{-g})^{-1}$ transforms the variation into a tensor, as we will see later. The factor -2 is a convention determining the coefficient of the action of matter.

The variational derivative on the right-hand side of (3.49) is defined as follows. We split the field $\psi^a(x)$ in the variation formula (3.32) into two parts, $\psi^a(x) = (g_{\mu\nu}(x), \phi^A(x))$, and substitute for the variations:

$$\delta x^\mu = 0, \quad \delta\phi^A(x) = 0 \,.$$

Here $\delta g_{\mu\nu}(x)$ is compactly supported in the (open) domain of integration. Then $\delta_* = \delta$, and the variation of the action becomes

$$\delta S = \int d^4x \, W^{\mu\nu}(x) \delta g_{\mu\nu}(x) \,.$$

The coefficients $W^{\mu\nu}(x)$ are not yet determined, as the metric is symmetric in the indices μ and ν. It is possible to add an arbitrary antisymmetric tensor to $W^{\mu\nu}(x)$. We remove this freedom by requiring that $W^{\mu\nu}$ be symmetric in μ and ν. With these conventions, the stress-energy tensor (3.49) becomes well defined, where

$$\frac{\delta S_M}{\delta g_{\mu\nu}(x)} := W^{\mu\nu}(x) \,.$$

From the variation formula (3.32), we infer

$$W^{\mu\nu}(x) = \frac{\partial \mathscr{L}_M}{\partial g_{\mu\nu}(x)} - \frac{\partial}{\partial x^\rho} \frac{\partial \mathscr{L}_M}{\partial g_{\mu\nu,\rho}(x)} \,. \tag{3.50}$$

Note that we defined only the stress-energy tensor for matter in this way—the method does not work for gravity. For instance, the variation of the total action with respect to $g_{\mu\nu}(x)$ vanishes, if the equations of motion are satisfied. Hence the method does not produce a quantity like a stress-energy tensor for the whole system, including gravity. Up to now no reasonable candidate for the stress-energy tensor of gravity has been defined. This is most likely related to the fact that such a tensor would have to determine the energy of gravity contained in an infinitesimal volume. On the other hand, the equivalence principle implies that a gravitational field in such a volume can be transformed away by an appropriate coordinate change.

3.4.2 Properties

The definition of the stress-energy tensor implies immediately that

1. $T^{\mu\nu}(x)$ is symmetric in the indices μ and ν,
   and furthermore,
2. $T^{\mu\nu}(x)$ is a tensor field of type $(2,0)$.

Proof Equation (3.49) implies

$$\delta S_M = -\frac{1}{2} \int d^4x \sqrt{-g} T^{\mu\nu}(x) \delta g_{\mu\nu}(x) .$$

δS_M is a scalar for each variation $\delta g_{\mu\nu}(x)$. It follows that we also have

$$\delta S_M = -\frac{1}{2} \int d^4x' \sqrt{-g'} T'^{\mu\nu}(x') \delta g'_{\mu\nu}(x') ,$$

where x'^μ are different coordinates in the domain of integration. In the second integral, we substitute $x'^\mu = x'^\mu(x)$:

$$\delta S_M = -\frac{1}{2} \int d^4x \left| \frac{\partial(x'^0, \dots, x'^3)}{\partial(x^0, \dots, x^3)} \right| \sqrt{-g'(x'(x))} T'^{\mu\nu}(x'(x)) X^\rho_{\mu'} X^\sigma_{\nu'} \delta g_{\rho\sigma}(x)$$

$$= -\frac{1}{2} \int d^4x \sqrt{-g(x)} \left(T'^{\mu\nu}(x'(x)) X^\rho_{\mu'} X^\sigma_{\nu'} \right) \delta g_{\rho\sigma}(x) .$$

Here we used (1.14) to transform the determinant of g. It follows that

$$\int d^4x \sqrt{-g(x)} \left(T'^{\mu\nu}(x'(x)) X^\rho_{\mu'} X^\sigma_{\nu'} - T^{\rho\sigma}(x) \right) \delta g_{\rho\sigma}(x) = 0$$

for all $\delta g_{\mu\nu}(x)$. Hence, the coefficient of $\delta g_{\mu\nu}(x)$ must be zero, and we obtain

$$T'^{\mu\nu}(x'(x)) X^\rho_{\mu'} X^\sigma_{\nu'} = T^{\rho\sigma}(x)$$

qed.

The following theorem states the most important property of $T^{\mu\nu}(x)$.

Theorem 10 *If S_M is a scalar and the equations of motion are satisfied for the matter fields,*

$$\frac{\partial \mathscr{L}_M}{\partial \varphi^A(x)} - \frac{\partial}{\partial x^\rho} \frac{\partial \mathscr{L}_M}{\partial \varphi^A_{,\rho}(x)} = 0, \tag{3.51}$$

then

$$\nabla_\mu T^{\mu\nu}(x) = 0. \tag{3.52}$$

Equation (3.52) is called *divergence formula* for stress-energy tensor.

Proof The coordinate invariance of S_M implies that the variation δS_M with respect to an infinitesimal coordinate transformation (3.24) vanishes. On the other hand, this variation can be computed by the variation formula (3.32). We demand that the variation $\delta x^\mu(x)$ be compactly supported in the domain of integration V for the action. The corresponding variation of the remaining fields can be computed from the vector field $\delta x^\mu(x)$. For example, for $g_{\mu\nu}(x)$ we get

$$g'_{\mu\nu}(x') = X^\rho_{\mu'} X^\sigma_{\nu'} g_{\rho\sigma}(x) ,$$

where

$$X^\rho_{\mu'} = \delta^\rho_\mu - \partial_\mu \delta x^\rho(x) .$$

Hence, we have

$$\delta g_{\mu\nu}(x) = -g_{\alpha\nu}(x)\partial_\mu \delta x^\alpha(x) - g_{\mu\alpha}(x)\partial_\nu \delta x^\alpha(x) ,$$

and the corresponding form-variation $\delta_* g_{\mu\nu}$ is (cf.(3.30)):

$$\delta_* g_{\mu\nu}(x) = -g_{\alpha\nu}(x)\partial_\mu \delta x^\alpha(x) - g_{\mu\alpha}(x)\partial_\nu \delta x^\alpha(x) - \partial_\alpha g_{\mu\nu}(x)\delta x^\alpha(x) .$$

The right-hand side can be recast to yield the following important formula (exercise):

$$\delta_* g_{\mu\nu}(x) = -\nabla_\mu \delta x_\nu - \nabla_\nu \delta x_\mu, \tag{3.53}$$

where $\delta x_\mu = g_{\mu\nu} \delta x^\nu$.

The variations $\delta \varphi^A(x)$ and $\delta_* \varphi^A(x)$ are also non-trivial, as the fields $\varphi^A(x)$ behave non-trivially under a change of coordinates. However, the explicit formula for these variations are not needed for the proof.

The variation formula (3.32) then implies

$$\delta S_M = \int_V d^4x \left[\frac{\partial}{\partial x^\mu}(\ldots)^\mu + \left(\frac{\partial \mathscr{L}_M(x)}{\partial \varphi^A(x)} - \frac{\partial}{\partial x^\mu} \frac{\partial \mathscr{L}_M(x)}{\partial \varphi^A_{,\mu}(x)} \right) \delta_* \varphi^A(x) \right.$$

$$\left. + \left(\frac{\partial \mathscr{L}_M(x)}{\partial g_{\rho\sigma}(x)} - \frac{\partial}{\partial x^\mu} \frac{\partial \mathscr{L}_M(x)}{\partial g_{\rho\sigma,\mu}(x)} \right) \delta_* g_{\rho\sigma}(x) \right] .$$

The total divergence vanishes since $\delta x^\mu(x)$ has compact support and the second term in parenthesis is zero due to (3.51). We substitute (3.50) and (3.53) into the remainder:

$$\delta S_{\mathrm{M}} = 1/2 \int_V d^4x \, \sqrt{-g} T^{\mu\nu}(x) \left(\nabla_\mu \delta x_\nu(x) + \nabla_\nu \delta x_\mu(x) \right) .$$

By symmetry of the stress-energy tensor, we can omit the factor 1/2 and the last term in parentheses. The rest can be written as follows:

$$\delta S_{\mathrm{M}} = \int_V d^4x \, \sqrt{-g} \nabla_\mu \left(T^{\mu\nu} \delta x_\nu \right) - \int_V d^4x \, \sqrt{-g} \left(\nabla_\mu T^{\mu\nu} \right) \delta x_\nu$$

$$= \int_V d^4x \, \partial_\mu \left(\sqrt{-g} T^{\mu\nu} \delta x_\nu \right) - \int_V d^4x \left(\sqrt{-g} g_{\nu\rho} \nabla_\mu T^{\mu\nu} \right) \delta x^\rho .$$

The first term vanishes, and the second only vanishes for all δx^ρ if $\sqrt{-g} g_{\nu\rho} \nabla_\mu T^{\mu\nu} = 0$, which is equivalent to (3.52), qed.

3.4.3 Interpretation of the Divergence Formula

Let p be a point in space–time and $\{x^\mu\}$ a coordinate system which is geodesic at p, that is $\Gamma^\mu_{\rho\sigma}(p) = 0$, and satisfies

$$g_{\mu\nu}(p) = \eta_{\mu\nu} .$$

Furthermore, we set $x^\mu(p) = 0$. In these special coordinates (3.52) at p becomes

$$\partial_\mu T^{\mu\nu}(p) = 0 .$$

Multiplying this equation with the invariant 4-volume element d^4x and expanding the sum yields

$$\partial_\mu T^{\mu\nu}(p) d^4x = \sum_\mu \left(\partial_\mu T^{\mu\nu}(0) dx^\mu \right) d^3_\mu x,$$

where

$$d^3_\mu x = \left(dx^1 dx^2 dx^3, dx^0 dx^2 dx^3, dx^0 dx^1 dx^3, dx^0 dx^1 dx^2 \right)$$

are the area elements of the four coordinate hyperplanes. These yield the correct measure of the 3-surfaces in the corresponding units (cm^3 or cm^{2}s, if $c \neq 1$) and the surface d^3_μ is perpendicular to the μ-axis.

The expression in parentheses can be interpreted as the difference of the field $T^{\mu\nu}(x)$ at the points p and p_μ, where p_μ are the vertices adjacent to p in the infinitesimal coordinate cube (Fig. 3.1):

$$x^\mu(p_0) = \left(dx^0, 0, 0, 0 \right), x^\mu(p_1) = \left(0, dx^1, 0, 0 \right) ,$$

$$x^\mu(p_2) = \left(0, 0, dx^2, 0 \right), x^\mu(p_3) = \left(0, 0, 0, dx^3 \right) ,$$

whence

$$\sum_\mu \left(\partial_\mu T^{\mu\nu}(0) dx^\mu \right) d^3_\mu x = \sum_\mu \left(T^{\mu\nu}(p_\mu) - T^{\mu\nu}(p) \right) d^3_\mu x .$$

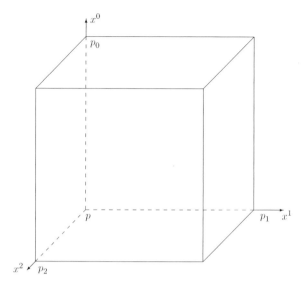

Fig. 3.1 Coordinate cube in a local inertial system

Each of the eight terms in the sum evidently equals the flow of current density (that is the current) of the ν-component of the 4-momentum through one face of the co-ordinate cube. As the sum has to vanish, the ν-component of the 4-momentum is *locally* conserved. This is the interpretation of the divergence (3.52). This corresponds to the result we inferred for particles

$$\frac{\nabla p^\mu}{\mathrm{d}s} = 0 \,.$$

In general, this conservation law *cannot* be integrated on a finite domain of integration so that a global conservation law follows as in Minkowski space. There we have a global, geodesic coordinate system in which (3.52) becomes

$$\partial_\mu T^{\mu\nu}(x) = 0 \,.$$

Integrating this equation over the volume between two hypersurfaces $x^0 = t_1$ and $x^0 = t_2$, we find, using Gauss' theorem, that the total momentum (respectively energy) of the field is conserved:

$$P^\nu = \int \mathrm{d}_0^3 x \, T^{0\nu} = \text{const.}$$

On the other hand, in curved space–time, there is no such global geodesic system, we have to use a new geodesic system for each point, and the coordinate cubes do not fit together. If we try to work in a general coordinate system, then the Γ-corrections to $\partial_\mu T^{\mu\nu}(x)$ in (3.52) do not vanish, and the volume integral cannot be transformed to a surface integral.

The fact that total energy and momentum are only conserved (or even well defined) in flat space–time can be understood as follows. A system of matter in a gravitational field is not isolated, it can exchange energy and momentum with the gravitational field. Equation (3.52) shows that this exchange happens in a *non-local* way. That is, the gravitational field does not have an stress-energy tensor such that the sum of this tensor and $T^{\mu\nu}$—a "total" stress-energy tensor—is locally conserved, as this is the case for two systems of matter in Minkowski space–time [5]. The *total* energy of all fields including gravity can be defined in asymptotically flat space-times as the monopole in the multipole expansion of gravitational field. An example is the Eddington–Robertson expansion for rotationally symmetric space-times. For general stationary fields in weak-field approximation, this will be done in Sect. 4.4.5, (4.66). A more general theory can be found in [6].

3.4.4 Ideal Fluids

Besides electrodynamics, relativistic hydrodynamics is another classical field theory which plays an important role in astrophysics. There are many different fluids, ranging from the cold dust of the galaxies to the hot plasma in the solar corona or accretion discs (magneto-hydrodynamics). The fluid approximation is sufficient in many cases, in particular, for all models in these Notes. More serious astrophysics [7, 8], however, often has to consider more sophisticated forms of gas dynamics, such as relativistic kinetic theory [9].

The equations of motion for fluids in special relativity is the divergence equation $\partial_\nu T^{\mu\nu} = 0$, valid in any inertial frame. The equivalence principle then leads to the postulate

$$\nabla_\nu T^{\mu\nu} = 0, \tag{3.54}$$

valid in all curved space–times and for all coordinates. Here $T^{\mu\nu}$ must be the stress-energy tensor of the whole system. The kind of matter considered is determined by the form of $T^{\mu\nu}$.

Here, we restrict our attention to the most basic fluid, namely the *ideal fluid* (ideal = no viscosity). We proceed by giving a precise definition.

The ideal fluid is characterized by two properties: in every point, there is a rest frame for the fluid (which is unique up to rotation of the spatial axes), and the fluid looks isotropic in this rest frame. A rest frame of a continuous system of matter with stress-energy tensor $T^{\mu\nu}$ at p is a local inertial frame $\bar{x}^\mu$ at p, such that the components of the energy current $T^{0\mu}$ have the form:

$$\bar{T}^{0\mu}(p) = (\rho, 0, 0, 0), \tag{3.55}$$

in these coordinates. Then $\bar{u}^\mu := (1, 0, 0, 0)$ is called the *4-velocity* of the ideal fluid at p. The real number ρ is called *mass density* (or *energy density* or simply "density") of the fluid. To be isotropic means that the tensor $\bar{T}^{\mu\nu}$ is invariant with respect to rotations of the three spatial axes, that is transformations of the form:

$$\bar{x}'^0 = \bar{x}^0 , \quad \bar{x}'^k = \sum_l O_{lk}\bar{x}^l , \tag{3.56}$$

where O_{lk} is an orthogonal matrix. Thus $\{\bar{x}'^\mu\}$ is also a local inertial frame and a rest frame at p, and (3.55) is not affected by the transformation (3.56).

The definition of an ideal fluid implies that the stress-energy tensor $\bar{T}^{\mu\nu}$ and the 4-velocity $\bar{u}^\mu$ have components (with respect to a rest frame) of the form:

$$\bar{T}^{\mu\nu} = \begin{pmatrix} \rho & 0 & 0 & 0 \\ 0 & p & 0 & 0 \\ 0 & 0 & p & 0 \\ 0 & 0 & 0 & p \end{pmatrix} , \quad \bar{u}^\mu = \begin{pmatrix} 1 \\ 0 \\ 0 \\ 0 \end{pmatrix} . \tag{3.57}$$

The number p is called *pressure* of the fluid (thus isotropy is just Pascal's law). In every point we have a rest frame, whence density ρ, pressure p, and 4-velocity u^μ are functions on the space-time M. We assume that these functions are of class C^∞. Therefore, $\rho(x)$ and $p(x)$ are scalar fields and $u^\mu(x)$ is a vector field on M with norm 1.

For each point, there is a different rest frame, but there are no coordinates in which $T^{\mu\nu}(x)$ has the form (3.57) globally. For some computations, it is important to have an expression for the stress-energy tensor in terms of ρ, p, and u^μ, which is valid in an arbitrary coordinate system $\{x^\mu\}$. To find such an expression, note that in a rest frame we have that

$$\bar{g}^{\mu\nu} = \begin{pmatrix} 1 & 0 & 0 & 0 \\ 0 & -1 & 0 & 0 \\ 0 & 0 & -1 & 0 \\ 0 & 0 & 0 & -1 \end{pmatrix} , \quad \bar{u}^\mu \bar{u}^\nu = \begin{pmatrix} 1 & 0 & 0 & 0 \\ 0 & 0 & 0 & 0 \\ 0 & 0 & 0 & 0 \\ 0 & 0 & 0 & 0 \end{pmatrix} ,$$

whence

$$\bar{T}^{\mu\nu} = \rho \bar{u}^\mu \bar{u}^\nu - p(\bar{g}^{\mu\nu} - \bar{u}^\mu \bar{u}^\nu) .$$

The quantities on both sides of this equation are tensors of type $(2,0)$, and the equation tells us that the components of these tensors agree in one coordinate system. As they transform in the same way, they have to agree in every coordinate system. Thus we find the desired relation:

$$T^{\mu\nu}(x) = (\rho(x) + p(x)) u^\mu(x) u^\nu(x) - p(x) g^{\mu\nu}(x) . \tag{3.58}$$

Let us investigate how the state of a fluid at a point can be described, and how equations of motion for the fluid can be derived. The pressure and the density are not independent of each other in actual systems. They are related by the so-called *equation of state*, for example $p = p(\rho)$. The state of an ideal fluid at a point x is determined if we fix $\rho(x)$, $u^1(x)$, $u^2(x)$, and $u^3(x)$. The pressure is then given by the equation of state, and $u^0(x)$ can be computed from the normalization for u^μ (a time-like unit vector). The evolution of the functions $\rho(x)$, $u^1(x)$, $u^2(x)$, and $u^3(x)$ from their initial values is given by (3.54). We want to rewrite the divergence equation as a system of differential equations for the fields ρ, p, and u^μ. Substitute (3.58) into the left-hand side of (3.52). This yields

$$\nabla_\mu [(\rho + p)u^\mu u^\nu - pg^{\mu\nu}]$$
$$= (\partial_\mu \rho + \partial_\mu p)\, u^\mu u^\nu + (\rho + p)u^\nu \nabla_\mu u^\mu + (\rho + p)u^\mu \nabla_\mu u^\nu - g^{\mu\nu}\partial_\mu p$$
$$= [u^\mu \partial_\mu \rho + (\rho + p)\nabla_\mu u^\mu]\, u^\nu + [(\rho + p)u^\mu \nabla_\mu u^\nu - (g^{\mu\nu} - u^\mu u^\nu)\partial_\mu p]\ .$$

The second term is a vector orthogonal to u^μ, which can be seen as follows. The tensor $P^\mu_\nu = g^\mu_\nu - u^\mu u_\nu$ satisfies the equations

$$P^\mu_\nu u^\nu = 0, \quad P^\mu_\nu s^\nu = s^\mu$$

for all vectors s^μ orthogonal to u^μ, that is $u_\nu s^\nu = 0$. Therefore, P^μ_ν is a projection operator onto the directions that are spatial with respect to an observer with 4-velocity u^μ. In addition, $u^\mu \nabla_\mu u^\nu$ is orthogonal to u^ν (exercise). Thus, the conclusion $u_\nu \nabla_\mu T^{\mu\nu} = 0$ of the divergence equation takes the form

$$u^\mu \partial_\mu \rho + (\rho + p)\nabla_\mu u^\mu = 0\ . \tag{3.59}$$

If we substitute this back into the divergence equation, all that remains is

$$(\rho + p)u^\mu \nabla_\mu u^\nu = (g^{\mu\nu} - u^\mu u^\nu)\partial_\mu p\ . \tag{3.60}$$

The first equation says that the time derivative of the energy density is proportional to the divergence of the integral curves of the vector field u^μ plus the work done by the pressure. This equation is called *energy equation*. The second equation has only three independent components. On the left-hand side, there is the 4-acceleration of the integral curves (streamline of the fluid) multiplied by the mass density plus a relativistic correction $pu^\mu \nabla_\mu u^\nu$. The right-hand side is the gradient of the pressure projected to the spatial directions. Thus (3.60) is the relativistic version of *Euler's equation* [10]. It describes the movement of an ideal fluid (no viscosity) in arbitrary gravitational fields.

3.5 Exercises

1. Let $V_\mu(x)$ and $W_{\mu\nu}(x)$ be covariant tensor fields, with $W_{\mu\nu}(x) = -W_{\nu\mu}(x)$ for all μ, ν, x. Show that the following equations, valid in any coordinate system, define tensor fields and that these tensor fields are anti-symmetric in all indices.

$$T_{\mu\nu} := \partial_\mu V_\nu - \partial_\nu V_\mu\ ,$$
$$U_{\mu\nu\rho} := \partial_\mu W_{\nu\rho} + \partial_\nu W_{\rho\mu} + \partial_\rho W_{\mu\nu}\ .$$

2. Let $\{x^\mu\}$ be coordinates, $g_{\mu\nu}(x)$ the components of the metric with respect to $\{x^\mu\}$, $g(x) := \det(g_{\mu\nu}(x))$ and d^4x the 4-volume element. Show that the expression $\sqrt{-g}\mathrm{d}^4x$ is invariant.

3. Show that

$$\mathrm{d}\sqrt{-g} = -(1/2)\sqrt{-g}\,g_{\mu\nu}\mathrm{d}g^{\mu\nu}$$
$$= (1/2)\sqrt{-g}\,g^{\mu\nu}\mathrm{d}g_{\mu\nu} .$$

Hint: Use the sub-determinant formula to express the inverse of a matrix in terms of the derivative of the determinant with respect to a matrix element.

4. Use the variation formula to compute the variation of the action

$$S = -\frac{1}{16\pi}\int \mathrm{d}^4x\, F^{\alpha\beta}F_{\alpha\beta}$$

for the Maxwell field $A_\mu(x)$ in Minkowski space-time with respect to infinitesimal Poincaré transformations. Assume that the field equations

$$\frac{\partial \mathcal{L}}{\partial A_\mu} - \frac{\mathrm{d}}{\mathrm{d}x^\rho}\frac{\partial \mathcal{L}}{\partial A_{\mu,\rho}}$$

are satisfied. Hint: first show that

$$\frac{\partial(F^{\alpha\beta}F_{\alpha\beta})}{\partial(\partial_\mu A_\nu)} = 4F^{\mu\nu} .$$

5. Prove the following three equations:

(a) $\nabla_\lambda\nabla_\kappa\,\varphi(x) = \nabla_\kappa\nabla_\lambda\,\varphi(x)$ for all scalar fields $\varphi(x)$.
(b) $\nabla_\lambda\nabla_\kappa V^\mu(x) - \nabla_\kappa\nabla_\lambda V^\mu(x) = R^\mu_{\nu\lambda\kappa}(x)V^\nu(x)$ for all vector fields $V^\mu(x)$.
  Use the following equations:

$$\nabla_\lambda V^\mu = \partial_\lambda V^\mu + \Gamma^\mu_{\rho\lambda} V^\rho ,$$
$$R^\mu_{\nu\lambda\kappa} = \partial_\lambda\Gamma^\mu_{\nu\kappa} - \partial_\kappa\Gamma^\mu_{\nu\lambda} + \Gamma^\mu_{\lambda\rho}\Gamma^\rho_{\nu\kappa} - \Gamma^\mu_{\kappa\rho}\Gamma^\rho_{\nu\lambda} .$$

(c) $\nabla_\lambda\nabla_\kappa U_\mu(x) - \nabla_\kappa\nabla_\lambda U_\mu(x) = -R^\nu_{\mu\lambda\kappa}U_\nu$ for all covector fields $U_\mu(x)$. Show that this follows directly from the two preceding results.

6. Prove the formula
$$\Gamma^\mu_{\nu\mu} = \partial_\nu \log\sqrt{|g|} .$$

7. Let $\xi^\mu(x)$ be a Killing vector field. Show that in this case

$$\nabla_\mu\xi_\nu + \nabla_\nu\xi_\mu = 0,$$

where $\xi_\mu = g_{\mu\nu}\xi^\nu$ (and the affine connection is metric).

8. Determine the stress-energy tensor of the Klein–Gordon scalar field $\varphi(t,x)$ with action

$$S_\varphi = \kappa \int d^4x \sqrt{-g} \left(g^{\mu\nu} \partial_\mu \varphi \partial_\nu \varphi - m^2 \varphi^2 \right) .$$

Which is the sign for κ that yields non-negative energy density?

9. Let $\xi^\mu(x)$ be a Killing vector field. Show that in this case the relation

$$\nabla_\mu (T^{\mu\nu} \xi_\nu) = 0$$

follows from the fact that $T^{\mu\nu}$ has vanishing divergence.

10. Let $u^\mu(x)$ be a vector field with constant norm: $g_{\mu\nu} u^\mu u^\nu = $ const for all x. Prove that in this case $u^\rho \nabla_\rho u^\mu$ is orthogonal to u^μ.

11. Let $u(x)$ be a function with light-like gradient: $g^{\mu\nu} \partial_\mu u \partial_\nu u = 0$, where $u^\mu = g^{\mu\nu} \partial_\nu u$. Show that $u^\rho \nabla_\rho u^\mu = 0$. This implies that the integral curves of u^μ are light-like autoparallels.

References

1. J. D. Jackson, *Classical Electrodynamics*, Wiley, New York, 1962.
2. K. R. Lang, *The Cambridge Encyclopedia of the Sun*, Cambridge University Press, Cambridge, UK, 2001.
3. R. Penrose and W. Rindler, *Spinors in Spacetime. Vol. I. Two-Spinors Calculus and Relativistic Fields*, Cambridge University Press, Cambridge, UK 1984.
4. H. Goldstein, *Classical Mechanics*, Addison-Wesley, Reading, MA, 1980.
5. C. W. Misner, K. S. Thorne and J. A. Wheeler, em Gravitation, Freeman, San Francisco, CA, 1973.
6. R. M. Wald, *General Relativity*, University of Chicago Press, Chicago, IL, 1984.
7. J. Frank, A. King and D. Raine, *Accretion Power in Astrophysics*, Cambridge University Press, Cambridge, UK, 2002.
8. J. H. Krolik, *Active Galactic Nuclei*, Princeton University Press, Princeton, NJ, 1999.
9. J. Ehlers, *General Relativity and Cosmology*, edited by R. K. Sachs, Academic Press, New York, 1971.
10. G. F.R. Ellis, *General Relativity and Cosmology*, edited by R. K. Sachs, Academic Press, New York, 1971.

Chapter 4
Dynamics of Gravity

So far, we considered several matter systems (particles, fields) in a prescribed gravitational field. The matter was influenced by the gravitational field—it was minimally coupled to gravity—but its own influence on the gravitational field was not discussed. But we know that the matter of the sun and the planets generates a measurable gravitational field. In Newton's theory, this effect is described by the Poisson equation. This section is devoted to establish the corresponding relativistic equations for gravity and to study some of their basic properties.

Finally, after long but necessary preliminary work, we arrive at the heart of Einstein's theory. As the Maxwell equations are the core of Maxwell's theory, so are the so-called Einstein equations central to general relativity. Generations of physicists have studied these equations and have understood them now to some extent. The reason for the slow progress is that these equations are difficult, non-linear, and rather different from other field equations in theoretical physics. The non-linearity can be treated as a perturbation in only few cases. The numerical treatment is not easy [1], either. We devote the rest of these Notes to study several important solutions of these equations.

4.1 The Action

The Lagrange function of gravity cannot be "derived", only postulated. To select reasonable candidates, we make simplifying and analogy-motivated assumptions.

First, we demand that the Lagrange function implies differential equations of second order for the metric. This is the common choice which proved to be fruitful for all other fields and dynamical systems. Then

$$L(x) = F_0\left(g_{\mu\nu}(x), g_{\mu\nu,\rho}(x)\right) . \tag{4.1}$$

Hájíček, P.: *Dynamics of Gravity.* Lect. Notes Phys. **750**, 125–157 (2008)
DOI 10.1007/978-3-540-78659-7_4   © Springer-Verlag Berlin Heidelberg 2008

Thus F_0 is a function of $10 + 40 = 50$ arguments. Second, we require that

$$\sqrt{-g}F\left(g_{\mu\nu}(x), g_{\mu\nu,\rho}(x), g_{\mu\nu,\rho\sigma}(x)\right)$$
$$= \sqrt{-g}F_0\left(g_{\mu\nu}(x), g_{\mu\nu,\rho}(x)\right) + \frac{\partial}{\partial x^\sigma}\sqrt{-g}F^\sigma\left(g_{\mu\nu}(x), g_{\mu\nu,\rho}(x)\right), \quad (4.2)$$

where F^σ are functions of $g_{\mu\nu}(x)$ and $g_{\mu\nu,\rho}(x)$, and $F(x)$ is a scalar field. That is, the value of F at a point p has to be independent of the choice of coordinates in which the components $g_{\mu\nu}(p)$ of the metric and its derivatives $g_{\mu\nu,\rho}(p)$ and $g_{\mu\nu,\rho\sigma}(p)$ are computed. That is

$$F\left(g'_{\mu\nu}(x'), g'_{\mu\nu,\rho}(x'), g'_{\mu\nu,\rho\sigma}(x')\right) = F\left(g_{\mu\nu}(x), g_{\mu\nu,\rho}(x), g_{\mu\nu,\rho\sigma}(x)\right)$$

(this is the same function on both sides—only its arguments are different) at each point x. Such functions are called *invariant*. This condition means that if the 10 functions $g_{\mu\nu}(x)$ form a solution of the equation, then so do the functions $g'_{\mu\nu}(x')$, where $g'_{\mu\nu}(x')$ arise from $g_{\mu\nu}(x)$ by a coordinate transformation. Dynamical field models with this property are called *generally covariant*. Note that the function F in (4.2) is a linear function of the variables $g_{\mu\nu,\rho\sigma}(x)$:

$$\frac{\partial}{\partial x^\sigma}F^\sigma\left(g_{\mu\nu}(x), g_{\mu\nu,\rho}(x)\right) = \frac{\partial F^\sigma}{\partial g_{\mu\nu}(x)}g_{\mu\nu,\sigma}(x) + \frac{\partial F^\sigma}{\partial g_{\mu\nu,\rho}(x)}g_{\mu\nu,\rho\sigma}(x).$$

About invariant functions, the following is well known:

Theorem 11 *Let $\mathcal{M}$ be a n-manifold, $g_{\mu\nu}(x)$ a metric on $\mathcal{M}$, and F an invariant function depending on the following arguments:*

1. $F(g_{\mu\nu}(x), g_{\mu\nu,\rho}(x))$,
2. $F(g_{\mu\nu}(x), g_{\mu\nu,\rho}(x), g_{\mu\nu,\rho\sigma}(x))$.

Then F must be of the following form:

1. $F = const$,
2. $F = \bar{F}(g_{\mu\nu}(x), R^\mu_{\nu\rho\sigma}(x))$.

Thus, F can only be non-trivial in the second case and can distinguish between different metrics. Furthermore, F can only depend on the arguments $g_{\mu\nu,\rho}(x)$ and $g_{\mu\nu,\rho\sigma}(x)$ via the curvature tensor at x. Part 1 of the theorem is evident. At every fixed point each metric can be changed to $\eta_{\mu\nu}$ by a coordinate transformation, and simultaneously all first derivatives can be made to vanish. Then

$$F\left(g_{\mu\nu}(x), g_{\mu\nu,\rho}(x)\right) = F(\eta_{\mu\nu}, 0) = const.$$

Part 2 is more difficult. We had to show that in a specific coordinate system (the so-called normal coordinates [2]) all non-zero first and second derivatives of the metric could be computed from the curvature tensor [2]. Then, it is easy and we need not go into detail.

Hence we must assume the form (4.2). The total sum F has to be invariant but not necessarily F_0 alone. Then F depends linearly on the variables $g_{\mu\nu,\rho\sigma}(x)$. The following theorem describes invariant functions of this kind:

Theorem 12 *All invariant functions F of type 2, depending linearly on $g_{\mu\nu,\rho\sigma}(x)$ have the form*

$$F = aR + b,$$

where a and b are constants and R is the scalar curvature of the metric $g_{\mu\nu}(x)$.

For each metric affine connection, the scalar curvature R is defined by

$$R := g^{\mu\nu}R_{\mu\nu},$$

where $R_{\mu\nu}$ is the so-called *Ricci tensor*

$$R_{\mu\nu} := R^{\rho}_{\mu\rho\nu}.$$

Thus, the constants a and b are the only freedom we have at this stage. For historical reasons, we express these constants in terms of two others, G and Λ:

$$a = -\frac{1}{16\pi G}, \qquad b = -\frac{\Lambda}{8\pi G}.$$

As it turns out, G agrees with Newton's constant, whereas Λ is the so-called *cosmological constant*. We prove this claim later and leave G and Λ undetermined for now. The action S_G of gravity is thus of the following form:

$$S_G = -\frac{1}{16\pi G}\int d^4x\sqrt{-g}(R + 2\Lambda).\tag{4.3}$$

4.2 The Einstein Equations

The field equations arise from variation of the action (4.3) with respect to the metric $g_{\mu\nu}(x)$. Let the variation $\delta g_{\mu\nu}(x)$ be C^{∞} with support in a compact region. We write the variation in the following form:

$$\delta S_G = \frac{1}{16\pi G}\int d^4x\sqrt{-g}A^{\mu\nu}(x)\delta g_{\mu\nu}(x),\tag{4.4}$$

where $A^{\mu\nu}(x)$ is a tensor field. In the presence of matter, the variation of the total action with respect to $g_{\mu\nu}(x)$ has to be computed and set to zero:

$$\delta S_G + \delta S_M = 0.$$

In combination with (4.4) and (3.49) this yields

$$A^{\mu\nu}(x) = 8\pi G T^{\mu\nu}(x).$$

The tensor field $A^{\mu\nu}(x)$ has to be computed directly. The variation formula (3.32) does not help us here, as it only applies to Lagrange functions depending on at most first derivatives of the fields, whereas S_G also contains second derivatives. Compute the variation:

$$
\begin{aligned}
\delta S_G &= -\frac{1}{16\pi G}\int d^4x\, \delta\left[\sqrt{-g}\left(g^{\mu\nu}R_{\mu\nu}+2\Lambda\right)\right]\\
&= -\frac{1}{16\pi G}\int d^4x\, \sqrt{-g}\left(R_{(\mu\nu)}-1/2\,Rg_{\mu\nu}-\Lambda g_{\mu\nu}\right)\delta g^{\mu\nu}\\
&\quad -\frac{1}{16\pi G}\int d^4x\, \sqrt{-g}g^{\mu\nu}\delta R_{\mu\nu}\,.
\end{aligned}
\tag{4.5}
$$

We have

$$
\delta g^{\mu\nu} = -g^{\mu\kappa}g^{\mu\lambda}\delta g_{\kappa\lambda}\,.
\tag{4.6}
$$

Substituting this for $\delta g^{\mu\nu}(x)$, we obtain

$$
\begin{aligned}
\delta S_G &= \frac{1}{16\pi G}\int d^4x\, \sqrt{-g}\left(R^{(\mu\nu)}-1/2\,Rg^{\mu\nu}-\Lambda g^{\mu\nu}\right)\delta g_{\mu\nu}\\
&\quad -\frac{1}{16\pi G}\int d^4x\, \sqrt{-g}g^{\mu\nu}\delta R_{\mu\nu}\,,
\end{aligned}
\tag{4.7}
$$

where $R^{(\mu\nu)}$ denotes the symmetric part of the Ricci tensor. (Later, we show that the Ricci tensor is symmetric.)

We compute the variation of the Ricci tensor as follows. We have

$$
R_{\mu\nu} = R^\rho_{\mu\rho\nu} = \partial_\rho\Gamma^\rho_{\mu\nu}-\partial_\nu\Gamma^\rho_{\mu\rho}+\Gamma^\rho_{\kappa\rho}\Gamma^\kappa_{\mu\nu}-\Gamma^\rho_{\kappa\nu}\Gamma^\kappa_{\mu\rho},
$$

thus,

$$
\delta R_{\mu\nu} = \partial_\rho\delta\Gamma^\rho_{\mu\nu}-\partial_\nu\delta\Gamma^\rho_{\mu\rho}+\delta\Gamma^\rho_{\kappa\rho}\Gamma^\kappa_{\mu\nu}+\Gamma^\rho_{\kappa\rho}\delta\Gamma^\kappa_{\mu\nu}-\delta\Gamma^\rho_{\kappa\nu}\Gamma^\kappa_{\mu\rho}-\Gamma^\rho_{\kappa\nu}\delta\Gamma^\kappa_{\mu\rho}\,.
$$

But what is $\delta\Gamma^\rho_{\mu\nu}$? This is the difference of two affine connections $\Gamma^\rho_{\mu\nu}[g_{\kappa\lambda}]$ and $\Gamma^\rho_{\mu\nu}[g_{\kappa\lambda}+\delta g_{\kappa\lambda}]$ in the same coordinate system. Then $\delta\Gamma^\rho_{\mu\nu}$ transforms as a tensor of type $(1,2)$, since the non-homogeneous term in transformation law for $\Gamma^\rho_{\mu\nu}$ (cf. (1.9)) is independent of $\Gamma^\rho_{\mu\nu}$ and these terms cancel in the difference. Thus the covariant derivative of $\delta\Gamma^\rho_{\mu\nu}$ makes sense, and we observe that

$$
\delta R_{\mu\nu} = \nabla_\rho\delta\Gamma^\rho_{\mu\nu}-\nabla_\nu\delta\Gamma^\rho_{\mu\rho}\,.
$$

Thus we arrive at

$$
\begin{aligned}
\int d^4x\, \sqrt{-g}g^{\mu\nu}\delta R_{\mu\nu} &= \int d^4x\, \sqrt{-g}g^{\mu\nu}\left(\nabla_\rho\delta\Gamma^\rho_{\mu\nu}-\nabla_\nu\delta\Gamma^\rho_{\mu\rho}\right)\\
&= \int d^4x\, \sqrt{-g}\left[\nabla_\rho\left(g^{\mu\nu}\delta\Gamma^\rho_{\mu\nu}\right)-\nabla_\nu\left(g^{\mu\nu}\delta\Gamma^\rho_{\mu\rho}\right)\right]\\
&= \int d^4x\, \sqrt{-g}\nabla_\rho\left(g^{\mu\nu}\delta\Gamma^\rho_{\mu\nu}-g^{\mu\rho}\delta\Gamma^\nu_{\mu\nu}\right)\,.
\end{aligned}
$$

The expression in parentheses,

$$\delta V^\rho = g^{\mu\nu}\delta\Gamma^\rho_{\mu\nu} - g^{\mu\rho}\delta\Gamma^\nu_{\mu\nu},$$

is a vector field. Equation (3.45) then yields:

$$\int d^4x\sqrt{-g}g^{\mu\nu}\delta R_{\mu\nu} = \int\left(\sqrt{-g}\delta V^\rho\right).$$

Hence the last term in (4.7), an integral of a total divergence, does not affect the field equations. We simply obtain

$$A^{\mu\nu} = R^{(\mu\nu)} - 1/2\,Rg^{\mu\nu} - \Lambda g^{\mu\nu} = G^{\mu\nu} - \Lambda g^{\mu\nu},$$

where $R^{(\mu\nu)}$ is the symmetric part of the Ricci tensor,

$$R^{(\mu\nu)} = 1/2\left(R^{\mu\nu} + R^{\nu\mu}\right)$$

and $G^{\mu\nu}$ is the so-called *Einstein tensor*,

$$G^{\mu\nu} = R^{(\mu\nu)} - 1/2\,Rg^{\mu\nu}.$$

This results in the field equations for gravity with matter:

$$G^{\mu\nu} - \Lambda g^{\mu\nu} = 8\pi G T^{\mu\nu}. \tag{4.8}$$

They are called the *Einstein equations*.

4.2.1 Properties of the Curvature Tensor

We can make two very general remarks about these equations. To this end, we need some properties of the curvature tensor.

Theorem 13 *The curvature tensor and Ricci tensor of a metric affine connection have the following symmetries:*

$$R_{\mu\nu\rho\sigma} = -R_{\nu\mu\rho\sigma}, \quad R_{\mu\nu\rho\sigma} = R_{\rho\sigma\mu\nu}, \tag{4.9}$$

$$R_{\mu\nu} = R_{\nu\mu}. \tag{4.10}$$

Proof At first, we need the following lemma:

Lemma 1 *The curvature of a metric affine connection satisfies*

$$R_{\mu\nu\rho\sigma} = 1/2\left(\partial_{\nu\rho}g_{\mu\sigma} + \partial_{\mu\sigma}g_{\nu\rho} - \partial_{\nu\sigma}g_{\mu\rho} - \partial_{\mu\rho}g_{\nu\sigma}\right) + \Delta R_{\mu\nu\rho\sigma}, \tag{4.11}$$

where all terms in $\Delta R_{\mu\nu\rho\sigma}$ are quadratic in the first derivatives of the metric, $\partial_\nu g_{\mu\sigma}$, and do not contain second derivatives.

(Proof: exercise). Now, choose a point p in space–time and a geodesic coordinate system $\{x^\mu\}$ at p. The result of Exercise 2 of Chap. 2, then yields that $\partial_\mu g_{\nu\rho} = 0$. Thus (4.11) implies for the components of the curvature tensor with respect to $\{x^\mu\}$ that

$$R_{\mu\nu\rho\sigma}(p) = 1/2 \left(\partial_{\nu\rho} g_{\mu\sigma} + \partial_{\mu\sigma} g_{\nu\rho} - \partial_{\nu\sigma} g_{\mu\rho} - \partial_{\mu\rho} g_{\nu\sigma} \right)\big|_p .$$

Hence, (4.9) holds at p and for the components with respect to $\{x^\mu\}$. But p is arbitrary, and symmetry properties of tensors are invariant—if they hold in one coordinate system, then they hold in each. For the Ricci tensor we have

$$R_{\mu\nu} = g^{\rho\sigma} R_{\mu\rho\nu\sigma} . \tag{4.12}$$

Using the second equation (4.9) and the symmetry of the metric, we obtain:

$$g^{\rho\sigma} R_{\mu\rho\nu\sigma} = g^{\rho\sigma} R_{\nu\sigma\mu\rho} = g^{\sigma\rho} R_{\nu\sigma\mu\rho},$$

and, by (4.12), this equals $R_{\nu\mu}$, qed.

Due to (4.10) we can subsequently omit the parentheses around the indices of the Ricci tensor. Further important properties of a general curvature tensor (not necessarily of a metric affine connection) are stated in the following theorem.

Theorem 14 *The curvature tensor satisfies the following identities:*

$$R^\mu_{\nu\rho\sigma} + R^\mu_{\rho\sigma\nu} + R^\mu_{\sigma\nu\rho} = 0 , \tag{4.13}$$

$$\nabla_\tau R^\mu_{\nu\rho\sigma} + \nabla_\rho R^\mu_{\nu\sigma\tau} + \nabla_\sigma R^\mu_{\nu\tau\rho} = 0 . \tag{4.14}$$

These are called the *first and second Bianchi identities*. However, (4.13) only holds if

$$\Gamma^\mu_{\rho\sigma} = \Gamma^\mu_{\sigma\rho} . \tag{4.15}$$

Proof Again, choose a point p and a geodesic coordinate system $\{x^\mu\}$ at p. Then

$$R^\mu_{\nu\rho\sigma}(p) = \partial_\rho \Gamma^\mu_{\nu\sigma}(p) - \partial_\sigma \Gamma^\mu_{\nu\rho}(p) .$$

($\Gamma^\mu_{\nu\sigma}(p)$ vanishes, but not its derivative $\partial_\tau \Gamma^\mu_{\nu\sigma}(p)$). Substituting this relation into the left-hand side of (4.13), equation (4.15) implies that (4.13) holds. As $\nabla_\tau = \partial_\tau$ at p and the fact that the rest in $R^\mu_{\nu\rho\sigma}$ is quadratic in Γ, the above relation also implies that

$$\nabla_\tau R^\mu_{\nu\rho\sigma}\big|_p = \partial_{\tau\rho} \Gamma^\mu_{\nu\sigma}(p) - \partial_{\tau\sigma} \Gamma^\mu_{\nu\rho}(p) .$$

The left-hand side of (4.14) then vanishes due to the symmetry of the expression $\partial_{\tau\rho} \Gamma^\mu_{\nu\sigma}(p)$ in τ and ρ, qed.

The symmetries of the curvature tensor and its covariant derivative we just proved are not independent. For example, the first equation of (4.9) clearly follows from the second equation of (4.9) and the antisymmetry in the last two indices. It can also be shown that the second equation of (4.9) follows from the antisymmetry in the first and second pair of indices together with the first Bianchi identity.

The second Bianchi identity implies an important property of the Einstein equations. Contracting equation (4.14) in ρ and μ yields

$$\nabla_\tau R_{\nu\sigma} - \nabla_\sigma R_{\nu\tau} + \nabla_\mu R^\mu_{\nu\sigma\tau} = 0 . \tag{4.16}$$

Contracting again in τ and ν implies

$$2\nabla_\tau R^\tau_\sigma - \nabla_\sigma R = 0 . \tag{4.17}$$

This equation can be written in the form

$$\nabla_\tau (R^\tau_\sigma - 1/2\, \delta^\tau_\sigma R) = 0 ,$$

which means

$$\nabla_\mu G^{\mu\nu} = 0 . \tag{4.18}$$

Thus the Einstein tensor is divergence free, regardless of the metric from which it is calculated, as (4.18) is an identity. That is, the Einstein equations imply that the stress-energy tensor $T^{\mu\nu}$ is divergence free without reference to the equations of motion of the matter.

4.3 General Covariance of the Einstein Equations

The form of the Einstein equations can symbolically be represented as:

$$g^{..}\partial_{..}g_{..} + g^{..}g^{..}\partial_. g_{..}\partial_. g_{..} - \Lambda g_{..} = 8\pi G T_{..} .$$

There are three types of terms on the left-hand side, and all have fixed numbers like $1/2$, -1, Λ, etc., as coefficients. That is, if the components $g_{\mu\nu}(x)$ of the metric are known in any coordinate system, the left-hand side can be computed from them. Two facts are remarkable:

1. The equations are formed in the same way, regardless of the chosen coordinate system x^μ.
2. The left-hand sides (there are 10 independent expression) computed in coordinates x'^μ equal certain linear combinations of the left-hand sides computed in x^μ (it is indeed a tensor transformation!).

Precisely this property is called general covariance.

The general covariance of the field equations has an unexpected consequence. This troubled Einstein so much that he conjectured that the field equations for the metric must *not* be generally covariant [3]. We want to understand this now. To this end, we need some more mathematics. In Sect. 2.6.2 we already introduced the notion of a diffeo and explained the action φ_* of a diffeo φ on tensor fields. We have to elaborate on this.

Let $\mathcal{M}$ be a n-manifold. Consider two diffeos from $\mathcal{M}$ into itself, $\varphi : \mathcal{M} \mapsto \mathcal{M}$. They are defined on the whole manifold, are onto and also invertible. Two diffeos φ_1 and φ_2 can thus be composed

$$\varphi := \varphi_2 \circ \varphi_1,$$

and the composition is again a diffeo. The set of all diffeos is a group with the composition as multiplication. Indeed, the composition of two maps is always associative, the identity map id is a diffeo, and the inverse, φ^{-1}, of φ has the property that $\varphi^{-1} \circ \varphi = $ id. This group is called Diff$\mathcal{M}$.

Diff$\mathcal{M}$ is a large group as its dimension is infinite [4]. Without proof we mention two interesting properties, demonstrating its size. Let $(p_1, \cdots, p_m)$ and $(q_1, \cdots, q_m)$ be two M-tuples of points in $\mathcal{M}$. Can we find $\varphi \in $ Diff$\mathcal{M}$, so that

$$\varphi(p_i) = q_i \quad \forall i = 1, \cdots, m?$$

The answer is yes for almost all m-tuples. In particular, Diff$\mathcal{M}$ can map any point in $\mathcal{M}$ to any other point in $\mathcal{M}$.

The second property is the following. Let $U \subset \mathcal{M}$ be an arbitrary neighborhood in $\mathcal{M}$ (a "hole"). There exists $\varphi \in $ Diff$\mathcal{M}$ such that

$$\varphi|_{\mathcal{M} \setminus U} = \text{id}$$

and

$$\varphi|_U \neq \text{id} .$$

Then it is clear that $\varphi(U) = U$. If there was $p \in U$ such that $\varphi(p) \notin U$, then φ^{-1} is non-trivial at $\varphi(p)$, but on the other hand the definition φ implies that it has to be trivial (id), a contradiction.

Let us return to the Einstein equations. We will show the following:

Theorem 15 *Let $g_{\mu\nu}(x)$ be a solution of Einstein's equations with source $T_{\mu\nu}(x)$ on a manifold $\mathcal{M}$ and let $\varphi \in $ Diff$\mathcal{M}$ be an arbitrary diffeo. Then $(\varphi_* g)_{\mu\nu}(x)$ is also a solution for the source $(\varphi_* T)_{\mu\nu}(x)$.*

The proof uses the general covariance. We know that the components for g and T in coordinates x^μ satisfy the equations around a point p. At $\varphi(p)$ we choose the coordinate system which is the φ-image of x^μ. In this system the components $(\varphi_* g)_{\mu\nu}(x)$ and $(\varphi_* T)_{\mu\nu}(x)$ have exactly the same form as those of g and T with respect to x^μ and have to solve an equation of the same form.

We actually proved that Diff$\mathcal{M}$ is a symmetry group of Einstein's equations, meaning that $(\varphi_* g)$ is a solution whenever g is. This gives a recipe to construct a relatively large number of new solutions from a given one. Indeed, $(\varphi_* g)$ is a tensor field on $\mathcal{M}$ which is different from g unless φ is an isometry of g, which is rarely the case.

This appears to have an inconvenient consequence. The Einstein equations seem to have too many solutions. Consider for example an asymptotically flat space–time with a bounded central source of mass μ and radius R. If we have this situation in Newtonian theory, then this source and boundary condition allows only one solution for the Poisson equation, which seems plausible for physical reasons.

Consider the same situation in general relativity, and assume that $g_{\mu\nu}(x)$ is a solution of the Einstein equations on $\mathcal{M}$ corresponding to the above situation. Fix a small neighborhood $U \subset \mathcal{M}$ outside the source. Then there exists $\varphi \in \mathrm{Diff}\,\mathcal{M}$ which does not change the source or the asymptotic region, but changes g non-trivially in U! It seems that the Einstein equations allow many different solutions for a physically unique situation. This roughly describes Einstein's famous *hole argument* [3].

The only way out of this difficulty is to assume that two metrics g and φ_*g for arbitrary $\varphi \in \mathrm{Diff}\,\mathcal{M}$ cannot be distinguished by any measurable properties. That is, $\mathrm{Diff}\,\mathcal{M}$ is a *gauge group* of general relativity.

At first this is difficult to believe. The two fields g and φ_*g are different on $\mathcal{M}$. Thus there exists at least one point $p \in \mathcal{M}$, where $g(p) \neq (\varphi_*g)(p)$. As the metric is measurable, we seem to have a contradiction.

If we look closely at our description of the measurement of the metric components in Sect. 2.4, we find that we needed a family of observers. In particular, the points of the manifold $\mathcal{M}$ were identified by these observers. If we apply a diffeo $\varphi \in \mathrm{Diff}\,\mathcal{M}$, then we not only have to change the metric to φ_*g, but also transport the observers. Then the measurement will yield the same components of the metric, namely those of φ_*g with respect to the φ-image of the observers.

This motivates the following postulate:

Postulate 4.1 *The bare points of the manifold $\mathcal{M}$ are not identifiable or distinguishable by measurements or observations.*

To determine a physical point, one has to describe a sufficient number of measurable properties of certain fields (including the geometry of a metric) at this point. Such physical points do not agree with the bare points of $\mathcal{M}$, as the bare points of $\mathcal{M}$ change under $\mathrm{Diff}\,\mathcal{M}$, whereas the physical points do not.

4.4 Weak Gravitational Field

We can get some idea on the rich physics behind the Einstein equations if we apply them to weak gravitational fields. In comparison with Newtonian theory, a number of new effects appear. Nevertheless, these effects are very weak and difficult to measure. More important consequences of the Einstein equations follow from their non-linearity and the particular coupling between matter and gravity they postulate. These lead to truly noticeable phenomena (somebody would even say "catastrophic") that can be observed far away in the cosmos. We shall consider them in the following chapters. However, the field in the solar system and, in particular, around Earth, is weak. By the way, we can also investigate the meaning of the constant G.

We have already defined weak fields for the spherically symmetric static case as "asymptotically flat" metric. A more general definition can be formulated as follows:

Definition 17 Let $(\mathscr{M},g)$ be a space–time. If there are coordinates x^μ with the properties

1. they cover the whole $\mathscr{M}$ and their range is $\mathbb{R}^4$;
2. the components $g_{\mu\nu}$ of the metric with respect to x^μ satisfy

$$g_{\mu\nu}(x) - \eta_{\mu\nu} = O(\varepsilon), \tag{4.19}$$

$$g_{\mu\nu,\rho}(x) = O(\varepsilon),$$

$$g_{\mu\nu,\rho\sigma}(x) = O(\varepsilon), \quad \forall x,\mu,\nu,\rho,\sigma,$$

where $\eta_{\mu\nu}$ is the matrix defined by (2.1) and ε is a small number (10^{-6}, say), then the field $g_{\mu\nu}$ is called *weak*.

In the weak-field theory, the metric is dimensionless and the coordinates have the dimension of length.

This definition could be formulated so that the "space–time" in it is only a part of the space–time that we consider. For example, the asymptotically flat region of a space–time with a strong field near the center is such a "space–time". On the other hand, any point p of any space–time has a neighborhood in which condition (4.19) holds: we just have to use a local inertial frame at p. Hence, the condition is only non-trivial if $\mathscr{M}$ is "sufficiently large". In the present section, we avoid these questions and assume that the coordinate range is the whole $\mathbb{R}^4$.

We consider a system of fields containing gravity and call all deviations from the flat geometry *disturbances*. An important assumption underlying the weak-field theory (with $\Lambda = 0$) is that any disturbance of Minkowski space–time that is small at some time remains so if evolved by the exact (non-linearized) field equations. This assumption, called *stability of Minkowski space–time*, is non-trivial because of the non-linearity of the Einstein equations. If Minkowski space–time would be unstable, then solutions to the linearized equations might start to deviate strongly from exact ones. Mathematicians have shown a number of theorems about the stability [5]. All of them require stronger fall-off conditions at infinity than those of Definition 17. The recent status is described in [6]. Numerical simulations show that there is a well-defined (model-dependent) threshold. Disturbances remain small if they start under the threshold and black holes and singularities evolve if they surpass it [7]; then the deviations from Minkowski space–time become large. In the present chapter, we adhere to the "naive" Definition 17 but we shall be careful to discuss only those cases in which the linearized theory is reliable. The mathematics needed for an adequate analysis of these problems lies beyond the scope of these Notes.

4.4.1 Auxiliary Metrics and Gauge Transformations

In Definition 17, $g_{\mu\nu}$ is the *physical metric* on the manifold $\mathcal{M}$. It is determined by observations and measurements described in Sect. 2.4 and it describes the gravitational field. Obviously, the definition uses another metric on $\mathcal{M}$, the Minkowski metric with components $\eta_{\mu\nu}$ with respect to the coordinates x^μ. This will be called the *auxiliary metric*.

We can easily see that the auxiliary metric is not uniquely determined. The measurements that determine metrics would anyway lead to the metric $g_{\mu\nu}$ and the only property that can be used for the definition of the auxiliary metric is condition (4.19). Let us study the freedom in this condition.

To this end, we define new coordinates x'^μ by the transformation

$$x'^\mu = x^\mu + \varepsilon X^\mu(x) , \qquad (4.20)$$

where $X^\mu(x)$ is an arbitrary vector field on $\mathcal{M}$ such that the components with respect to x^μ and their derivatives are everywhere of order 1 or lower. The inverse transformation to the first order of ε is

$$x^\mu = x'^\mu - \varepsilon X^\mu(x') , \qquad (4.21)$$

where $X^\mu(x')$ is to be understood as the same four functions of four variables x'^μ as $X^\mu(x)$ are of x^μ. Then, the components of the physical metric with respect to the new coordinates are

$$g'_{\mu\nu}(x') = \frac{\partial x^\rho}{\partial x'^\mu} \frac{\partial x^\sigma}{\partial x'^\nu} g_{\rho\sigma} \left(x'^\kappa - \varepsilon X^\kappa(x')\right) .$$

To first order in ε, we obtain

$$g'_{\mu\nu}(x') = g_{\mu\nu} - \varepsilon \left(X_{\mu,\nu} + X_{\nu,\mu}\right) ,$$

where $X_\mu = \eta_{\mu\nu} X^\nu$ and again the values x'^μ have to be substituted for the arguments of all functions of x^μ on the right-hand side. It follows immediately

$$g'_{\mu\nu} - \eta_{\mu\nu} = g_{\mu\nu} - \eta_{\mu\nu} - \varepsilon \left(X_{\mu,\nu} + X_{\nu,\mu}\right) = O(\varepsilon) . \qquad (4.22)$$

Hence, for the purpose of the Definition 17, the coordinates x'^μ are as good as x^μ.

However, the auxiliary metrics determined by their components $\eta_{\mu\nu}$ with respect to either x^μ or x'^μ are different because the transformation (4.20) need not be a Poincaré transformation even to first order. We can recognize the origin for this non-uniqueness in the freedom in how the flat metric can be put on a given manifold, as it has been shown in Sect. 2.2. Strictly speaking, the auxiliary metric has not the physical meaning of a space–time metric. In particular, the light cones are determined by the physical metric.

Define the field $h_{\mu\nu}(x)$ by its components with respect to x^μ,

$$\varepsilon h_{\mu\nu}(x) = g_{\mu\nu}(x) - \eta_{\mu\nu} \,, \tag{4.23}$$

and the field $\bar{h}'_{\mu\nu}(x')$ by its components with respect to x'^μ,

$$\varepsilon \bar{h}'_{\mu\nu}(x') = g'_{\mu\nu}(x') - \eta_{\mu\nu} \,.$$

Both are defined as differences of two tensor fields of type $(0,2)$, so they are tensors of the same type. From (4.22), we can calculate the difference of these two sets of components:

$$\bar{h}'_{\mu\nu}(x) = h_{\mu\nu}(x) - X_{\mu,\nu}(x) - X_{\nu,\mu}(x) + \mathrm{O}(\varepsilon) \,. \tag{4.24}$$

This is a very practical equation, but it is *not* a tensor equation. The left-hand side is a component of the tensor field $\bar{h}'_{\mu\nu}(x')$ with respect to the coordinates x'^μ and the arguments of it are written as x^0, x^1, x^2, and x^3, that is, without the primes. The right-hand side is a sum of components of the fields $h_{\mu\nu}(x)$ and $X_{\mu,\nu}(x)$ with respect to the coordinates x^μ at the same value of their arguments as the left-hand side. Thus, the equation does not compare components of two tensors in one coordinate system and at one point!

We obtain a tensor equation, if we transform the left-hand side of (4.24) to coordinates x^μ by the inverse to the transformation

$$\bar{h}'_{\mu\nu}(x') = \frac{\partial x^\rho}{\partial x'^\mu} \frac{\partial x^\sigma}{\partial x'^\nu} \bar{h}_{\rho\sigma}\left(x'^\kappa - \varepsilon X^\kappa(x')\right) \,. \tag{4.25}$$

To zeroth order in ε, this yields

$$\bar{h}_{\mu\nu}(x) = h_{\mu\nu}(x) - X_{\mu,\nu}(x) - X_{\nu,\mu}(x) + \mathrm{O}(\varepsilon) \,. \tag{4.26}$$

Here, components of two fields with respect to the same coordinate system are compared at the same point. The form is still the same as in (4.24) because all changes on the left-hand side are of the order of ε.

Equations (4.25) and (4.26) decompose the transformation from $h_{\mu\nu}(x)$ to $\bar{h}'_{\mu\nu}(x')$ into two steps. The first, (4.26), corresponds to the change from $h_{\mu\nu}(x)$ to $\bar{h}_{\mu\nu}(x)$ keeping the coordinates fixed. The second, (4.25), is a transformation of tensor representation functions due to a coordinate transformation (4.20). It is the change (4.26) of the field that is called *gauge transformation* in the weak-field theory. All measurable quantities such as the physical metric, must be independent of the choice, that is, they must be gauge invariant.

4.4.2 Affine Connection and Curvature

If we expand all equations and quantities in powers of ε and throw away all terms of order higher than linear, we obtain the so-called *linearized theory of gravity*. If the field is weak, then this is a very practical tool to obtain a lot of insight in a relatively quick way. Moreover, the freedom in the choice of the auxiliary metric can be exploited so that the linearized equations simplify further.

Assume that (4.19) is valid and let the tensor field h be defined by (4.23). Let us calculate all important quantities with respect to x^μ in the *linear approximation*, that is, we neglect all terms of higher order than linear in ε.

We easily see that the physical contravariant metric in this approximation is

$$g^{\mu\nu} = \eta^{\mu\nu} - \varepsilon h^{\mu\nu} , \qquad (4.27)$$

where the indices on the right-hand side have been raised by the auxiliary metric

$$h^{\mu\nu} = \eta^{\mu\rho}\eta^{\nu\sigma}h_{\rho\sigma} .$$

Similarly, the physical connection is

$$\Gamma^\mu_{\rho\sigma} = \frac{\varepsilon}{2}\eta^{\mu\nu}\left(h_{\nu\rho,\sigma} + h_{\nu\sigma,\rho} - h_{\rho\sigma,\nu}\right) , \qquad (4.28)$$

because all derivatives of $\eta_{\mu\nu}$ vanish. Test particles move along autoparallels of the physical affine connection. The autoparallel equation in the linear approximation reads

$$\ddot{x}^\mu + \frac{\varepsilon}{2}\eta^{\mu\nu}\left(h_{\nu\rho,\sigma} + h_{\nu\sigma,\rho} - h_{\rho\sigma,\nu}\right)\dot{x}^\rho\dot{x}^\sigma = 0 . \qquad (4.29)$$

The components of the physical curvature tensor in the linear approximation are easily calculated from (4.28) to be

$$R_{\mu\nu\rho\sigma} = \frac{\varepsilon}{2}\left(h_{\mu\sigma,\nu\rho} + h_{\nu\rho,\mu\sigma} - h_{\mu\rho,\nu\sigma} - h_{\nu\sigma,\mu\rho}\right) . \qquad (4.30)$$

In this approximation, we also have

$$R_{\mu\nu} = \eta^{\rho\sigma}R_{\mu\rho\nu\sigma} , \quad R = \eta^{\mu\nu}\eta^{\rho\sigma}R_{\mu\rho\nu\sigma} .$$

Thus, we obtain the Ricci tensor in the linear approximation with respect to coordinates x^μ (exercise),

$$R_{\mu\nu} = \frac{\varepsilon}{2}\left[-\Box h_{\mu\nu} + \left(h^\rho_\mu - \frac{1}{2}\delta^\rho_\mu h\right)_{,\rho\nu} + \left(h^\rho_\nu - \frac{1}{2}\delta^\rho_\nu h\right)_{,\rho\mu}\right] , \qquad (4.31)$$

where the "wave operator" is defined by

$$\Box = \eta^{\mu\nu}\frac{\partial}{\partial x^\mu}\frac{\partial}{\partial x^\nu} ,$$

and

$$h = \eta^{\mu\nu}h_{\mu\nu} .$$

We see that $R_{\mu\nu}$ consists of the wave operator term plus divergence terms. This is analogous to the Maxwell equations expressed in terms of the potential A_ρ, except that our divergence terms are more complicated. Can these terms be removed by a gauge transformation, as in electrodynamics?

If we make a gauge transformation

$$\bar{h}_{\mu\nu}(x) = h_{\mu\nu}(x) + X_{\mu,\nu}(x) + X_{\nu,\mu}(x) , \qquad (4.32)$$

the divergence term for the new field is (exercise)

$$\left(\bar{h}_{\mu}^{\rho} - \frac{1}{2}\delta_{\mu}^{\rho}\bar{h}\right)_{,\rho} = \left(h_{\mu}^{\rho} - \frac{1}{2}\delta_{\mu}^{\rho}h\right)_{,\rho} + \Box X_{\mu} .$$

Hence, if we choose X^{μ} so that

$$\Box X_{\mu} = -\left(h_{\mu}^{\rho} - \frac{1}{2}\delta_{\mu}^{\rho}h\right)_{,\rho} , \qquad (4.33)$$

then the new field $\bar{h}_{\mu\nu}(x)$ satisfies

$$\left(\bar{h}_{\mu}^{\rho} - \frac{1}{2}\delta_{\mu}^{\rho}\bar{h}\right)_{,\rho} = 0 . \qquad (4.34)$$

From the mathematical point of view, (4.33) is just the wave equation (with a source) on Minkowski space–time. This equation has many different solutions. We shall make use of this additional freedom later. Hence, we *can* choose X^{μ} so that (4.34) holds.

The remaining question is: has the Ricci tensor $\bar{R}_{\mu\nu}$ the same form, if expressed by means of $\bar{h}_{\mu\nu}(x)$ as $R_{\mu\nu}$ in terms of $h_{\mu\nu}(x)$? To study how the connection and the curvature change under gauge transformations, we have to express $h_{\mu\nu}(x)$ in terms of $\bar{h}_{\mu\nu}(x)$ and $X^{\mu}(x)$ and plug it into (4.28), (4.29), (4.30) and (4.31). We find that

$$\bar{\Gamma}_{\rho\sigma}^{\mu} = \frac{\varepsilon}{2}\eta^{\mu\nu}\left(\bar{h}_{\nu\rho,\sigma} + \bar{h}_{\nu\sigma,\rho} - \bar{h}_{\rho\sigma,\nu}\right) - \varepsilon X_{,\rho\sigma}^{\mu} .$$

The curvature tensor and all tensors constructed from it have exactly the same form if expressed in terms of $\bar{h}_{\mu\nu}(x)$ as equations (4.30) and (4.31) in terms of $h_{\mu\nu}(x)$. The reason is that the two terms resulting from $\varepsilon X_{,\rho\sigma}^{\mu}$ in Γ cancel out in the formula for the curvature. Thus, we arrive at the important theorem:

Theorem 16 *The curvature tensor is gauge invariant.*

The Ricci tensor $\bar{R}_{\mu\nu}$ *has* the same form and is therefore given by

$$\bar{R}_{\mu\nu} = -\frac{\varepsilon}{2}\Box\bar{h}_{\mu\nu}$$

in view of (4.34).

We can even remove the additional term in Γ by transforming the coordinates,

$$x'^{\mu} = x^{\mu} + \varepsilon X^{\mu}(x) , \qquad (4.35)$$

as the inhomogeneous term in the transformation law for Γ is of first order in ε. The transformation of the curvature tensor and the Ricci tensor is trivial under (4.35) since their transformation laws are homogeneous and they are already of first order in ε. This also follows directly from the fact that the new auxiliary metric has components $\eta_{\mu\nu}$ with respect to coordinates x'^{μ}.

To summarize, by a suitable choice of auxiliary metric and coordinates, the basic equations of the linearized theory can be brought to the form (written without bars and primes):

The physical metric is

$$g_{\mu\nu}(x) = \eta_{\mu\nu} + \varepsilon h_{\mu\nu}(x) , \tag{4.36}$$

where the disturbance field $h_{\mu\nu}(x)$ satisfies

$$\left(h_\mu^\rho - \frac{1}{2}\delta_\mu^\rho h \right)_{,\rho} = 0 . \tag{4.37}$$

The connection (the force of gravity in linearized theory) is

$$\Gamma_{\rho\sigma}^\mu = \frac{\varepsilon}{2}\eta^{\mu\nu}\left(h_{\nu\rho,\sigma} + h_{\nu\sigma,\rho} - h_{\rho\sigma,\nu} \right) . \tag{4.38}$$

The curvature tensor is

$$R_{\mu\nu\rho\sigma} = \frac{\varepsilon}{2}\left(h_{\mu\sigma,\nu\rho} + h_{\nu\rho,\mu\sigma} - h_{\mu\rho,\nu\sigma} - h_{\nu\sigma,\mu\rho} \right) \tag{4.39}$$

and the Ricci tensor is

$$R_{\mu\nu} = -\frac{\varepsilon}{2}\Box h_{\mu\nu} . \tag{4.40}$$

This enables us to write Einstein equations

$$R_{\mu\nu} - \frac{1}{2}Rg_{\mu\nu} - \Lambda g_{\mu\nu} = 8\pi G T_{\mu\nu} \tag{4.41}$$

in the linear approximation as follows. First, we remove the curvature scalar from the equation by taking the trace,

$$R = -4\Lambda - 8\pi G g^{\mu\nu}T_{\mu\nu}$$

and substituting this back:

$$R_{\mu\nu} + \Lambda g_{\mu\nu} = 8\pi G\left(T_{\mu\nu} - \frac{1}{2}g_{\mu\nu}g^{\rho\sigma}T_{\rho\sigma} \right) . \tag{4.42}$$

Second, we linearize (4.42) using (4.36) and (4.40):

$$-\Box h_{\mu\nu} + 2\Lambda h_{\mu\nu} = -\frac{2\Lambda}{\varepsilon}\eta_{\mu\nu} + \frac{16\pi G}{\varepsilon}\left(T_{\mu\nu} - \frac{1}{2}\eta_{\mu\nu}T \right) , \tag{4.43}$$

where

$$T = \eta^{\mu\nu} T_{\mu\nu} .$$

4.4.3 The Cosmological Constant

Let us consider the Einstein equations (4.43) in the case of vanishing stress-energy tensor:

$$-\Box h_{\mu\nu} + 2\Lambda h_{\mu\nu} + \frac{2\Lambda}{\varepsilon} \eta_{\mu\nu} = 0 . \tag{4.44}$$

If we compare, e.g., with the Klein–Gordon equation,

$$-\Box\Phi - m^2\Phi = 0 ,$$

we see that the second term on the left-hand side of (4.44) has the form of a mass term.

The third term on the left-hand side of (4.44) is large and does not contain $h_{\mu\nu}$, it has the form of a source. This source drives the solution away from flat space–time. As we shall see in Chap. 5, evolution over distances comparable with $|\Lambda|^{-1/2}$ results in large deviations from the Minkowski metric.

The reason why such a source appears here is that we have expanded a solution of the equation

$$G_{\mu\nu} - \Lambda g_{\mu\nu} = 0$$

around a space–time that itself is not a solution to it. If we expand around de Sitter ($\Lambda > 0$) or anti-de Sitter ($\Lambda < 0$) space–times, which are solutions of the above equation, the source term disappears and the equation becomes homogeneous. The cosmological constant would then only introduce a mass term of a slightly different form.

Can we say that Λ has the meaning of mass of the gravitational field? This is difficult because first, Λ does much more than a humble mass term could do: it changes the background space–time. Second, even if we accept the new background, some problems remain. Mass is usually defined as one of two parameters (mass and spin) that distinguish the irreducible representations of the Lorentz group and as such has only relevance to fields in Minkowski background. One can say at most that Λ is a characteristic of the gravitational field similarly as mass is a characteristic of the electron–positron field, but it is not really a mass.

There is an altogether different interpretation of Λ, which seems to have been invented by Pauli. Instead of trying to attribute it to the gravitational field, it can be attributed to matter. Thus, the Λ term can be taken from the left-hand side to the right-hand side of the Einstein equations. There, it can be written in the form $8\pi G T_{\mu\nu}^{\Lambda}$, where

$$T_{\mu\nu}^{\Lambda} = \frac{\Lambda}{8\pi G} \eta_{\mu\nu} . \tag{4.45}$$

However, this $T^\Lambda_{\mu\nu}$ is the stress-energy tensor of a rather strange kind of matter, it is homogeneous ($\Lambda = $ const) and Lorentz invariant (any multiple of $\eta_{\mu\nu}$ is). This matter seems to be covered by the definition of an ideal fluid with density and pressure

$$\rho_\Lambda = \frac{\Lambda}{8\pi G} , \quad p_\Lambda = -\frac{\Lambda}{8\pi G} , \tag{4.46}$$

but then either ρ or p must be negative. Now, the ground state of any quantum field has these properties if its energy density is non-zero. Such states exist in quantum field theory and the idea has been very fruitful in cosmology.

How can Λ be measured and what is its value? We shall study this question in the cosmology chapter. The value is about

$$\Lambda \approx 10^{-52}\mathrm{m}^{-2} ,$$

so small that it can be neglected in all processes except the cosmological evolution.

To summarize, the Λ-term can be considered to be a property of either the gravitational field or the matter. What it is and why it has the observed value is an enigma. In any case, any linear theory with a non-vanishing cosmological constant must start from an expansion around the space–time with constant curvature Λ. In this chapter, we neglect Λ and expand only around Minkowski space–time.

4.4.4 The Linearized Einstein Equations

If we drop the cosmological constant, the linearized Einstein equations become

$$-\varepsilon\Box h_{\mu\nu} = 16\pi G \left(T_{\mu\nu} - \frac{1}{2}\eta_{\mu\nu}T \right) , \tag{4.47}$$

$$\left(h^\rho_\mu - \frac{1}{2}\delta^\rho_\mu h \right)_{,\rho} = 0 . \tag{4.48}$$

An immediate consequence of these equations is that $GT_{\mu\nu}$ is of order ε. The linearized theory is only applicable if the source is sufficiently weak. Another consequence is the conservation of energy and momentum in the linearized form

$$T^\rho_{\mu,\rho} = 0 . \tag{4.49}$$

Indeed, (4.47) implies

$$-\varepsilon\Box h = -16\pi GT ,$$

whence

$$-\varepsilon\Box \left(h^\rho_\mu - \frac{1}{2}\delta^\rho_\mu h \right) = 16\pi GT^\rho_\mu . \tag{4.50}$$

Taking the divergence of both sides and applying (4.48), we obtain (4.49).

If (4.49) were exactly valid, it would mean that the matter (for instance an ideal fluid) moves on the auxiliary space–time as if no gravity was present. The influence of the gravitational field on the motion of matter appears only in higher orders of approximation. Indeed, if written out, the exact equation reads

$$T^\rho_{\mu,\rho} + \Gamma^\rho_{\nu\rho}T^\nu_\mu - \Gamma^\nu_{\mu\rho}T^\rho_\nu = 0 \,.$$

The last two terms are of higher order in ε than the first one. We may still use this equation as well as the autoparallel equation when calculating the influence of the gravitational field on the motion of matter. However, we must keep in mind that when doing so, we already work in a higher approximation than the linear one.

The linearized Einstein equations have a lot of properties in common with the Maxwell equations and we shall use the methods well known from electrodynamics [8]. First, the equations are invariant with respect to all Poincaré transformations. Therefore, we can adapt our frame to the particular source or to a particular observer (if there are some in the problem to solve) to simplify the algebra. Second, there is a residual gauge freedom. We require that the analogue of the Lorentz condition, (4.33), is satisfied. Similarly as in electrodynamics, this does not determine the field $X^\mu(x)$ uniquely but only up to addition of an arbitrary field $X^\mu_0(x)$,

$$X'^\mu(x) = X^\mu(x) + X^\mu_0(x) \,,$$

that satisfies the condition

$$\Box X^\mu_0 = 0 \,.$$

The most important property of the system (4.47) and (4.48) is its linearity. Hence, the general solution is a sum of a particular solution and a general solution to the homogeneous system. The particular solution to an arbitrary source can again be written as a linear combination of solutions corresponding to point sources. The solution $G(\vec{x}, \vec{x}')$ to the source at the point $\vec{x}'$ is called Green's function and satisfies the equation

$$\Box_x G\left(\vec{x}, \vec{x}'\right) = 4\pi\delta\left(\vec{x} - \vec{x}'\right) \,.$$

The index x at the wave operator indicates that the derivatives act on the variables $\vec{x}$, not $\vec{x}'$. The Green's function $G(\vec{x}, \vec{x}')$ depends only on the difference $\vec{x} - \vec{x}'$ because of the Poincaré translation invariance. Moreover, from the physical point of view, it is most advantageous to choose the retarded form of the Green's function:

$$G\left(\vec{x}\right) = \frac{\delta\left(x^0 - |\mathbf{x}|\right)}{|\mathbf{x}|} \,.$$

The resulting particular solution corresponds to zero incoming radiation, and we can write it in the form

$$\varepsilon h_{\mu\nu}\left(\vec{x}\right) = -4G \int d^3x'\, G\left(\vec{x} - \vec{x}'\right)\left[T_{\mu\nu}\left(\vec{x}'\right) - \frac{1}{2}\eta_{\mu\nu}T\left(\vec{x}'\right)\right] \,. \tag{4.51}$$

After carrying out the integration in x^0, we obtain

$$\varepsilon h_{\mu\nu}\left(x^0, \mathbf{x}\right) = -4G \int d^3x' \frac{T_{\mu\nu}\left(x^0 - |\mathbf{x}'|, \mathbf{x}'\right) - \frac{1}{2}\eta_{\mu\nu}T\left(x^0 - |\mathbf{x}'|, \mathbf{x}'\right)}{|\mathbf{x} - \mathbf{x}'|}. \quad (4.52)$$

This is in any case a particular solution to (4.47), but does it also solve (4.48)? To answer this question, we rewrite (4.51):

$$\varepsilon\left[h_\mu^\rho(\vec{x}) - \frac{1}{2}\delta_\mu^\rho h(\vec{x})\right] = -4G \int d^3x'\, G\left(\vec{x} - \vec{x}'\right) T_{\mu\nu}(\vec{x}')$$

and obtain step by step

$$\varepsilon\left[h_\mu^\rho(\vec{x}) - \frac{1}{2}\delta_\mu^\rho h(\vec{x})\right]_{,\rho} = -4G \int d^3x'\, \frac{\partial G(\vec{x} - \vec{x}')}{\partial x^\rho}\, T_\mu^\rho(\vec{x}')$$

$$= 4G \int d^3x'\, \frac{\partial G(\vec{x} - \vec{x}')}{\partial x'^\rho}\, T_\mu^\rho(\vec{x}') = -4G \int d^3x'\, G(\vec{x} - \vec{x}')T_{\mu,\rho}^\rho(\vec{x}') = 0.$$

We have used the property of Green's function that

$$\frac{\partial G(\vec{x} - \vec{x}')}{\partial x^\rho} = -\frac{\partial G(\vec{x} - \vec{x}')}{\partial x'^\rho}$$

as well as (4.49). Hence, we have a particular solution to the whole system.

The general solution to the system (4.47) and (4.48) with $T_{\mu\nu} = 0$ can be decomposed into plane waves (Fourier transformation). Plane waves will be studied in Sect. 4.4.7.

4.4.5 Stationary Fields

Suppose that the matter source and the gravitational field are stationary. More precisely, set

$$T_{\mu\nu,0} = 0, \quad h_{\mu\nu,0} = 0. \quad (4.53)$$

The divergence equation becomes

$$\sum_k T_{\mu k,k} = 0 \quad (4.54)$$

and the retarded integral (4.52) simplifies to

$$\varepsilon h_{\mu\nu}(\mathbf{x}) = -4G \int d^3x' \frac{T_{\mu\nu}(\mathbf{x}') - \frac{1}{2}\eta_{\mu\nu}T(\mathbf{x}')}{|\mathbf{x} - \mathbf{x}'|}. \quad (4.55)$$

Assume further that the source is spatially bounded, that is the stress-energy tensor $T_{\mu\nu}(\mathbf{x})$ is only non-zero in a 3-volume V with finite boundary ∂V. In addition,

$T_{\mu\nu}(\mathbf{x})$ is required to be smooth so that it vanishes together with its derivatives at the boundary ∂V. The assumptions are analogous to those made in electrodynamics at the starting point of the multipole expansion [8], and we proceed similarly.

Let us calculate the gravitational field far away from V. We choose the origin of the coordinates x^1, x^2, and x^3 somewhere inside V and expand everything in powers of the small parameter $|\mathbf{x}'|/|\mathbf{x}|$. Thus,

$$\frac{1}{|\mathbf{x}-\mathbf{x}'|} = \frac{1}{|\mathbf{x}|}\left(1+\sum_k \frac{x^k}{|\mathbf{x}|}\frac{x'^k}{|\mathbf{x}|}+\sum_{k,l}\frac{x^k x^l}{|\mathbf{x}|^2}\frac{3x'^k x'^l-|\mathbf{x}'|^2\delta^{kl}}{2|\mathbf{x}|^2}+\cdots\right).$$

Here we used the expansion

$$(1+a)^{-1/2}=1-\frac{1}{2}a+\frac{3}{8}a^2-\frac{5}{16}a^3+\cdots,$$

valid for small a, and collected all terms with common factors

$$\frac{x^k}{|\mathbf{x}|},\quad \frac{x^k x^l}{|\mathbf{x}|^2},\quad \cdots$$

We calculate, retaining only terms up to the second order in $|\mathbf{x}'|/|\mathbf{x}|$:

$$\varepsilon h_{\mu\nu}(\mathbf{x}) = -\frac{4G}{|\mathbf{x}|}\left\{\int d^3x'\left[T_{\mu\nu}(\mathbf{x}')-\frac{1}{2}\eta_{\mu\nu}T(\mathbf{x}')\right]\right.$$
$$\left.+\sum_k\frac{x^k}{|\mathbf{x}|^2}\int d^3x' \, x'^k\left[T_{\mu\nu}(\mathbf{x}')-\frac{1}{2}\eta_{\mu\nu}T(\mathbf{x}')\right]+\cdots\right\}.$$

In this way, we obtain

$$\varepsilon h_{00}(\mathbf{x}) = -\frac{2G}{|\mathbf{x}|}\left[M+M\sum_k\frac{x^k}{|\mathbf{x}|^2}X^k+\int d^3x'\sum_k T_{kk}\right.$$
$$\left.+\sum_k\frac{x^k}{|\mathbf{x}|^2}\int d^3x' \, x'^k\sum_l T_{ll}+\cdots\right], \qquad (4.56)$$

where

$$M = \int_V d^3x \, T_{00}(\mathbf{x}) \qquad (4.57)$$

is the total mass and

$$X^k = \frac{1}{M}\int_V d^3x \, x^k T_{00}(\mathbf{x}) \qquad (4.58)$$

are the center-of-mass coordinates. The last two terms in the square brackets of (4.56) can be dealt with as follows. From (4.54), it follows that

$$T_{kl} = \sum_m\left(T^{mk}x^l+T^{ml}x^k\right)_{,m}-\frac{1}{2}\sum_{m,n}\left(T^{mn}x^k x^l\right)_{,mn} \qquad (4.59)$$

and

$$\sum_l T_{ll} x^k = \sum_{m,l} \left(T^{ml} x^m x^k - \frac{1}{2} T^{lk} x^m x^m \right)_{,l} . \tag{4.60}$$

Hence, the integrands are divergences and can be turned into surface integrals along ∂V. These integrals vanish because T_{kl} is zero there. Thus, we have

$$\varepsilon h_{00} = -\frac{2GM}{|\mathbf{x}|} \left(1 + \frac{1}{|\mathbf{x}|^2} \sum_k X^k x^k + \cdots \right) . \tag{4.61}$$

Next, we calculate h_{0k}:

$$\varepsilon h_{0k}(\mathbf{x}) = -\frac{4G}{|\mathbf{x}|} \left[\int_V d^3 x'\, T_{0k}(\mathbf{x}') + \sum_l \frac{x^l}{|\mathbf{x}|^2} \int_V d^3 x'\, x'^l T_{0k}(\mathbf{x}') + \cdots \right] .$$

From the identity

$$T_{0k} = -\sum_l \left(T_0^l x^k \right)_{,l} , \tag{4.62}$$

we conclude that the first integral is zero. It is nothing but the total 3-momentum P_k of the source. It vanishes as the source would not be stationary with $P^k \neq 0$. Indeed, (4.62) follows from the conservation, $T_{0,\mu}^\mu = 0$ together with the stationarity, $T_{0,0}^0 = 0$. We can write further

$$-T_0^k x^l = -\frac{1}{2} \left(T_0^k x^l + T_0^l x^k \right) - \frac{1}{2} \left(T_0^k x^l - T_0^l x^k \right)$$

$$= -\sum_m \left(T_0^m x^k x^l \right)_{,m} + \frac{1}{2} \sum_m \varepsilon^{klm} \varepsilon_{mij} T_0^i x^j .$$

Here ε_{klm} is the totally antisymmetric quantity satisfying $\varepsilon_{123} = 1$ (ε^{klm} is obtained from ε_{klm} by raising its indices with $\eta^{\mu\nu}$), and we have used the identity

$$T_0^k x^l + T_0^l x^k = \sum_m \left(T_0^m x^k x^l \right)_{,m} .$$

This divergence does not contribute to the integral and we obtain

$$\varepsilon h_{0k}(\mathbf{x}) = \frac{2G}{|\mathbf{x}|^3} \sum_{l,m} \varepsilon_{kml} J^m x^l , \tag{4.63}$$

where

$$J_m = \varepsilon_{mkl} \int_V d^3 x\, x^k T_0^l(\mathbf{x}) \tag{4.64}$$

is the total angular momentum (observe that $J^m = -J_m$).

Finally,

$$\varepsilon h_{kl}(\mathbf{x}) = -\frac{2G}{|(\mathbf{x})|} \left[M\delta_{kl} + M\delta_{kl} \sum_m \frac{x'^m}{|(\mathbf{x})|^2} X^m + \int_V d^3x' \left(2T_{kl} - \delta_{kl} \sum_m T_{mm} \right) \right.$$
$$\left. + \sum_m \frac{x'^m}{|(\mathbf{x})|^2} \int_V d^3x' \, x'^m \left(2T_{kl} - \delta_{kl} \sum_m T_{mm} \right) + \cdots \right]. \tag{4.65}$$

We can turn the volume integrals into vanishing surface integrals in view of (4.59), (4.60) and

$$\sum_l T^{kl} x^l = \frac{1}{2} \sum_l \left(T^{kl} |(\mathbf{x})|^2 \right)_{,l}.$$

Hence,

$$\varepsilon h_{kl}(\mathbf{x}) = -\frac{2GM}{|\mathbf{x}|} \delta_{kl} \left(1 + \frac{1}{|(\mathbf{x})|^2} \sum_m X^m x^n + \cdots \right). \tag{4.66}$$

The comparison to electrodynamics shows that the analogue of the electric monopole is the total mass M and that of the electric dipole is MX^k, where X^k is the position vector of the center of mass. The analogue of the electric current is the total momentum P^k and that of the magnetic dipole is the angular momentum J^k.

The metric can be simplified if the coordinates x^k are transformed by a shift $-X^k$ so that the center of mass is at the origin. Such a coordinate system is called *co-moving* and *mass centered*. The final form of the metric is

$$ds^2 = \left(1 - \frac{2GM}{|\mathbf{x}|} \right) (dx^0)^2 + \frac{4G}{|\mathbf{x}|^3} \varepsilon_{kmn} J^m x^n \, dx^0 \, dx^k - \left(1 + \frac{2GM}{|\mathbf{x}|} \right) \delta_{kl} \, dx^k \, dx^l. \tag{4.67}$$

Keep in mind that J_m are the components of the angular momentum as it is usually defined, having the same direction as the vector of angular velocity.

Now, we can understand the meaning of the constant G in the Einstein equations. Consider the leading terms in the metric (4.67),

$$ds^2 = \left(1 - \frac{2GM}{|\mathbf{x}|} \right) (dx^0)^2 - \delta_{kl} \, dx^k dx^l,$$

and compare this to (2.54) in which we expand the exponential,

$$e^{2\Phi} \approx 1 + 2\Phi$$

and write out Φ,

$$\Phi = -\frac{G'M}{|x|},$$

where G' is Newton's constant. Equality can only be obtained if $G = G'$. That is, G must be Newton's constant to ensure that Newtonian theory arises as the first approximation of the Einstein equations.

4.4.6 Gravitomagnetic Phenomena

The gravitomagnetic effect that is mentioned most frequently is the *dragging of inertial frames* or *Lense–Thirring effect*. An orthonormal tetrad that is parallel transported along any time-like trajectory near a stationary rotating body rotates with respect to the stationary coordinates. One part of the tetrad rotation is proportional to the angular momentum of the body and its angular velocity is roughly aligned with that of the body, as if "dragged" by it.

Let us study this effect using the metric (4.67) for a trajectory of a stationary observer,

$$x^0 = \lambda , \quad x^m = x_0^m .$$

A vector field $V^\mu(\lambda)$ that is parallel along this curve satisfies the equation

$$\dot{V}^\mu + \Gamma^\mu_{\rho 0} V^\rho = 0 .$$

Equation (4.38) implies that

$$\Gamma^0_{m0} = \Gamma^m_{00} = \frac{\varepsilon}{2} h_{00,m} ,$$

and

$$\Gamma^m_{n0} = -\frac{\varepsilon}{2} \left(h_{0m,n} - h_{0n,m} \right) .$$

Substituting for $\frac{\varepsilon}{2} h_{\mu\nu}$ from (4.67) leads after some calculation to

$$\Gamma^0_{m0} = \Gamma^m_{00} = \frac{GM}{|\mathbf{x}|^3} x^m ,$$

and

$$\Gamma^m_{n0} = \Omega^k \varepsilon_{kmn}$$

with

$$\Omega^k = \frac{G}{|\mathbf{x}|^3} \left(J_k - 3 J_l \frac{x^l x^k}{|\mathbf{x}|^2} \right) .$$

Now, the equations of parallel transport become

$$\dot{V}^0 + \frac{GM}{|\mathbf{x}|^3} x^m V^m = 0$$

and

$$\dot{V}^m + \frac{GM}{|\mathbf{x}|^3} x^m V^0 + \Omega^k \varepsilon_{kmn} V^n = 0 .$$

Thus, the tetrad suffers a boost associated to the velocity

$$v^m = \frac{GM}{|\mathbf{x}|^3} x^m ,$$

and a rotation with angular velocity Ω^k—this is the dragging. Similar expressions are present in the parallel transport along any curve that has a non-zero component $\dot{x}^0$ of its tangent vector. We can also see from the above formula that $J_k x^k = 0$ in the equatorial plane of the rotating body and hence $\Omega^k = GJ_k/|\mathbf{x}|^3$ there.

The dragging of inertial frames can be measured by observing, say, gyroscopes in a satellite on a circular trajectory around Earth. As is easily understood by invoking the equivalence principle, the rotational axis of an ideal gyroscope is parallel transported. Since the effect has $|\mathbf{x}|^{-3}$ dependence, it is extremely tiny. The first measurement of the dragging is just on the way.

Another important effect is the difference in the attraction that a rotating body exerts on co-rotating and counter-rotating satellites in its equatorial plane. To study it, let us transform the metric (4.67) to spherical coordinates corresponding to the frame such that $J_k = J\delta_k^3$, where J is a number:

$$ds^2 = \left(1 - \frac{2GM}{r}\right) dt^2 - \frac{4GJ}{r} \sin^2 \vartheta \, dt \, d\varphi$$

$$- \left(1 + \frac{2GM}{r}\right) \left(dr^2 + r^2 \, d\vartheta^2 + r^2 \sin^2 \vartheta \, d\varphi^2\right) . \qquad (4.68)$$

Its Killing vectors are the time translation δ_0^μ and the rotation along the z-axis δ_3^μ. Let us restrict ourselves to the equatorial plane $\vartheta = \pi/2$ and write the conservation laws

$$e = \left(1 - \frac{2GM}{r}\right) \dot{t} , \qquad (4.69)$$

$$j = \left(1 + \frac{2GM}{r}\right) r^2 \dot{\varphi} , \qquad (4.70)$$

$$1 = \left(1 - \frac{2GM}{r}\right) \dot{t}^2 - \frac{4GJ}{r} \dot{t}\dot{\varphi} - \left(1 + \frac{2GM}{r}\right) \left(\dot{r}^2 + r^2\dot{\varphi}^2\right) . \qquad (4.71)$$

Substitute the expressions for $\dot{t}$ and $\dot{\varphi}$ from (4.69) and (4.70) into (4.71) and calculate only the linear terms in ε. Thus, the so-called radial equation results:

$$\dot{r}^2 + V(r) = 0 , \qquad (4.72)$$

where the effective potential is (exercise)

$$V(r) = 1 - e^2 - \frac{2GM}{r} + \frac{j^2}{r^2} + \frac{4GJ\,je}{r^3} - \frac{2GM\,j^2}{r^3} . \qquad (4.73)$$

The interpretation of its different terms is the following. The third and fourth terms coincide with the well-known Newtonian effective potential. The fifth one describes the influence of the angular momentum J and the sixth one is a correction to the Newtonian term j^2/r^2.

Differentiating (4.72) with respect to λ yields

$$\ddot{r} = -\frac{V'(r)}{2} \, .$$

Hence, the additional radial acceleration due to J is

$$\ddot{r} = \frac{6GJje}{r^4} \, .$$

We see that this term is repulsive for $Jj > 0$ and attractive for $Jj < 0$. This means that co-rotating satellites ($Jj > 0$) are attracted less by the central body than the counter-rotating ones ($Jj < 0$). Indeed, the angular momentum term is only a small correction to the gravity pull given by the third term in $V(r)$. This effect plays a role in the energetics of black holes.

4.4.7 Plane Waves

We mentioned that any general solution of the homogeneous linearized Einstein equations can be written as a linear combination of *monochromatic plane waves*,

$$h_{\mu\nu}(\vec{x}) = A_{\mu\nu} \exp\left(ik_\rho x^\rho\right) \, ,$$

where $A_{\mu\nu}$ is a constant tensor and k_ρ a constant null vector satisfying the transversality relation in the form

$$\eta^{\rho\sigma}\left(A_{\rho\mu}k_\sigma - A_{\rho\sigma}k_\mu\right) = 0 \, .$$

The monochromatic plane wave alone is not a good approximation to any exact solution of the full Einstein equations as it does not satisfy suitable fall-off conditions. However, any field $h_{\mu\nu}$ satisfying such conditions allows a decomposition into plane waves, which may be useful to study its properties.

A general (not monochromatic) plane wave also depends on the space–time coordinates only through the linear function $k_\rho x^\rho$, but each component $h_{\mu\nu}$ is a different function of the variable $k_\rho x^\rho$. Such waves are idealizations that approximately describe local properties of gravitational waves far away from their sources.

Given a plane wave, we can rotate the space coordinates so that the wave travels in the z direction: $k_t = -k_z = \omega$ and $k_x = k_y = 0$. Then, the field has the form

$$h_{\mu\nu}(t - z)$$

(the constant ω has been incorporated into $h_{\mu\nu}$). Our study of this wave is simplified by using coordinates u, x, y, and v instead of t, x, y, and z,

$$u = t - z, \quad v = t + z \, .$$

The auxiliary metric then has components $\tilde{\eta}_{\mu\nu}$, given by the matrix

$$
\begin{pmatrix}
0 & 0 & 0 & 1/2 \\
0 & -1 & 0 & 0 \\
0 & 0 & -1 & 0 \\
1/2 & 0 & 0 & 0
\end{pmatrix}.
$$

Its inverse $\tilde{\eta}^{\mu\nu}$ is

$$
\begin{pmatrix}
0 & 0 & 0 & 2 \\
0 & -1 & 0 & 0 \\
0 & 0 & -1 & 0 \\
2 & 0 & 0 & 0
\end{pmatrix}.
$$

Observe that the change from inertial coordinates to null coordinates is not a gauge transformation, but a pure coordinate transformation so that the auxiliary metric must change components in order to remain the same tensor field.

The field $h_{\mu\nu}(u)$ satisfies (4.47) with $T_{\mu\nu} = 0$, but not necessarily the transversality condition (4.48), which now reads

$$\left(h^\mu_\mu - 1/2\delta^u_\mu h \right)_{,u} , \tag{4.74}$$

where

$$h^\mu_\rho = \tilde{\eta}^{\mu\nu} h_{\nu\rho} , \quad h = \tilde{\eta}^{\mu\nu} h_{\mu\nu} ,$$

so that $h^u_u = h^v_v$. Thus, (4.74) implies that

$$\left(h^x_x + h^y_y \right)_{,u} = 0 , \tag{4.75}$$

and

$$\left(h^u_x \right)_{,u} = \left(h^u_y \right)_{,u} = \left(h^u_v \right)_{,u} = 0 . \tag{4.76}$$

As we know, condition (4.74) does not determine the gauge uniquely. We can still change the gauge by a field X_μ which satisfies the wave equation. Any such field that preserves the form $h_{\mu\nu}(u)$ (that is the fact that $h_{\mu\nu}(u)$ depends only on u) can be written as

$$X_\mu(u) + X^0_\mu(x, y, v) ,$$

with four arbitrary functions $X_\mu(u)$ of one variable and four arbitrary linear functions $X^0_\mu(x, y, v)$ of three variables. This leads to the change of components of $h_{u\mu}(u)$,

$$h_{u\mu}(u) \longmapsto h_{u\mu}(u) + \left(1 + \delta^u_\mu \right) X_{\mu,u}(u)$$

for any μ. Hence, all of them can be transformed to zero by a suitable choice of $X_\mu(u)$. According to (4.75) and (4.76), $h_{xx} + h_{yy}$, h_{vx}, h_{vy}, and h_{vv} are constant. We can transform these constants to zero by a suitable choice of the linear functions $X^0_\mu(x, y, v)$ (exercise). The final result is that the only non-zero components of $h_{\mu\nu}(u)$ are

$$h_{xx} = -h_{yy} \tag{4.77}$$

and

$$h_{xy} = h_{yx} . \tag{4.78}$$

This gauge is called *transverse traceless* or TT-gauge. The only gauge freedom left is (exercise)

$$h_{xx} \mapsto h_{xx} + c , \quad h_{yy} \mapsto h_{yy} - c ,$$
$$h_{xy} \mapsto h_{xy} ,$$

where c is a constant. Hence, the value of h_{xx} is not gauge invariant (unless some fall-off conditions are used).

The non-zero components (4.77) and (4.78) can be considered as components of a tensor in the (x, y)-plane for each u. This tensor can always be written in the form

$$\mathbf{e}_+ f_+(u) + \mathbf{e}_\times f_\times(u)$$

with $f_+(u) = h_{xx}(u)$, $f_\times(u) = h_{xy}(u)$, and

$$\mathbf{e}_+ = \begin{pmatrix} 1 & 0 \\ 0 & -1 \end{pmatrix} , \quad \mathbf{e}_\times = \begin{pmatrix} 0 & 1 \\ 1 & 0 \end{pmatrix} , \tag{4.79}$$

where $\mathbf{e}_+$ and $\mathbf{e}_\times$ are two *polarization tensors* of gravitational waves. This is analogous to the two polarization vectors of electromagnetic waves, that is two vectors orthogonal to the plane wave propagation direction. The components $h_{\mu\nu}$ satisfying (4.77) and (4.78) transform under the rotation in the (x, y)-plane

$$x = x' \cos\phi - y' \sin\phi ,$$
$$y = x' \sin\phi + y' \cos\phi$$

as follows

$$h'_{11} = h_{11} \cos 2\phi + h_{12} \sin 2\phi ,$$
$$h'_{12} = -h_{11} \sin 2\phi + h_{12} \cos 2\phi .$$

In particular, $\mathbf{e}_+$ is mapped to $-\mathbf{e}_\times$ and $\mathbf{e}_\times$ to $\mathbf{e}_+$ if $\phi = \pi/4$.[1]

Any trace-free symmetric tensor in the plane can be diagonalized by a rotation. To obtain $h'_{12} = 0$ we need

$$\tan 2\phi = \frac{h_{12}}{h_{11}} .$$

Thus, any $h_{\mu\nu}$ is a multiple of a rotated $\mathbf{e}_+$. The x-axis can be aligned with the eigenvector of $h_{\mu\nu}$ corresponding to its positive eigenvalue at some fixed point. In

---

[1] In general, to transform between the polarization states of a spin-s-zero-rest-mass field, one needs the rotation by $\pi/2s$, [9]. Accordingly, gravity has spin 2.

this way, the polarization determines certain directions in the (x, y)-plane. A plane wave is called *linearly polarized* if these directions do not depend on u, that is if there is $\alpha \in [0, 2\pi)$ and a function $f(u)$ such that

$$f_+(u) = f(u) \cos \alpha, \quad f_\times(u) = f(u) \sin \alpha.$$

4.4.8 Measurable Properties of Plane Waves

We showed that the changes in the amplitude $f(u)$ and the polarization of a linearly polarized plane wave are gauge invariant. How can we measure them? We describe a simple measurement that works in principle. A real measurement can be based on this idea but must be technically much subtler as the effect is very tiny.

Let two observers Alice (A) and Bob (B) move along auto-parallels. Let A be steadily sending a light signal of a fixed frequency ν_A and B receiving it and measuring the frequency $\nu_B(t)$ at each value t of his proper time.

To see how both amplitude and polarization of a wave can be measured, we assume that the space–time contains a linearly polarized plane wave. That is,

$$ds^2 = du\, dv - [1 + \varepsilon f(u)]dx^2 - [1 - \varepsilon f(u)]dy^2, \tag{4.80}$$

where $f(u)$ is a smooth function for $u \in \mathbb{R}$.

We need two time-like autoparallels for the observers and a null one for the signal. The general autoparallel for the metric (4.80) is easily found as the geometry has the typical plane wave symmetry[2]: there are three mutually orthogonal Killing fields, two of them space-like, δ_x^μ and δ_y^μ, and one null, δ_u^μ. The conservation laws then read

$$P_x = [1 + \varepsilon f(u)]\dot{x}, \quad P_y = [1 - \varepsilon f(u)]\dot{y}, \tag{4.81}$$

and

$$P = \dot{u}. \tag{4.82}$$

If we require that the square norm of the tangent vector to the autoparallel is σ, and if we substitute the components $\dot{x}$, $\dot{y}$, and $\dot{u}$ from (4.81) and (4.82) into the metric, we obtain

$$P\dot{v} - [1 + \varepsilon f(u)]^{-1}P_x^2 - [1 - \varepsilon f(u)]^{-1}P_y^2 = \sigma.$$

For a time-like autoparallel parameterized by the proper time we have $\sigma = 1$ and for a light-like autoparallel $\sigma = 0$. We calculate everything just to first order in ε.

---

[2] The Einstein equations can be solved exactly for this symmetry. The result is the so-called exact plane wave, found by Rosen, by Bondi, by Ehlers and Kundt (see [10], p. 957). It also contains two arbitrary functions of u.

Thus, we obtain

$$\dot{u} = P ,$$ (4.83)

$$\dot{x} = P_x[1 - \varepsilon f(u)] ,$$ (4.84)

$$\dot{y} = P_y[1 + \varepsilon f(u)] ,$$ (4.85)

$$\dot{v} = P^{-1}\left\{\sigma + P_x^2[1 - \varepsilon f(u)] + P_y^2[1 + \varepsilon f(u)]\right\} .$$ (4.86)

Integration of these differential equations is straightforward:

$$x(u) = P_x P^{-1} u - \varepsilon P_x P^{-1} F(u) + X ,$$ (4.87)

$$y(u) = P_y P^{-1} u + \varepsilon P_y P^{-1} F(u) + Y ,$$ (4.88)

$$v(u) = \frac{\sigma + P_x^2 + P_y^2}{P} u - \varepsilon \frac{P_x^2 - P_y^2}{P} F(u) + V ,$$ (4.89)

where

$$F(u) = \int_0^u dx\, f(x) ,$$

X, Y, and V are constants and $u = P\lambda$.

Let the observers follow autoparallels with $P_x = P_y = 0$ and $\sigma = P = 1$. It follows that they stay at fixed values of the original coordinates x, y, and z because of (4.84), (4.85), and $\dot{z} = \dot{v} - \dot{u} = P^{-1} - P$ due to (4.83) and (4.86). Further, we choose the values $X_{A,B}$, $Y_{A,B}$, and $V_{A,B} = V$ of the constants X, Y, and V for the observers. Thus, they both lie at the same value of z and their proper time coincides with u.

The light-like autoparallel that joins them starts at the value $u = u_A$ from A. The constant P is determined by the physical parameter λ of the autoparallel. It will be fixed but arbitrary. However, for the sake of simplicity, we shall carry out all calculations with the affine parameter u along the null autoparallel. When we need the 4-momentum p^μ, we calculate it via $p^\mu = P^{-1}\, dx^\mu/du$. The autoparallel then has the form

$$x(u) = P_x(u - u_A) - \varepsilon P_x[F(u) - F(u_A)] + X_A ,$$ (4.90)

$$y(u) = P_y(u - u_A) + \varepsilon P_y[F(u) - F(u_A)] + Y_A ,$$ (4.91)

$$v(u) = \left(P_x^2 + P_y^2\right)(u - u_A) - \varepsilon\left(P_x^2 - P_y^2\right)[F(u) - F(u_A)] + u_A + V .$$ (4.92)

We are looking for the values of P_x, P_y, and u_B such that

$$x(u_B) = X_B , \quad y(u_B) = Y_B , \quad v(u_B) = u_B + V .$$ (4.93)

Let us expand this

$$P_x = P_x^{(0)} + \varepsilon P_x^{(1)} , \quad P_y = P_y^{(0)} + \varepsilon P_y^{(1)} , \quad u_B = u_B^{(0)} + \varepsilon u_B^{(1)} ,$$

and collect the terms up to linear order in ε from (4.93). The zero-order equations give

$$P_x^{(0)} = \frac{X_B - X_A}{D_0} \,, \quad P_y^{(0)} = \frac{Y_B - Y_A}{D_0} \,, \quad u_B^{(0)} - u_A = D_0 \,, \tag{4.94}$$

where

$$D_0 = \sqrt{(X_B - X_A)^2 + (Y_B - Y_A)^2}$$

is the distance between the observers in the $\varepsilon = 0$ approximation. In fact, we can define a distance in the (x,y)-surface at any "time" u using the metric (4.80):

$$D(u) = \sqrt{[1 + \varepsilon f(u)](X_B - X_A)^2 + [1 - \varepsilon f(u)](Y_B - Y_A)^2}$$

and write down $D(u)$ with the accuracy to first order in ε:

$$D(u) = D_0 \left[1 + \frac{\varepsilon}{2} \frac{(X_B - X_A)^2 - (Y_B - Y_A)^2}{(X_B - X_A)^2 + (Y_B - Y_A)^2} f(u) \right] \,. \tag{4.95}$$

The first-order equations in which $P_x^{(0)}$, $P_y^{(0)}$, and $u_B^{(0)}$ are expressed with the help of (4.94) can be solved for $P_x^{(1)}$, $P_y^{(1)}$, and $u_B^{(1)}$:

$$P_x^{(1)} = \frac{X_B - X_A}{D_0} \left[1 + 2\frac{(Y_B - Y_A)^2}{D_0^2} \right] \frac{F(u_A + D_0) - F(u_A)}{2D_0} \,, \tag{4.96}$$

$$P_y^{(1)} = -\frac{Y_B - Y_A}{D_0} \left[1 + 2\frac{(X_B - X_A)^2}{D_0^2} \right] \frac{F(u_A + D_0) - F(u_A)}{2D_0} \,, \tag{4.97}$$

$$u_B^{(1)} = \frac{(X_B - X_A)^2 - (Y_B - Y_A)^2}{2D_0^2} [F(u_A + D_0) - F(u_A)] \,. \tag{4.98}$$

To calculate the redshift, we use formula (2.32). Hence, we need the 4-momentum p^μ of the null autoparallel at its intersections with the observers and the tangent vectors $\dot{x}_{A,B}^\mu$ of the observers there. We obtain, after some calculation (exercise)

$$\frac{\lambda_B - \lambda_A}{\lambda_A} = \frac{\varepsilon}{2} \frac{(X_B - X_A)^2 - (Y_B - Y_A)^2}{(X_B - X_A)^2 + (Y_B - Y_A)^2} [f(u_A + D_0) - f(u_A)] \tag{4.99}$$

the final formula for the redshift.

It is interesting to compare the redshift formula with a formula for the change in the relative distance between the observers from the time u_A to the time u_B. Equation (4.95) gives, to first order in ε,

$$\frac{D(u_B) - D(u_A)}{D(u_A)} = \frac{\varepsilon}{2} \frac{(X_B - X_A)^2 - (Y_B - Y_A)^2}{(X_B - X_A)^2 + (Y_B - Y_A)^2} [f(u_A + D_0) - f(u_A)] \,. \tag{4.100}$$

Hence, the redshift determines the relative change of distance. We shall see in our treatment of cosmology that the cosmological redshift has the same property. There are some differences, however. First, the wave formula holds only in the linear approximation while the cosmological formula is exact. Second, the cosmological result is completely isotropic, while the wave one is not.

Thus, we arrive at the effect of the polarization of the wave. The polarization tensor of our wave is $\mathbf{e}_+$. It has two orthogonal eigenvectors, δ_x^μ and δ_y^μ, and the corresponding eigenvalues, $+1$ and -1, respectively. The redshift or relative distance changes are depending uniformly on the angle ζ between the separation vector $(X_B - X_A, Y_B - Y_A)$ of the observers and the $+1$-eigenvector. We have

$$\cos \zeta = \frac{X_B - X_A}{D_0} ,$$

and the non-isotropy factor in both formulas (4.99) and (4.100) is simply $\cos 2\zeta$. For example, if the observers are aligned along the $+1$-axis, the factor is 1, along the -1-axis, it is -1, and if they are arranged diagonally ($\alpha = \pi/4$), then the effect is zero, etc. This angle dependence gives an observable meaning to the polarization of the wave.

Modern detectors of gravitational waves [11] are utilizing the relative change of distances in two mutually orthogonal directions. An arrangement similar to the Michelson–Morley experiment can in principle detect changes in the interference pattern if a gravitational wave is passing.

There is much more to the theory of gravitational waves than we have just described. The questions of energy that the waves carry, that they take away from their sources (back reaction), and that they transfer to matter systems are fascinating and very difficult (they cannot be answered within the linearized theory). On the other hand, the action of gravitational waves on matter is extremely weak: all techniques that were in use at the time of writing these Notes (2006) were not enough to detect the waves directly. That is why we now turn to more conspicuous effects of gravity.

4.5 Exercises

1. Show that

$$\delta \left\{ {}^{\mu}_{\rho\sigma} \right\} = \frac{1}{2} g^{\mu\nu} \left(\nabla_\rho \delta g_{\nu\sigma} + \nabla_\sigma \delta g_{\nu\rho} - \nabla_\nu \delta g_{\rho\sigma} \right) ,$$
$$\delta R^{\mu}_{\nu\rho\sigma} = \nabla_\rho \delta \Gamma^{\mu}_{\nu\sigma} - \nabla_\sigma \delta \Gamma^{\mu}_{\nu\rho} .$$

Show that $\delta g^{\mu\nu} = -g^{\mu\kappa} g^{\nu\lambda} \delta g_{\kappa\lambda}$. Use this equation and the fact that the variation commutes with the derivative.

2. Prove that the curvature of a metric connection satisfies

$$R_{\mu\nu\rho\sigma} = \frac{1}{2} \left(\partial_\nu \partial_\rho g_{\mu\sigma} + \partial_\mu \partial_\sigma g_{\nu\rho} - \partial_\mu \partial_\rho g_{\nu\sigma} - \partial_\nu \partial_\sigma g_{\mu\rho} \right) + \Delta R_{\mu\nu\rho\sigma},$$

where all terms in $\Delta R_{\mu\nu\rho\sigma}$ are quadratic in the first derivatives of the metric, $\partial_\rho g_{\mu\nu}$, and do not contain second derivatives.

3. Prove that the three relations

$$R_{\mu\nu\rho\sigma} = -R_{\nu\mu\rho\sigma}, \tag{4.101}$$

$$R_{\mu\nu\rho\sigma} = -R_{\mu\nu\sigma\rho}, \tag{4.102}$$

$$R_{\mu\nu\rho\sigma} + R_{\mu\rho\sigma\nu} + R_{\mu\sigma\nu\rho} = 0, \tag{4.103}$$

imply the symmetry $R_{\mu\nu\rho\sigma} = R_{\rho\sigma\mu\nu}$.

4. Find the number of independent components of $R_{\mu\nu\rho\sigma}$ in a n-dimensional manifold, subject to the assumption that all independent symmetries are described by the (4.101), (4.102), and (4.103). Hint: consider how many independent components there are for a totally anti-symmetric tensor of arbitrary type in a n-dimensional manifold.

5. Compare (4.26) for the gauge transformation with (2.78) for the Lie derivative of a $(0,2)$ symmetric tensor and explain the relation between them (including the sign!)

6. For many calculations with the (non-linearized) Einstein equations, it is advantageous to choose coordinates such that the condition

$$\left(\sqrt{-g}\,g^{\mu\nu}\right)_{,\nu} = 0$$

is satisfied. These are called *harmonic* coordinates or *de Donder* gauge condition. Show that (1) Equation (4.37) is the linearized form of the de Donder condition, and that (2) harmonic coordinates are harmonic functions in the sense that they satisfy the covariant Laplace equation.

7. Find the orders of magnitude of the corrections $2GMR^{-1}$ and $2GJR^{-2}$ to the flat metric at the surface of Earth, that is, $R = R_{\text{Earth}}$, $M = M_{\text{Earth}}$ and $J = J_{\text{Earth}}$. Calculate J_{Earth} within Newtonian mechanics assuming the Earth to be a homogeneous perfect sphere.

8. Find the error in the following argument. Suppose that (1) metric determines all lengths and time intervals and (2) that components of the metric in a given space–time are time–dependent. Then *all* lengths (wavelengths of light signals, distance between observers, etc.) and time intervals in the space–time change with time in a *universal* way given by the space–time dependence of the metric. (Hint: the components of a metric depend not only on the geometry but also on the coordinates ...)

References

1. B. Schutz, *Gravity from the Ground up*, Cambridge University Press, Cambridge, UK, 2003.
2. J. L. Synge, *Relativity: The General Theory*. North-Holland, Amsterdam, 1960.

3. J. Stachel, *Einstein and the History of General Relativity, Einstein Studies*, Vol. 1, P. 63, Birkhauser, Boston, MA, 1989.

4. J. Milnor, *Relativity, Groups and Topology II*, edited by B. C. DeWitt and R. Stora, Elsevier, Amsterdam, 1984.

5. D. Christodoulou and S. Klainerman, *The Global Nonlinear Stability of Minkowski Spacetime*, Princeton University Press, Princeton, NJ, 1993.

6. H. Lindblad and J. Rodnianski, arXiv, math.AP/0411109.

7. M. W. Choptuik, Phys. Rev. Lett. **70** (1993) 9.

8. J. D. Jackson, *Classical Electrodynamics*, Wiley, New York, 1962.

9. S. Weinberg, *The Quantum Theory of Fields* Vol. I, Cambridge University Press, Cambridge, UK, 1995.

10. C. W. Misner, K. S. Thorne and J. A. Wheeler, em Gravitation, Freeman, San Francisco, CA, 1973.

11. I. Ciufolini and J. A. Wheeler, *Gravitation and Inertia*. Princeton University Press, Princeton, NJ, 1995.

Chapter 5
Cosmological Models

In this chapter we study the question which solutions of Einstein's equations can describe the universe that we observe around us.

We consider the details of space–time geometry, such as the gravitational field of single galaxies, stars, or planes, as small perturbations of a smooth background space-time. We make some assumptions about this coarse geometry that seem reasonable today. We shall see that the consequences of these hypotheses are in remarkable agreement with many observations.

Modern cosmology is a field that uses all kinds of physics and influences our view of the whole world. In this section we discuss some general-relativistic aspects of cosmology, such as the basics of the dynamics of matter subject to gravity, as well as the geometry of the space-times in question, the so-called *cosmological models*. We shall not consider the thermal evolution of the universe, the baryon synthesis, the formation of the elements, or the origin of galaxies.

5.1 Homogeneous Isotropic 3-Spaces

5.1.1 The Cosmological Principle

The starting hypothesis of cosmology is that from a mathematical point of view, the cosmological models are the solutions of Einstein's equations with the largest possible spatial symmetry. This assumption of symmetry is aesthetically appealing. The universe, as a whole, should be symmetric, and all structures perturbing the symmetry should be relatively small. Besides that, there is observational evidence for this assumption, namely the *isotropy* of the *cosmic microwave background radiation* (CMB).

The CMB was discovered in 1965 by Penzias and Wilson [1]. These scientists constructed a radio receiver for wavelength $\lambda = 7.25$ cm with a very low noise level. They were receiving a noise signal roughly corresponding to the temperature

Hájíček, P.: *Cosmological Models*. Lect. Notes Phys. **750**, 159–208 (2008)
DOI 10.1007/978-3-540-78659-7_5   © Springer-Verlag Berlin Heidelberg 2008

$T = 2.5 - 4.5$ K. This discovery was rewarded with the Nobel Prize. Since then, this thermal radiation was measured in all wavelengths and in all directions. The result is the spectrum of a nearly perfect black-body radiation of temperature 2.7 K with almost perfect isotropy and no polarization. This radiation is interpreted as a remainder from the early times of the universe, when densities and temperatures were large, and radiation was in thermal equilibrium with and dominated matter ("thermal history of the universe", [2, 3]).

In general, we assume that we are not at a distinguished point, or center, of the universe. This is the so-called *Copernican principle*. Then the CMB is isotropic everywhere else in the universe, too. This is only possible if the spatial geometry of the universe, averaged over large distances, is homogeneous and isotropic. This is the so-called *cosmological principle*. We will study examples of such 3-spaces.

5.1.2 Euclidean Space

The simplest example is $\mathbb{E}^3$, that is the manifold $\mathbb{R}^3$ with the metric

$$d_3s^2 = \left(dx^1\right)^2 + \left(dx^2\right)^2 + \left(dx^3\right)^2$$

in Cartesian coordinates x^1, x^2, x^3, and

$$d_3s^2 = dr^2 + r^2 d\vartheta^2 + r^2 \sin^2 \vartheta \, d\varphi^2 \tag{5.1}$$

in spherical coordinates.

Every two points p and q in $\mathbb{E}^3$ define a translation $\phi : \mathbb{E}^3 \mapsto \mathbb{E}^3$, which maps p to q, $\phi(p) = q$, and does not change the metric. This is the mathematical description of the fact that each point is equivalent to every other point, that is the homogeneity of space. Similarly, at each point p, we can represent two arbitrary directions by unit vectors u^k and v^k and find a rotation $\phi : \mathbb{E}^3 \mapsto \mathbb{E}^3$, which fixes p, $\phi(p) = p$ but transforms u^k to v^k, $\phi(u^k) = v^k$. The metric is also invariant under rotations. This property is called the isotropy of space at p. If space is isotropic at all points, we simply call it isotropic. We remark that the isometry group $\mathscr{E}^3$ of $\mathbb{E}^3$ is six-dimensional [4]. It is generated by three independent infinitesimal translations, and three independent infinitesimal rotations.

5.1.3 The Sphere S^3

Another 3-space with as much symmetry is the 3-sphere S^3. We can deduce many properties of S^3 by regarding S^3 as a sub-manifold of $\mathscr{E}^4$. However, this is just a method to simplify calculations and to help imagination. All properties which are of interest to us also follow directly from the metric of S^3. Therefore, consider $\mathbb{E}^4$ with Cartesian coordinates $\{X^\mu\}$ and the metric

$$dS^2 = \sum_{\mu} (dX^{\mu})^2 \ . \tag{5.2}$$

S^3 consists of the points satisfying the equation

$$\sum_{\mu} (X^{\mu})^2 = R^2 \ , \tag{5.3}$$

where $R > 0$ is the radius of the sphere. The S^3 constructed in this way is invariant with respect to all rotations about the origin in $\mathbb{E}^4$. This is the group SO(4), generated by six independent infinitesimal rotations [4] (in $\mathbb{E}^4$ there are six distinct pairs of coordinate axes). In addition, we need a metric on S^3 in some coordinates. We choose coordinates as follows (Fig. 5.1):

Let $r(p)$ be the distance of the point p from the X^4-axis. All points with the same distance r from the X^4-axis are contained in a 2-sphere, which can be projected onto the hyperplane $X^4 = 0$. For this projection, we choose the usual spherical coordinates ϑ and φ. Then the coordinates X^{μ} and r, ϑ, and φ on S^3 are related as follows:

$$X^4 = \pm\sqrt{R^2 - r^2}, \quad X^k = rn^k \ , \tag{5.4}$$

where

$$n^1 := \sin\vartheta\cos\varphi, \quad n^2 := \sin\vartheta\sin\varphi, \quad n^3 := \cos\vartheta \ . \tag{5.5}$$

Some computations simplify a lot if we directly employ the properties of the three functions $n^k(\vartheta, \varphi)$ (exercise). If the range of r is the interval $[0, 1]$ and the range of ϑ and φ are the usual intervals, then we obtain the upper (lower) hemisphere for the upper (lower) sign. Clearly, the coordinates X^{μ}, given by the above embedding,

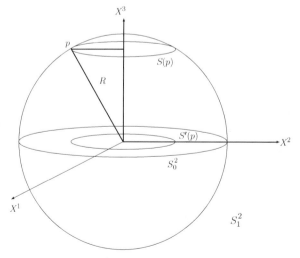

Fig. 5.1 The sphere S^2 is represented by its intersection S_0^2 with the (X^1, X^2)-plane and by S_1^2 with the (X^2, X^3)-plane. The circle $S(p)$ through p is projected to $S'(p)$ in the (X^1, X^2)-plane

satisfy (5.3) for all values of r, ϑ, and φ. We obtain the metric of the sphere by expressing the differentials of X^μ in terms of the differentials of r, ϑ, and φ and substituting into (5.2). This results in

$$d_3 s^2 = \frac{dr^2}{1 - R^{-2} r^2} + r^2 \left(d\vartheta^2 + \sin^2 \vartheta \, d\varphi^2 \right) . \tag{5.6}$$

5.1.4 The Pseudo-sphere P^3

The next example is the 3-space P^3, which is sometimes called pseudo-sphere, hyperbolic plane, or Lobachevsky space. We again use the method of embedding. Consider Minkowski space in an inertial frame $\{X^\mu\}$ and the metric

$$dS^2 = \left(dX^0 \right)^2 - \left(dX^1 \right)^2 - \left(dX^2 \right)^2 - \left(dX^3 \right)^2 .$$

P^3 is defined by the equation

$$\left(x^0 \right)^2 - \left(x^1 \right)^2 - \left(x^2 \right)^2 - \left(x^3 \right)^2 = R^2 \tag{5.7}$$

and $X^0 > 0$. As the Minkowski interval to the origin is invariant under all Lorentz transformations, this hypersurface is also invariant. Its isometry group is the Lorentz group $SO(1,3)$, a six-dimensional group generated by three infinitesimal boosts, and three infinitesimal rotations [4]. The construction of the coordinates on P^3 is analogous to the case of the sphere (Fig. 5.2).

We denote by r the Minkowski interval to the X^0-axis and by ϑ and φ the spherical coordinates of the 2-sphere in the $X^0 = 0$ plane. Then the embedding relations are

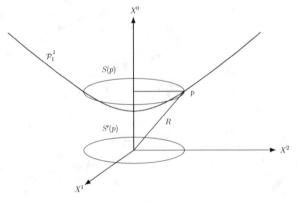

Fig. 5.2 The pseudo-sphere $\mathscr{P}^2$ represented by its intersection $\mathscr{P}^2_1$ with the (X^0, X^2)-plane. The circle $S(p)$ through p is projected to $S'(p)$ at the (X^1, X^2)-plane

$$X^k = rn^k,$$
$$X^0 = \sqrt{R^2 + r^2}.$$

Clearly these relations satisfy (5.7). All points in the pseudo-sphere can be obtained by taking r in the interval $[0, \infty]$. The metric is

$$d_3 s^2 = \frac{dr^2}{1 + R^{-2} r^2} + r^2 \left(d\vartheta^2 + \sin^2 \vartheta \, d\varphi^2 \right) . \tag{5.8}$$

All three metrics, (5.1), (5.6), and (5.8), can be expressed by one formula:

$$d_3 s^2 = \frac{dr^2}{1 - K r^2} + r^2 \left(d\vartheta^2 + \sin^2 \vartheta \, d\varphi^2 \right) . \tag{5.9}$$

where $K = R^{-2}$ results in (5.6), $K = -R^{-2}$ in (5.8), and $K = 0$ in (5.1). We shall need the curvature tensor of these spaces. A long but straightforward calculation yields (exercise)

$$R_{klmn} = K(g_{km} g_{ln} - g_{kn} g_{lm}) , \tag{5.10}$$

where g_{km} is the metric (5.9). K is the so-called *Gauss curvature* of the space, or briefly space curvature.

It can be shown that the previous three examples are (locally) the only homogeneous and isotropic 3-spaces with positive-definite metric. The corresponding mathematical field is called the "theory of symmetric spaces" [2, 5].

5.2 Robertson–Walker Space-Times

5.2.1 Metric

In the previous section, we studied the geometry of the 3-spaces. As a next step, we shall describe a metric such that the space-time can be foliated by space-like hypersurfaces with geometry given by (5.9). One possibility is

$$ds^2 = dt^2 - a^2(t) \left[\frac{dr^2}{1 - K r^2} + r^2 \left(d\vartheta^2 + \sin^2 \vartheta \, d\varphi^2 \right) \right] . \tag{5.11}$$

Space-times with the metric (5.11) are called *Robertson–Walker space-times*. The arbitrary function $a(t)$ is called *scale factor*, and K is a constant.

To see that the hypersurfaces $t = \text{const}$ of the metric (5.11) have the geometry (5.9), we change coordinates via $r = kr'$ with $k > 0$. The result is again a metric of the form (5.11),

$$ds^2 = dt^2 - a'^2(t) \left[\frac{dr'^2}{1 - K' r'^2} + r'^2 \left(d\vartheta^2 + \sin^2 \vartheta \, d\varphi^2 \right) \right] ,$$

where

$$a'(t) = ka(t), \quad K' = k^2 K .$$ (5.12)

Let $t = t_1$ be an arbitrary, but fixed time. Choosing $k = a^{-1}(t_1)$, we have $a'(t_1) = 1$, and the metric of the hypersurface $t = t_1$ agrees with (5.9), provided

$$K' = \frac{K}{a^2(t_1)} .$$ (5.13)

This yields the claim as t_1 is arbitrary. The above argument also shows that K is not the curvature of space, which is given by (5.13) at time t_1. The corresponding radius is

$$R' = \frac{a(t_1)}{\sqrt{|K|}} .$$

Hence, in the Robertson–Walker space-time, the geometry of space is time dependent.

The function $a(t)$ and the constant K do not have any direct physical meaning, as their values can be changed by a coordinate transformation. This is somewhat annoying. In many textbooks, this freedom is excluded by restricting K to its three standard values -1, 0, and $+1$. However, this breaks the freedom (5.12) only for $K \neq 0$. The description of the cosmological models is then discontinuous at $K = 0$. This is particularly unfortunate if it should turn out that our universe is described by a model with $K = 0$. But today this seems likely to be the case.

Thus we want to restrict the freedom differently, in the same way for all cases. We proceed similarly as Traschen and Eardley [6] and normalize the scale factor so that

$$a(t_0) = 1$$ (5.14)

for the time t_0 today. Then K has the meaning of today's space curvature. In principle, this is a measurable quantity and cannot be changed by a coordinate transformation. The transformation (5.12) is no longer just a coordinate transformation; it translates "today" to where $a' = 1$. This preserves the geometry of space-time. The method of Traschen and Eardley is based on the fact that the Robertson–Walker space-times are time dependent and look differently to cosmic observers at different times.

The space curvature K can be an arbitrary real number. If $K > 0$, the model is called *closed*, if $K = 0$ *spatially flat* (or flat), if $K < 0$ *hyperbolic*, and $K \leq 0$ *open*. The scale factor $a(t)$ is an arbitrary function of t which satisfies condition (5.14). Thus space can either expand or contract and the evolution is determined by the equations of motion (Einstein's equations).

What mathematical properties characterize the metric (5.11)? Let $\bar{\xi}^k(r, \vartheta, \varphi)$ be a Killing vector field of the metric (5.9) and define

$$\xi^0(t, r, \vartheta, \varphi) = 0, \quad \xi^k(t, r, \vartheta, \varphi) = \bar{\xi}^k(r, \vartheta, \varphi) .$$

Then $\xi^\mu(t, r, \vartheta, \varphi)$ is a Killing vector field of (5.11) (exercise). Thus the Robertson–Walker space-times have at least as much symmetry as the 3-spaces. One can show that (5.11) describes all spaces characterized in this way [2]. However, the metric (5.11) may have more symmetry. One example is Minkowski space-time, which results from setting $K = 0$ and $a = $ const in (5.11).

Le us make one remark about dimensions. In cosmology, t and r have the dimension of length ($c = 1$!), whereas a, ϑ, and φ are dimensionless.

5.2.2 Cosmic Rest System

For general functions $a(t)$, the isometry group of the space-time (5.11) agrees with the isometry group of the 3-spaces $t = $ const. Thus it is smaller than that of Minkowski space-time. This can be expressed in the following way. For each point p in space-time, the images of p with respect to all transformations in the isometry group of (5.11) yield the hypersurface $t = $ const containing p. Hence, these hypersurfaces are the orbits of the isometry group and thus determined uniquely (and geometrically). In Minkowski space, the $x^0 = $ const hypersurfaces also have six-dimensional symmetry, but the family $x^0 = $ const is not uniquely determined, as a boost will change it to another one.

This implies that Robertson–Walker space-times with isometry group no larger than six-dimensional symmetry have a *cosmic rest system* $\{e_a(p)\}$ in every point p, the frame with time axis orthogonal to this hypersurface. In coordinates t, r, ϑ, φ, we have

$$e_0^\mu (t, r, \vartheta, \varphi) = (1, 0, 0, 0) .$$

The observers with 4-velocity e_0 are called the *cosmological observers*. It is only these observers who perceive the universe as homogeneous and isotropic.

By what we said about the CMB before, our universe may not have more symmetry than homogeneity and isotropy. Such a thermal radiation defines a rest frame in each point (it is only isotropic in this frame). Thus the rest frame has to agree with the cosmic rest system. This opens up the possibility to determine the cosmic rest system. Indeed, this has been undertaken. In 1986, it was found that the CMB deviates weakly from perfect isotropy and that this deviation has the form of a dipole ($\cos \alpha$), with respect to the rest frame of the Earth. This deviation corresponds to a motion of Earth with speed $v_{\text{Earth}} \approx 361 \, \text{km s}^{-1}$ relative to the preferred frame. Subtracting the velocity of the Earth within our galaxy, one obtains

$$v_{\text{Galaxy}} \approx 500 \, \text{km s}^{-1} .$$

This is a surprisingly large speed.

Besides this, the CMB is isotropic and thermal to a very high accuracy. In 1992, the satellite COBE (COsmic Background Explorer) measured the CMB [3]. The results of these measurements give that the fluctuations about isotropy, subtracting out Earth's motion, are merely of the relative magnitude of

$$\frac{\Delta T}{T} \approx 10^{-5}$$

and that the temperature is

$$T = 2.726 \pm 0.01 \,\text{K} \,,$$

with a spectrum closer to that of black-body radiation than one could generate in a laboratory. The fluctuations of isotropy can be measured much more accurately. Its structure contains information about the early universe.

5.2.3 Cosmological Redshift

If $a' \neq 0$, then cosmological observers have a non-vanishing relative velocity. This relative motion exhibits a (positive or negative) redshift of their light signals. Let us calculate this redshift.

To this end, consider a light signal traveling along a light-like geodesic, parameterized by the functions $t(\kappa)$, $r(\kappa)$, $\vartheta(\kappa)$, and $\varphi(\kappa)$. Assume that light is emitted at κ_1 and received and analyzed at κ_0 (Fig. 5.3). Let κ be a physical parameter (cf. page 44), so that the vector $(\dot{t}, \dot{r}, \dot{\vartheta}, \dot{\varphi})$ is the 4-momentum of light. Thus

$$\dot{t} = h\nu \,,$$

where ν is the frequency of light with respect to the preferred frame at each point on the ray, in particular at the beginning and the end. The redshift z is defined by

$$z = \frac{\lambda_0 - \lambda_1}{\lambda_1} \,,$$

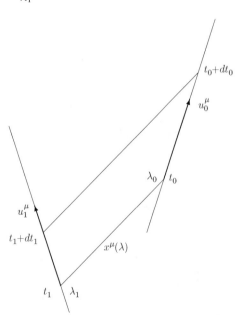

Fig. 5.3 Redshift and time dilation between two cosmological observers

or

$$z = \frac{v_1 - v_0}{v_0} = \frac{\dot{t}_1 - \dot{t}_0}{\dot{t}_0} . \tag{5.15}$$

We can calculate the relation of $\dot{t}_1$ to $\dot{t}_0$ from the geodesic equation. This is the Euler–Lagrange equation for the Lagrangian (2.15)

$$L = (1/2)\dot{t}^2 - (1/2)a^2\tilde{g}_{kl}\dot{x}^k\dot{x}^l ,$$

where $\tilde{g}_{kl}$ is the metric in square brackets on the right-hand side of (5.11). Note that here x^k corresponds to the coordinates r, ϑ, and φ. The corresponding Euler–Lagrange equations are

$$\ddot{t} + aa'\tilde{g}_{kl}\dot{x}^k\dot{x}^l = 0 , \tag{5.16}$$

$$\ddot{x}^m + 2a'a^{-1}\dot{t}\dot{x}^m + \tilde{\Gamma}^m_{kl}\dot{x}^k\dot{x}^l = 0, \tag{5.17}$$

where $\tilde{\Gamma}^m_{kl}$ are the Christoffel symbols of the metric $\tilde{g}_{kl}$. In addition, a light-like geodesic satisfies

$$\dot{t}^2 - a^2\tilde{g}_{kl}\dot{x}^k\dot{x}^l = 0 . \tag{5.18}$$

Substituting this into (5.16), we get

$$\ddot{t} + a'a^{-1}\dot{t}^2 = 0 ,$$

which simply implies that

$$a\dot{t} = \text{const} .$$

Thus (5.15) becomes

$$z = \frac{1 - a_1}{a_1} . \tag{5.19}$$

This is the general and exact formula for the cosmological redshift. Thus the redshift is given by the quotient of the scale factors $a_0 = 1$ at the time of receiving, and a_1 at the time of emission, and independent of the history in between. From (5.19) we infer that the redshift is positive in an expanding universe (and negative in a contracting one), as expected. Equation (5.19) can be used to measure the scale factor at time t_1. We have

$$a_1 = \frac{1}{1+z} .$$

The index 0, used in this calculation, has a special meaning in cosmology. It labels the quantities which refer to the present. We will adopt this convention subsequently.

Another important property of the Robertson–Walker metric is the connection of the redshift and the distance. For small distances, we can deduce a simple rule. Equation (5.18) implies the following relation between the spatial distance d, defined by

$$ds^2 = a^2\tilde{g}_{kl}dx^k dx^l ,$$

and the time dt, needed by the light signal to traverse this distance

$$ds = dt .$$

For infinitesimal distances, we thus obtain from (5.19) that

$$dz = \frac{d\lambda}{\lambda} = \frac{da}{a} = \frac{a'}{a}dt = \frac{a'}{a}ds .$$

Defining the *Hubble constant H* by

$$H = \left(\frac{1}{a}\frac{da}{dt}\right)_{t_0} = \left(\frac{da}{dt}\right)_{t_0} ,$$

we obtain *Hubble's law*:

$$\frac{d\lambda}{\lambda} = H\,ds . \tag{5.20}$$

Its content is that the redshift is proportional to the distance. As it is written, (5.20) only holds for infinitesimal distances, but the implied relation is approximately true for small finite distances. Here "small" can be understood as "small compared to the cosmological radius $a/\sqrt{|K|}$". This holds in particular for our observations. To measure the Hubble constant, we plot the redshift z of galaxies as a function of their distance, yielding the so-called *Hubble diagram*, cf. Fig. 5.4. The measurements [7, 8] give

$$H = 65 \pm 10 \text{ km s}^{-1}\text{Mpc}^{-1}$$

(Pc is "Parsec" $= 3.26$ light years; Mpc $= 10^6$ pc is the typical distance between galaxies). The difficulties are the motion of the galaxies (cf. the speed of our local group of galaxies is $500\,\text{km s}^{-1}$) and the measurement of the distances.

5.2.4 Cosmological Horizons

In the first part of these Notes, we mostly considered the local structure of space-time. Small enough neighborhoods are sufficiently similar to Minkowski space. We

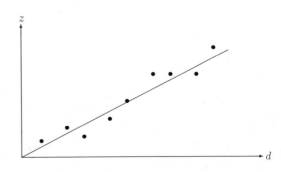

Fig. 5.4 The idea of Hubble diagram. The blobs represent values of individual galaxies. The tangent of the best-fit straight line is the Hubble constant. For a plot of a real measurement, see [7, 8]

now start to consider space-time as a whole. This so-called global structure will be surprisingly different from Minkowski space. One of the key aspects is what is called the *causal structure*, that is which events can be influenced by which other events. For example, which parts of the universe are visible for a cosmological observer? Into which parts of the universe can a cosmological observer send signals?

Consider an observer B. Of which events can he know when he reaches the point p on his trajectory? These events are obviously constrained by the past light cone at p. On the other hand, an observer starting his life at p can only influence the events in space-time contained in the future light cone of p (Fig. 5.5).

Let us calculate the location of the light cones in Robertson–Walker space-time. To this end, it is best to rewrite the Robertson–Walker metric in the following way:

$$ds^2 = a^2(t) \left(d\eta^2 - \frac{dr^2}{1 - Kr^2} - r^2 d\vartheta^2 - r^2 \sin^2 \vartheta \, d\varphi^2 \right),$$

where

$$d\eta = \frac{dt}{a(t)}.$$

We say that two metrics $g_{\mu\nu}$ and $g'_{\mu\nu}$ are *conformally related* (or that $g'_{\mu\nu}$ is a *conformal deformation* of $g_{\mu\nu}$) if $g'_{\mu\nu}(x) = F(x)g_{\mu\nu}(x)$ for a positive function $F(x)$. Thus the Robertson–Walker metric is conformally related to the metric in parentheses above. One can show that the light-like geodesics agree for two conformally related metrics (not the autoparallels as the affine parameter is different). Thus the path of light with respect to the coordinates η, r, ϑ, and φ is independent of $a(t)$. The metric in parenthesis is that of a static space-time. For $K > 0$, we have topology $\mathbb{R} \times S^3$, for the topology is $K \leq 0$ $\mathbb{R}^4$, and $K = 0$ gives Minkowski space.

Consider the observer with trajectory $r = 0$. This observer represents all possible cosmological observers due to complete homogeneity. The radial light rays through the point $\eta = \eta_0$ on its trajectory satisfy the equations

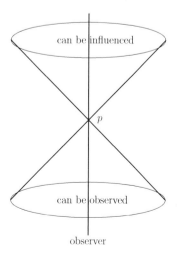

Fig. 5.5 The region that can be observed before point p is reached, and the region that can be influenced starting from point p

$$\vartheta = \vartheta_0, \quad \varphi = \varphi_0,$$

$$\dot{\eta}^2 - \frac{\dot{r}^2}{1 - Kr^2} = 0.$$

The latter equation can be solved readily:

$$r = \pm S_K(\eta - \eta_0), \tag{5.21}$$

where $S_K(\eta) = K^{-1/2} \sin(K^{1/2}\eta)$ for $K > 0$, $S_K(\eta) = (-K)^{-1/2}\sinh[(-K)^{1/2}\eta]$ for $K < 0$, and $S_K(\eta) = \eta$ for $K = 0$. Thus the light cone of the point with $\eta = \eta_0$ and $r = 0$ is determined. The upper sign holds for the future and the lower sign for the past light cone.

These light cones are obviously not very important boundaries. The observer can wait a little longer until his trajectory passes the point p, and his "horizon" will expand. Similarly, the observer could send a signal before p, and thus influence a larger region. It is an interesting question, whether such extensions can be made without restriction. Consider the simple case $K = 0$ (the other cases are similar). In this case, we have the usual light cones of Minkowski space with coordinates η, r, ϑ, and φ. The t-dependence of η is given by

$$\eta(t) = \eta_0 \pm \int_{t_0}^{t} \frac{d\tau}{a(\tau)}.$$

The crucial point is that this integral can converge if t approaches a "boundary" of the universe. This can happen in three different cases. First, $a(t)$ could remain regular and positive in the whole interval $t \in (-\infty, \infty)$, but

$$|\eta_s| := \lim_{t = \pm\infty} \left| \int_{t_0}^{t} \frac{d\tau}{a(\tau)} \right| < \infty,$$

or second, $a(t_s) = 0$ for a finite t_s and

$$|\eta_s| := \lim_{t = t_s} \left| \int_{t_0}^{t} \frac{d\tau}{a(\tau)} \right| < \infty,$$

or third $a(t_s) = \infty$ for a finite t_s, in which case certainly $|\eta_s| < \infty$. Then the Robertson–Walker space-time is represented by a subset of Minkowski space, determined either by $\eta > \eta_s$ or by $\eta < \eta_s$ or both. In the first case, we have a well-defined future light cone of the point $\eta = \eta_s$, $r = 0$ in space-time. Its surface is called *particle horizon* of the particle at the trajectory $r = 0$. In the second case, there is a well-defined past light cone of the point $\eta = \eta_s$, $r = 0$ in the space-time. Its surface is called *event horizon* of the observer on the trajectory $r = 0$ (cf. Fig. 5.6).

The particle horizons can be a serious problem in cosmology: vast parts of the universe cannot be influenced by certain particles. Thus large portions of the universe cannot interact with each other. But how can the same density, the same temperature, etc. be established, given these circumstances? On the other hand, the event horizons imply that an observer cannot observe certain parts of the universe. Then large parts of nature might be beyond our perception.

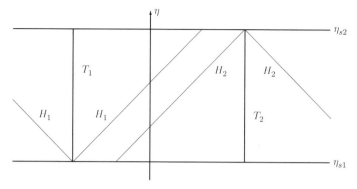

Fig. 5.6 Particle horizon H_1 of trajectory T_1 and event horizon H_2 of trajectory T_2

Whether such horizons exist is determined only by the behavior of the function $a(t)$, which results as a solution of the Einstein equations.

5.2.5 Einstein Tensor of the Robertson–Walker Space-Time

We can read the components of the affine connection for the metric (5.11), which we write in the form

$$ds^2 = dt^2 - a^2 \tilde{g}_{kl} dx^k dx^l \, ,$$

from (5.16) and (5.17):

$$\Gamma^0_{kl} = aa' \tilde{g}_{kl}, \quad \Gamma^k_{0l} = \frac{a'}{a} \delta^k_l, \quad \Gamma^m_{kl} = \tilde{\Gamma}^m_{kl} \, , \tag{5.22}$$

where the $\tilde{\Gamma}^m_{kl}$ are the Christoffel symbols of the metric $\tilde{g}_{kl}$. This implies (exercise):

$$G_{00} = 3a^{-2} \left(a'^2 + K \right) , \tag{5.23}$$

$$G_{kl} = - \left(2aa'' + a'^2 + K \right) \tilde{g}_{kl} \, . \tag{5.24}$$

We shall need these expressions to write down the Einstein equations. For $K = 0$, we see that although the 3-spaces are flat, the space-time itself can be curved.

5.3 Cosmic Dynamics

5.3.1 Friedmann–Lemaître Equations

In this section we shall write the Einstein equations for the Robertson–Walker metric and discuss the resulting dynamics. Before we do so, we want to make some general remarks about the task of solving the Einstein equations. We best compare it to solving the Maxwell equations. The Maxwell equations are linear and thus one can find a standard solution once and for all, preferably in the form of a Green's function. Then, given a source in Minkowski space, the solution is given by an integral of the product of the source and the Green's function. In contrast, first, solutions of the Einstein equations cannot be decomposed into a linear combination of other solutions. Second, we cannot determine a source without knowing the metric, as any distribution of mass needs the information on how far the individual mass elements are away from each other. The method for solving the Einstein equations, that is often used [9], is to assume a symmetry of the solution, and then solve the equations for both fields—metric and stress-energy tensor—simultaneously. Of course, that works only for large symmetry groups, but this will be always the case in these Notes. Another tack has also been successful recently: numerical methods [10].

To apply our solution method, we have to characterize the stress-energy tensor of matter in more detail. We assume that it has the form (3.58) of an ideal fluid. Furthermore, matter should have the same symmetry as the metric, that is the rest frame of the fluid should agree with the cosmic reference frame, or $u^\mu = (1,0,0,0)$ (with respect to the coordinates t, r, ϑ, φ), and the scalar fields ρ and p should be constant along the $t = $ const. hypersurfaces, $\rho = \rho(t)$, $p = p(t)$. Substituting these values into (5.23) and (5.24) and invoking the Einstein equations (4.8) yields two independent equations, best written in the following form:

$$a'' = -\frac{4\pi G}{3}(\rho + 3p)a + \frac{\Lambda}{3}a \,, \tag{5.25}$$

$$a'^2 + K = \frac{8\pi G}{3}\rho a^2 + \frac{\Lambda}{3}a^2 \,. \tag{5.26}$$

Another important equation is the energy equation (3.60). In our special case, it can be written as

$$\left(a^3\rho\right)' + p\left(a^3\right)' = 0 \,. \tag{5.27}$$

If $a' \neq 0$, then only two of these equations are independent. Equations (5.25), (5.26), and (5.27) are called *Friedmann–Lemaître equations*; in the case $\Lambda = 0$, they are called *Friedmann equations*. They determine the equations of motion, provided we specify an equation of state $p = p(\rho)$.

5.3.2 Cosmic Acceleration

Let us study (5.25), and set $\Lambda = 0$ for now. On the left-hand side, we have the acceleration a'' of cosmic expansion. On the right-hand side, there is the source term $\rho + 3p$ multiplied by a negative number. If the source is positive, then expansion decelerates, or contraction accelerates. An interpretation of this fact is that gravitation is attractive for a positive source and repulsive for a negative one. The new feature is that the pressure enters the source term. This is necessary in relativity, since pressure and mass density can be mixed by a boost. Observe the following interesting fact.

Assume that the system is contracting, $a' < 0$, with $\rho > 0$ and $p > 0$. The positivity of the term $\rho + 3p$ then implies that a' becomes even more negative and the contraction accelerates. Furthermore, the contraction has the tendency to increase the values of ρ and p. Here, mass density partly increases directly by contraction and partly indirectly by the work done by the contraction against p. These changes overweight the simultaneous decrease of a in (5.25). Apparently, the energy is transformed into mass by gravity, and the energy of gravity becomes more and more negative. This results in an instability, which can lead to collapse.

For normal states of matter, we have the inequalities

$$\rho \ge 0, \quad p \ge 0,$$

as the existence of negative mass would allow the construction of a "perpetuum mobile", and negative pressures are thermodynamically unstable. (The construction: form a mass-dipole from a positive and a negative mass. Simple calculations based on Newtonian theory imply that the dipole moves with an acceleration that agrees with the direction of the dipole.)

The assumption that matter is always in such a normal state has an interesting implication. In this case, the graph of the function $a(t)$ is convex (Fig. 5.7). Thus if $a' > 0$ today, the graph must intersect the t-axis in the past. There we have $a = 0$, that is a singularity (infinite curvature and density). This *Big Bang* could not have happened before the time $\Delta t = a'^{-1}(t_0) = H_0^{-1}$. Δt is known as the *Hubble time* and gives an upper bound for the age of the universe (under the given assumptions). The Hubble time is approximately 15×10^9 for the Hubble constant $65\,\mathrm{km\,s^{-1}\,Mpc^{-1}}$. It is an interesting fact that the known ages of rocks or astronomic systems (e.g., globular clusters of galaxies) are smaller or comparable to the Hubble time [2].

Consider the Λ-term in (5.25). If $\Lambda > 0$, then the deceleration of the universe is decreased, and if it dominates, we have acceleration. Then the above argument

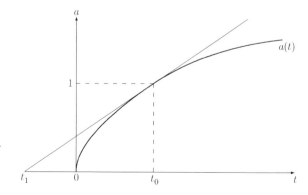

Fig. 5.7 A convex curve $a(t)$. The Hubble time $\Delta t = t_0 - t_1$ is upper bound to the age of the Universe t_0

is invalid. Indeed, this idea was suggested by Lemaître many years ago, as first observations gave a value for the Hubble constant so small that not even Earth's age of 4.5 billion years would fit into the corresponding Hubble time. Today, the value of Λ is determined by an independent measurement. We will return to this later.

The cosmological constant Λ is frequently expressed by the corresponding energy density ρ_Λ. We can move the Λ-term to the right-hand side of the Einstein equations and consider it as the effective stress-energy tensor $T_\Lambda^{\mu\nu}$. Then in the cosmological rest frame (and any other orthonormal frame):

$$T_\Lambda^{\mu\nu} = \frac{\Lambda}{8\pi G} \begin{pmatrix} 1 & 0 & 0 & 0 \\ 0 & -1 & 0 & 0 \\ 0 & 0 & -1 & 0 \\ 0 & 0 & 0 & -1 \end{pmatrix}.$$

Thus $T_\Lambda^{\mu\nu}$ has the form of an ideal fluid (3.57) with

$$\rho_\Lambda = \frac{\Lambda}{8\pi G}, \quad p_\Lambda = -\frac{\Lambda}{8\pi G}, \tag{5.28}$$

and it is called *dark energy* today. Whether this notion is just a mathematical analogy, or describes something real, is not clear at the moment (2007).

5.3.3 Linear Equations of State

The simplest equation of state is $p = 0$. Matter with this equation of state is called *dust* or *incoherent matter*. Sometimes we say that the "cold" matter has this equation of state. This means that single pieces or particles of the matter are massive, have small (non-relativistic) velocities, and massless particles can be neglected ("cold non-relativistic matter"). For example, today's galaxies can be viewed as this kind of dust.

We can also imagine that matter may be "hot". The massive particles have relativistic velocities and their dynamics is practically the same as for massless particles, or there are only massless particles. The equation of state of hot matter is

$$p = \frac{1}{3}\rho .$$

Both equations of state are linear and can be written in the general form

$$p = (\gamma - 1)\rho , \tag{5.29}$$

where $\gamma = 1$ for dust and $\gamma = 4/3$ for radiation. We can choose γ in the range between 1 and 2, so that $0 \leq p \leq \rho$. It can be shown that the sound speed of matter with

$p > \rho$ is greater than the speed of light, hence the upper bound. Equation (5.27) then becomes

$$\rho' a^3 + 3\gamma \rho a^2 a' = 0$$

or

$$\rho' a^{3\gamma} + 3\gamma \rho a^{3\gamma - 1} a' = 0$$

with the solution $\rho a^{3\gamma} = \text{const}$. Obviously, the meaning of the constant on the right is today's matter density ρ_0, and the result can be written in the following form:

$$\rho = \frac{3M}{8\pi G} a^{-3\gamma} . \tag{5.30}$$

Inserting into (5.26) yields that

$$a'^2 - Ma^{-3\gamma + 2} - La^2 = -K \tag{5.31}$$

where $L = \Lambda/3$.

The cosmological models that are usually studied, consist of three components with negligible mutual interaction. Then the stress-energy tensor of each component is conserved by itself and (5.30) holds separately for each component with the corresponding values for M and γ. The first of these components is dark energy. Its energy density ρ_Λ is independent of the scale factor a (at least during the cosmological epoch). Non-relativistic matter is the second component. Its energy density satisfies (5.30) with $\gamma = 1$ and some constant M, as if it was the only matter in the universe. Finally there is the radiation with density ρ, described by $\gamma = 4/3$.

Observe that the dependence of the three densities on a is different. This can lead to different roles for the three sorts of matter for very small or very large values of a. For small enough values of a, radiation dominates, and for very large values of a, the dark energy dominates. This holds independently of the values of the constants Λ and M.

5.3.3.1 Horizons

Let us consider the question of horizons. We will keep in mind that for $a \to 0$ radiation dominates, whereas in the case $a \to \infty$, the dark energy dominates. For the radiation model ($\gamma = 4/3$), (5.31) takes the form

$$a'^2 = Ma^{-2} - K + La^2 .$$

If $a \to 0$, then on the right-hand side, the first term dominates:

$$a'^2 \approx \frac{M}{a^2} .$$

This implies

$$a = (4M)^{1/4} \sqrt{t} ,$$

and

$$\eta(t) = \left(\frac{4}{M}\right)^{1/4} \sqrt{t} \to 0.$$

Hence there are horizons, particle horizons at the Big Bang, and event horizons at the Big Crunch.

In the other case, if $a \to \infty$, then the last term dominates. We obtain $a \sim \exp(\sqrt{L}t)$ and the integral will converge for $\Lambda > 0$. A detailed study of exact solutions will confirm these estimates.

5.4 Parameterization of Physically Distinct Models

We study the properties of solutions to (5.31). These properties represent predictions from general relativity about cosmology. Equation (5.31) uniquely determines the Robertson–Walker geometry for given constants γ, M, K, and L, as the constant of integration depends on the origin of the time scale and is meaningless. To see this, we use separation of variables to write the equation in the form

$$dt = \frac{da}{\sqrt{Ma^{-3\gamma+2} + La^2 - K}}. \tag{5.32}$$

Thus the constant of integration represents a translation of the time coordinate t. This is simply a coordinate transformation which does not affect physical properties. Hence the space of physically distinct solutions is four-dimensional. The ranges of the four parameters are $\gamma \in [1,2]$, $M \in (0,\infty)$, $K \in (-\infty,\infty)$, and $L \in (-\infty,+\infty)$.

The integral on the right-hand side of (5.32) can be expressed in terms of elementary functions, if $K = 0$ for all γ, if $L = 0$ for $\gamma = 1$ or $\gamma = 4/3$, and if $\gamma = 4/3$. However, all important qualitative properties of the integral can be determined in the general case in a different way.

5.4.1 Qualitative Discussion of the Dynamics

Let us define that the zeros, the divergences, and the extrema of a function are its qualitative properties. We shall study these properties subsequently of $a(t)$. To this end, rewrite radial equation (5.30) in the following form:

$$\left(\frac{da}{dt}\right)^2 + V(a) = -K, \tag{5.33}$$

where

$$V(a) := -Ma^{-3\gamma+2} - La^2$$

is the so-called *effective potential*. Equation (5.33) has the well-known form of the energy integral of the dynamical equation for non-relativistic motion of a point mass with mass 2 and total energy $-K$ along the a-axis subject to the action of the potential $V(a)$. The qualitative discussion is based on this Newtonian analogy (cf. the discussion of the perihelion shift, Sect. 2.9.2). Consider the following cases.

1. $\Lambda < 0$

In this case, Fig. 5.8 shows the diagram of $V(a)$. V is monotone increasing from $V = -\infty$ at $a = 0$ to $V = \infty$ at $a = \infty$, since $3\gamma - 2 > 0$. The possible orbits are determined by the part of the horizontal curves $E = -K$ which lie above the potential curve. All of these curves start at $a = 0$ (Big Bang), reach a maximal a, the so-called turning point, at a_U, where

$$V(a_U) = -K,$$

then turn around and return to $a = 0$. The time line in Fig. 5.9 is symmetric around the turning point. That is, the function $a(t)$ satisfies the equation

$$a(t - t_U) = a(t_U - t) . \tag{5.34}$$

In this case, the result is that all models with $K \in (-\infty, \infty)$ evolve from Big Bang to Big Crunch in finite time. Such solutions are called *recollapsing*.

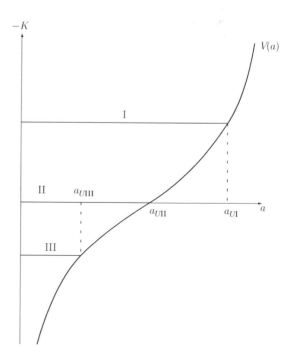

Fig. 5.8 Effective potential $V(a)$ in the case of $\Lambda < 0$. Trajectory I corresponds to $K < 0$, II to $K = 0$, and III to $K > 0$

Fig. 5.9 A typical
t-dependence of the scale
factor a in the case $\Lambda < 0$

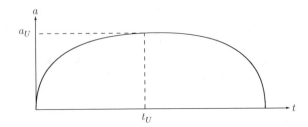

2. $\Lambda = 0$
Then

$$V(a) = -\frac{M}{a^{3\gamma-2}} \, .$$

The diagram in Fig. 5.10 shows that there are two sub-cases:

  $\underline{K > 0}$: the model is closed, recollapses as above, and satisfies (5.34).
  $\underline{K \leq 0}$: the model is open and expands for all times to $a \to \infty$ (cf. Fig. 5.11).
  The results in this case are the following:

1. Closed models recollapse,
2. Open models start at the Big Bang and expand for all times to $a \to \infty$. These
   models are called *eternally expanding*.

3. $\Lambda > 0$
The V-diagram is completely different (Fig. 5.12). The V-curve has a maximum
V_{Max} at a_{Max}. A simple calculation yields

$$a_{\text{Max}} = \left[\frac{M(3\gamma - 2)}{2L}\right]^{\frac{1}{3\gamma}} , \quad V_{\text{Max}} = -\frac{3\gamma L}{3\gamma - 2}\left[\frac{M(3\gamma - 2)}{2L}\right]^{\frac{2}{3\gamma}} . \tag{5.35}$$

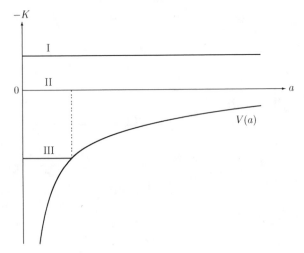

Fig. 5.10 Effective potential $V(a)$ in the case of $\Lambda = 0$. Trajectory I corresponds to $K < 0$, II to
$K = 0$, and III to $K > 0$

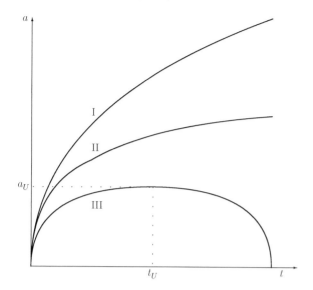

Fig. 5.11 The t-dependence of the scale factor a in the case $\Lambda = 0$. The meaning of the curves I, II, and III is explained by Fig. 5.10

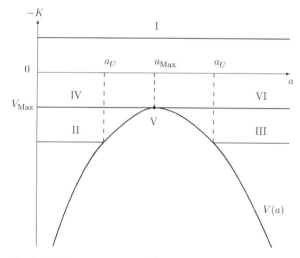

Fig. 5.12 Effective potential $V(a)$ in the case of $\Lambda > 0$. Trajectory I corresponds to $-K > V_{\text{Max}}$, II and III to $-K < V_{\text{Max}}$, IV, V, and VI to $-K = V_{\text{Max}}$

We have to distinguish three cases.

$-K > V_{\text{Max}}$. Then the trajectory with $E = -K$ does not meet the potential curve. These models start at the Big Bang and expand for all times to $a \to \infty$. They are of the eternally expanding kind.

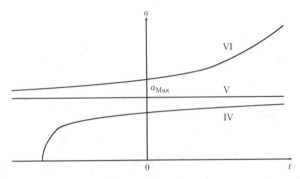

Fig. 5.13 The t-dependence of the scale factor a in the sub-case $-K = V_{\text{Max}}$ of $\Lambda > 0$. The meaning of the curves IV, V, and VI is explained by Fig. 5.12

$-K = V_{\text{Max}}$. We have to separate into three sub-cases:

1. The model starts at the Big Bang and expands for all times not to infinity, but to the value $a = a_{\text{Max}}$ (cf. Fig. 5.13).
2. The model is static with $a = a_{\text{Max}}$. It is unstable, as it lies at a maximum of the potential curve.
3. The model starts at $a = a_{\text{Max}}$ for $t \to -\infty$ and expands for all times to $a \to \infty$.

Thus the former two cases are models *without Big Bang*, but they do not seem to be important, as they are limit cases and not generic.

$-K < V_{\text{Max}}$. We have two sub-cases:

1. $a < a_{\text{Max}}$ and all models recollapse and satisfy (5.34).
2. $a > a_{\text{Max}}$ and all model contract from $a \to \infty$ as $\tau \to -\infty$, reach a turning point, and expand to $a \to \infty$. These are also models without Big Bang, but this time they are generic (Fig. 5.14).

In summary, the parameter space of K, L, and M is divided into three open domains. The first contains the recollapsing models, the second one the eternally expanding models, and the third one the models without Big Bang. On the boundaries of these domains, less interesting models lie.

5.4.2 Density Parameters

Let us now assume that γ is given. The parameters K, L, and M refer to a whole solution in the sense that each value of the 3-tuple determines a solution, and each solution a value of the 3-tuple. As the solutions are not static, we cannot observe the parameters directly. They are only accessible via measurable momentary quantities. Here, the "moment" can very well last 1000 years. This is why we want to express the global parameters in terms of others, which are simpler to measure.

One example for such a parameter is the Hubble constant H. We need two additional parameters. The Hubble constant, or its absolute value,

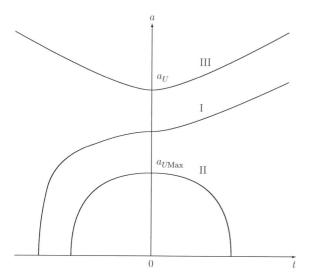

Fig. 5.14 The t-dependence of the scale factor a in the sub-case $-K \neq V_{\text{Max}}$ of $\Lambda > 0$. The meaning of the curves I, II, and III is explained by Fig. 5.12

$$|H| = |a'| \, ,$$

has the dimension of length^{-1} (since $c = 1$). Other momentary parameters can thus be made dimensionless. It turns out that it is favorable to choose dimensionless parameters. We multiply each quantity with non-trivial dimensions by powers of G and the Hubble constant to produce a dimensionless quantity.[1] These relative quantities and their properties are calculated from the theoretical model. The momentary values are then multiplied by the suitable power of the Hubble constant, before comparing them to the observations. This is a common procedure, which has been used for a long time in cosmology. The "complete" Hubble constant H contains another information, which is the sign of a':

$$\text{sgn}(a') = \text{sgn}(H) \, .$$

If a blueshift of the galaxies is observed, then this means that we are in the contracting phase of cosmological history with $a' < 0$.

A popular method is based on the so-called relative (momentary) densities Ω_M and Ω_Λ, which we define and describe in the following. To this end, we rewrite the Friedmann–Lemaître equation (5.26) for $t = t_0$ as follows:

$$K = \frac{8\pi G}{3} \left(\rho_M + \rho_\Lambda - \frac{3}{8\pi G} a'^2 \right) \, . \tag{5.36}$$

---

[1] This only works if $H \neq 0$, but these cases are the interesting ones.

All terms in parenthesis have the dimensions of mass density. This allows us to define the so-called *critical density*:

$$\rho_c := \frac{3}{8\pi G} a'^2 .$$ (5.37)

Why is ρ_c called "critical"? Rewrite (5.36) using the critical density:

$$K = \frac{8\pi G}{3}[(\rho_M + \rho_\Lambda) - \rho_c] .$$

The term in parenthesis is the total density, including dark energy. If this is larger than the critical density, then $K > 0$ and space is positively curved. If it is equal to the critical density, then space is flat, and if it is smaller than the critical density, then space is negatively curved. Thus the critical density is the value of the total density which forces the space to be closed.

Now we can introduce dimensionless parameters, by relating all densities to the critical one:

$$\Omega_M := \frac{\rho_M}{\rho_c}, \quad \Omega_\Lambda := \frac{\rho_\Lambda}{\rho_c} .$$

These are called *density parameters* or simply Ω-parameters. We see that they are an example of the method outlined above. Division by the critical density is little more than multiplication by the Hubble constant to the power -2 and with G to the power 1, as (5.37) shows. The parameters H, Ω_M, and Ω_Λ have one property in common. They depend on cosmic time t and are not constant as K, L, and M.

Expressed in Ω-parameters, the Friedmann–Lemaître equations for $t = t_0$ have a simple form. Equation (5.26) can be written in the form

$$K = a'^2(\Omega_M + \Omega_\Lambda - 1) ,$$

whereas (5.25) becomes

$$q = \frac{3\gamma - 2}{2}\Omega_M - \Omega_\Lambda ,$$

with

$$q := -\frac{a''}{a'^2} .$$

The current value of q is called *deceleration parameter*. In cosmology, it is common practice to express cosmic acceleration in terms of this dimensionless parameter.

We can compute all other quantities from Ω_M, Ω_Λ, and H. From the definition of the parameters and the densities, we infer that

$$\Omega_M = \frac{M}{H^2}, \quad \Omega_\Lambda = \frac{\Lambda}{3H^2} .$$ (5.38)

Conversely, from γ, H, Ω_M, and Ω_Λ, we can compute the parameters K, L, and M, which distinguish the solutions, as well as the momentary values of a and a':

$$K = (\Omega_M + \Omega_\Lambda - 1)H^2, \quad L = \Omega_\Lambda H^2, \quad M = \Omega_M H^2 ,$$ (5.39)

or

$$\Lambda = 3\Omega_\Lambda H^2 , \qquad (5.40)$$

and, for the sake of completeness,

$$a = 1 , \quad a' = H . \qquad (5.41)$$

The dimensions of these quantities can now be deduced from the powers of H in the previous equations.

5.4.3 The Ω-Diagram

In this section, we construct a map, the regions of which represents different momentary data. This is a two-dimensional diagram with axes Ω_M and Ω_Λ. It is increasingly popular to present certain properties of these models in form of a Ω-diagram, as most of them are independent of H.

We start by adding the lines corresponding to the values $K = 0$ and $q = 0$, Fig. 5.15. It is clear that these curves are boundaries of important subsets. Above the first curve, there are the closed models as $K > 0$ there, and the open models are below this curve. Similarly, above the second curve, there are the models which accelerate ($q < 0$), and below the curve are the decelerating ones.

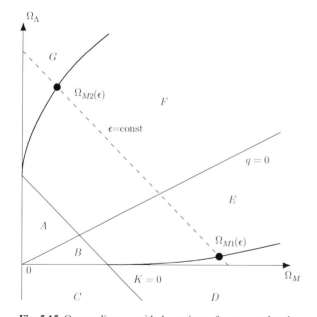

Fig. 5.15 Omega diagram with the regions of open, accelerating, eternally expanding models (A), open, decelerating, eternally expanding ones (B), open, decelerating, recollapsing ones (C), closed, decelerating, recollapsing ones (D), closed, decelerating, eternally expanding ones (E), closed, accelerating, eternally expanding ones (F) and models without Big Bang (G)

It turns out that each dynamical type can be found in a certain region of the Ω-diagram. In other words, the dynamical type is uniquely determined by the values of Ω_M and Ω_Λ and independent of H. The underlying reason is the behavior of the Einstein equations with respect to scaling (exercise).

At first, $\Omega_\Lambda < 0$ is equivalent to $\Lambda < 0$, and thus all points there correspond to the momentary states of the recollapsing models. Let us divide the region $\Omega_\Lambda > 0$ into $K < 0$ and $K \geq 0$. The first subset contains only states of eternally expanding models. The second one contains three generic types: eternally expanding models for $-K > V_{\text{Max}}$, recollapsing ones for $-K < V_{\text{Max}}$ and $1 < a_{\text{Max}}$, and models without Big Bang for $-K < V_{\text{Max}}$ and $1 > a_{\text{Max}}$.

Substituting for V_{Max} and a_{Max} from (5.35) and using (5.41) and (5.39) for a and K, we obtain

$$\frac{-K}{V_{\text{Max}}} = \left[A \frac{(\Omega_M + \Omega_\Lambda - 1)^{3\gamma}}{\Omega_M^2 \Omega_\Lambda^{3\gamma-2}} \right]^{\frac{1}{3\gamma}} , \tag{5.42}$$

with

$$A = \left(\frac{3\gamma - 2}{3\gamma} \right)^{3\gamma} \left(\frac{2}{3\gamma - 2} \right)^2 ,$$

and

$$\frac{1}{a_{\text{Max}}} = \left(\frac{2}{3\gamma - 2} \frac{\Omega_\Lambda}{\Omega_M} \right)^{\frac{1}{3\gamma}} . \tag{5.43}$$

We see that H cancels everywhere. Thus the above conditions determine well-defined regions in the Ω-diagram.

The eternally expanding models in the region $\Omega_\Lambda > 0$ and $K > 0$ satisfy the condition $-K > V_{\text{Max}}$ in the form

$$\frac{V_{\text{Max}}}{-K} > 1 .$$

This yields

$$\Omega_M^2 \Omega_\Lambda^{3\gamma-2} > A\varepsilon^{3\gamma} , \tag{5.44}$$

with

$$\varepsilon = \Omega_M + \Omega_\Lambda - 1 > 0 .$$

Consider the left-hand side of (5.44) along the line $\varepsilon = \text{const}$ between the points $\Omega_M = 0$ and $\Omega_\Lambda = 0$. There we have

$$\Omega_\Lambda = \varepsilon + 1 - \Omega_M , \quad \Omega_M \in (0, \varepsilon + 1)$$

and we turn to study the function

$$f(\Omega_M) = \Omega_M^2 (\varepsilon + 1 - \Omega_M)^{3\gamma-2}$$

in this interval. At the endpoints, we have

$$f(0) = f(\varepsilon + 1) = 0 ,$$

and in the interior of the interval, f is positive. There, it has a single maximum with

$$\Omega_{\text{Max}} = \frac{2}{3\gamma}(\varepsilon + 1) < \varepsilon + 1 ,$$

and the maximal value is

$$f_{\text{Max}} = A(\varepsilon + 1)^{3\gamma} .$$

Obviously we have

$$0 < A\varepsilon^{3\gamma} < f_{\text{Max}} ,$$

and thus the equation

$$f(\Omega_M) = A\varepsilon^{3\gamma}$$

always has two solutions, which we denote by $\Omega_{M1}(\varepsilon)$ and $\Omega_{M2}(\varepsilon)$. The eternally expanding models are in the interval $(\Omega_{M1}(\varepsilon), \Omega_{M2}(\varepsilon))$. For the recollapsing models and the models without Big Bang, we are left with the intervals $(0, \Omega_{M1}(\varepsilon))$ and $(\Omega_{M2}(\varepsilon), \varepsilon + 1)$. The former have to satisfy the inequality $1 < a_{\text{Max}}$, that is, in view of (5.43):

$$\Omega_\Lambda < \frac{3\gamma - 2}{2}\Omega_M ,$$

which obviously holds in the interval $(\Omega_{M1}(\varepsilon), \varepsilon + 1)$.

To get some idea of how the curves run near their endpoints, we study its behavior along the line $\varepsilon = 0$. The boundary curve starts there at the point $p_+ \equiv (\Omega_M = 0, \Omega_\Lambda = 1)$ and $p_- \equiv (\Omega_M = 1, \Omega_\Lambda = 0)$, as the equation for the boundary curve is

$$\Omega_M^2 \Omega_\Lambda^{3\gamma - 2} = A\varepsilon^{3\gamma} , \tag{5.45}$$

and these two points satisfy this equation for $\varepsilon = 0$. We try to capture the behavior of the boundary curve at p_+ by the power series

$$\Omega_M = x(\Omega_\Lambda - 1)^y + \dots$$

where $x > 0$ and $y > 0$. Substituting into (5.45) yields for the leading terms:

$$\Omega_\Lambda^{3\gamma - 2}(\Omega_\Lambda - 1)^{2y} = A\left[1 + (\Omega_\Lambda - 1)^{y-1}\right]^{3\gamma}(\Omega_\Lambda - 1)^{3\gamma} .$$

We thus have to set

$$2y = 3\gamma .$$

Then the limit of both sides is non-trivial and yields

$$x^2 = A .$$

Finally,
$$\Omega_M = \sqrt{A}(\Omega_\Lambda - 1)^{3\gamma/2} \,,$$
and since $3\gamma > 2$, the boundary is tangential to the axis $\Omega_M = 0$ at p_+.

Similarly, near the point p_- we set
$$\Omega_\Lambda = x(\Omega_M - 1)^y$$
and obtain
$$\Omega_\Lambda = A^{\frac{1}{3\gamma-2}}(\Omega_M - 1)^{\frac{3\gamma}{3\gamma-2}} \,.$$

Hence the boundary is tangential to the axis $\Omega_\Lambda = 0$ at p_-. The three generic models lie in the three distinguished regions of the Ω-diagram, whereas the non-generic ones lie on the boundaries.

5.4.3.1 The Age of the Universe in the Ω-Diagram

For all models with Big Bang, the age T_A can be defined as the proper time from the Big Bang $a = 0$ to the present $a = 1$.

To compute T_A, we return to (5.32) and define the function $T(K,L,M,a)$ (γ is considered fixed) by

$$T(K,L,M,a) := \int_0^a \frac{d\alpha}{\sqrt{M\alpha^{-3\gamma+2} + L\alpha^2 - K}} \,. \tag{5.46}$$

The models with Big Bang either expand for all times or recollapse. For the eternally expanding models, we simply have

$$T_A = T(M,K,L,1) \,.$$

For the recollapsing models, the situation is more complicated. The turning point a_U is the smallest zero of the function $f(a) = M\alpha^{-3\gamma+2} + L\alpha^2 - K$, so that $f(a) > 0$ for all $a \in (0, a_U)$. The graph of the solution is symmetric around a_U, and the value of a does not exceed a_U. Thus we have

$$1 \le a_U$$

and
$$T_U(M,K,L) \ge T(M,K,L,1) \,,$$

where
$$T_U(M,K,L) = T(K,L,M,a_U)$$

is the time from the Big Bang to the turning point.

We have to distinguish two cases. Either $H > 0$, then $a = 1$ is in the first half and again

$$T_A = T(M,K,L,1)$$

or $H < 0$ and $a = 1$ lies in the second half. By symmetry around the turning point, we have

$$T_A = 2T_U(M,K,L) - T(M,K,L,1) .$$

The information whether present day is in the interval $(0, T_U)$ or $(T_U, 2T_U)$ is given by $\mathrm{sgn}(H)$. Using the sign function, we can write the following general formula:

$$T_A = [1 - \mathrm{sgn}(H)]T_U + \mathrm{sgn}(H)T(M,K,L,1) .$$

The age has dimensions of length, and a handy dimensionless quantity arises by multiplication with the Hubble constant H. The physical interpretation is that we measure the real age in the corresponding Hubble time. Thus define

$$\chi := |H|T_A .$$

A simple calculation yields

$$\chi(M,K,L,H) = [1 - \mathrm{sgn}(H)]|H|T_U(M,K,L) + \mathrm{sgn}(H)|H|T(M,K,L,1) .$$

Substituting for M, K, and L in view of (5.39), we infer

$$|H|T(M,K,L,1) = \int_0^1 \frac{d\alpha}{\sqrt{\Omega_M \alpha^{-3\gamma+2} + \Omega_\Lambda \alpha^2 - \Omega_M - \Omega_\Lambda + 1}} \tag{5.47}$$

and

$$|H|T_U(M,K,L) = \int_0^{a_U} \frac{d\alpha}{\sqrt{\Omega_M \alpha^{-3\gamma+2} + \Omega_\Lambda \alpha^2 - \Omega_M - \Omega_\Lambda + 1}} .$$

The value of a_U, the root of the equation $f(a) = 0$, is independent of H, since

$$f(a) = H^2 \left(\Omega_M \alpha^{-3\gamma+2} + \Omega_\Lambda \alpha^2 - \Omega_M - \Omega_\Lambda + 1\right) .$$

Thus, we can draw the curves

$$\chi_\pm = \mathrm{const}$$

in the Ω-diagram, these are the so-called *isochrones*. In recollapsing models, there are two sorts of isochrones, one for each of the two signs of H.

Under the assumptions that γ is fixed, that we are at the point $(\Omega_M, \Omega_\Lambda)$, and that the Hubble constant has a certain value, we can determine the age of this point. A figure showing isochrones with the respective times for $H = 63$ can be found in [7, 8].

5.4.4 Luminosity Distance and the Measurement of Λ

Many measurements of cosmological parameters, like the Hubble constant or the deceleration parameter q, rely on distance measurements. What we mean by "distance" has to be carefully defined. In most cases, "distance" means *luminosity distance*. It can be defined as follows. The definition here is slightly simplified, first in comparison to astronomic conventions, but also compared to some cosmologists (who built curvature corrections into the definition [2]).

Assume that space (and space-time) is flat, and that a source of radiation has total power dE/dt. In distance R of the source, the radiation illuminates the area $4\pi R^2$. Let $\mathscr{L}_s$ denote the *apparent luminosity*, that is the corresponding current of energy per area element:

$$\mathscr{L}_s = \frac{dE/dt}{4\pi R^2} \ .$$

On the other hand, we can define the *absolute luminosity* $\mathscr{L}_a$ as the energy current through a sphere with radius 1 (astronomers use 10 Parsec here) around the source:

$$\mathscr{L}_a = \frac{dE/dt}{4\pi} \ .$$

Then

$$R = \sqrt{\frac{\mathscr{L}_a}{\mathscr{L}_s}} \ .$$

For our purposes, the luminosity distance $D_{\mathscr{L}}$ can be defined by the same formula in general. In particular, we can use this definition in curved and time-dependent cosmological space-times:

$$D_{\mathscr{L}} := \sqrt{\frac{\mathscr{L}_a}{\mathscr{L}_s}} \ .$$

For sources with known absolute luminosity, we can simply measure the apparent luminosity and compute the luminosity distance with the formula above. Roughly speaking, this is the common procedure. It does not make much sense to introduce curvature corrections for the Euclidean formula, since there is no unique way to define distances in Robertson–Walker space-time. In any case, it is not necessary, as any distance function, the values of which can be measured in the real cosmos on the one hand and computed from the model on the other, will serve our purposes.

The sources with known absolute luminosity which are bright enough to be observed at large distances are relatively sparse. Thus, in astronomy, they are called *standard candles*. The so-called Cepheid variables and supernovae of class IA are standard candles, which are presently used with some confidence. The Cepheid variables are periodically varying stars. There are two kinds of them, and only one of them is a reliable source [11, 12]. We will not discuss any further details here. The supernovae IA originate as a type of the so-called white dwarfs. These are stars of mass less than 1.2 solar masses, which used up their fuel. As their mass is relatively small, they can withstand gravitational collapse. If such a white dwarf is a component of a binary star system, and accretes mass from the second star, the critical value of 1.2 solar masses can eventually be exceeded, and the star begins to collapse. As the white dwarfs in the supernovae IA primarily consist of carbon and oxygen, the collapse ignites a nuclear reaction at once throughout the star. The result is a gigantic explosion—the supernova [10].

Let us calculate the "theoretical" luminosity distance. It will depend on the model, but also on the time of measurement in the model and the redshift of the source. Consider the Friedmann–Lemaître model with parameters γ, Ω_M, Ω_Λ, and H. Assume that the source is at $r = 0$ and $t = t_1$, follows the trajectory of a cosmological observer B_1, and radiates the power dE_1/dt_1. The energy per second measured at $r = r_0$ and time t_0 by a cosmological observer B_0 is subject to two redshifts. First, the redshift of the energy of each photon, which leads to the equation

$$dE_0 = dE_1 a_1 \ .$$

Second, the time of arrival of the photons at the cosmological observer B_0 is blueshift. If the times t_0 and t_1 are connected by light rays, so that $t_0(t_1)$ is the time of arrival of a light ray at B_0, where t_1 is the emission time at B_1 (Fig. 2.4), then we have

$$dt_0 = dt_1/a_1 \ .$$

The values of the scale factor at the moment of reception and emission are 1 and a_1, respectively (exercise). We infer

$$dE_0/dt_0 = a_1^2(dE_1/dt_1) \ .$$

This energy is distributed uniformly on the area $4\pi r_0^2$. Thus we obtain for the apparent luminosity that

$$\mathscr{L}_s = a_1^2 \frac{dE_1/dt_1}{4\pi r_0^2} \ .$$

This yields the simple formula:

$$D_{\mathscr{L}} = r_0/a_1 \ . \tag{5.48}$$

We see that $D_{\mathscr{L}}$ has the dimension of length. The corresponding dimensionless quantity $d_{\mathscr{L}}$ is defined by (assuming $H > 0$)

$$d_{\mathscr{L}} := HD_{\mathscr{L}} \ ,$$

or

$$d_{\mathscr{L}} = \frac{Hr_0}{a_1} \ .$$

We want to express the right-hand side in terms of the Ω-parameters and z. In general, this can only be done numerically. What we can do however is to show that the right-hand side does not depend on H. Indeed, a_1 can be expressed in terms of the redshift,

$$\alpha_1 = \frac{1}{1+z}$$

and the value of r_0 can be computed from (5.21):

$$Hr_0 = HS_K(\eta_0 - \eta_1),$$

where

$$\eta_0 - \eta_1 = \int_{t_1}^{t_0} \frac{dt}{a(t)} = \frac{1}{H} \int_{\tau_1}^{\tau_0} \frac{d\tau}{\beta(\tau)},$$

$\tau = Ht$ and $\beta(\tau) = a(\tau/H)$. The function $S_K(\eta)$ satisfies the identity:

$$xS_K(\eta/x) = S_{K/x^2}(\eta)$$

for arbitrary $x > 0$. Thus we infer

$$Hr_0 = S_{K/H^2} \left(\int_{\tau_1}^{\tau_0} \frac{d\tau}{\beta(\tau)} \right).$$

This can depend on H only via the value of τ. The times τ_0 and τ_1 are given by the function $T(K,L,M,a)$,

$$\tau_0 = HT(K,L,M,1), \quad \tau_1 = HT\left(K,L,M,\frac{1}{1+z}\right),$$

and $\beta(\tau)$ solves the equation

$$\tau = HT(K,L,M,\beta)$$

for β. Therefore (5.47) shows that the three functions on the right-hand side of the equations are independent of H.

This implies that the dimensionless luminosity distance $d_{\mathscr{L}}$ is a function of Ω_M, Ω_Λ, and z only (again γ is assumed to be fixed),

$$d_{\mathscr{L}}(\Omega_M, \Omega_\Lambda, z).$$

If we measure one source, we obtain the values of z and $d_{\mathscr{L}}$. The theoretical function $d_{\mathscr{L}}(\Omega_M, \Omega_\Lambda, z)$ then yields a relation between Ω_M and Ω_Λ, which can be drawn as a curve in the Ω-diagram. Multiple measurements yield multiple curves, which determine a region in the diagram [7, 8].

Further independent measurements and assumptions about the surface of the last scattering yield that $K \approx 0$. This and the measurements described previously result in the following values:

$$\Omega_{M0} \approx \frac{1}{3}, \quad \Omega_{\Lambda 0} \approx \frac{2}{3}.$$

Today we assume these values as valid.

This implies that we are currently in the Λ-dominated phase of an eternally expanding model and that the universe has been in a phase of accelerated expansion for some time.

5.4.5 The Friedmann Models

In this section we discuss the special case $\Lambda = 0$, and we consider only models dominated by dust ($\gamma = 1$), or radiation ($\gamma = 4/3$). They describe the cosmological evolution in the case that the contribution of Λ is negligible compared to M. These models were formerly used as the standard models, but they are not only historically relevant. In this case, (5.31) has the following exact solutions:

1. Dust ($\gamma = 1$):

$$a = \frac{M}{2K} \left(1 - \cos \sqrt{K}\eta \right) , \tag{5.49}$$

$$t = \frac{M}{2K^{3/2}} \left(\sqrt{K}\eta - \sin \sqrt{K}\eta \right) . \tag{5.50}$$

This solution is valid for all $K \in (-\infty, \infty)$, provided the expressions on the right-hand side for $K = 0$ are interpreted as limits $K \to 0$, and if $\sqrt{K} = i\sqrt{|K|}$ is used for $K < 0$.

2. Radiation ($\gamma = 4/3$):

$$a = \sqrt{t \left(2\sqrt{M} - Kt \right)} . \tag{5.51}$$

The solutions for $K = 0$ are sometimes called Einstein–de Sitter models.

5.5 Space-Times with Maximal Symmetry (10 Killing Fields)

Another important special case is $M = 0$. In this case, $\rho = p = 0$, (5.27) is satisfied, and (5.25) and (5.26) simplify to

$$a'' = La , \tag{5.52}$$
$$a'^2 - La^2 = -K . \tag{5.53}$$

We use the freedom (5.12), to transform K to $+1$, 0, or -1. In the subsequent discussion, we allow only these values for K.

Depending on the value of Λ, we distinguish three cases.

5.5.1 Minkowski Space-Time

Set $\Lambda = 0$. Then (5.53) implies that $K = 0$ or $K = -1$. For $K = 0$, we obtain $a = $ const. The corresponding Robertson–Walker metric becomes

$$ds^2 = dt^2 - a^2 \left(dr^2 + r^2 d\Omega^2 \right) .$$

This is Minkowski space, with inertial coordinates x^μ transformed in the following way

$$x^0 = t, \quad x^k = arn^k ,$$

where n^k is given by (5.5).

For $K = -1$, (5.53) yields that $a' = \pm 1$ or $a = \pm t$, and the Robertson–Walker metric is

$$ds^2 = dt^2 - t^2 \left(\frac{dr^2}{1+r^2} + r^2 d\Omega^2 \right) .$$

This is the metric in the interior of the light cone of the origin in Minkowski space. To see this, transform the inertial coordinates x^μ as follows

$$x^0 = t\sqrt{1+r^2}, \quad x^k = trn^k .$$

For $t > 0$ ($t < 0$) the coordinates t, r, ϑ, and φ only cover the interior of the future (past) light cone since

$$(x^0)^2 - \vec{x} \cdot \vec{x} = t^2 > 0 .$$

Hence, the pseudo-spheres are the surfaces of constant Minkowski interval from the origin. This space-time is sometimes called *Milne model* (Fig. 5.16).

5.5.2 de Sitter Space-Time

de Sitter space-time plays a prominent role in the inflation era of cosmology. Furthermore, it is likely the geometry that our universe approaches at late times and a popular toy of theoretical physics. Thus we will study its geometry in some detail.

For $\Lambda \neq 0$, we obtain from (5.52) that

$$a = c_+ \exp\left(\sqrt{L} t \right) + c_- \exp\left(-\sqrt{L} t \right) , \tag{5.54}$$

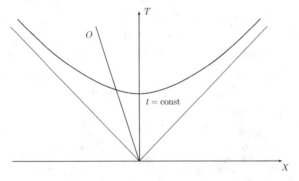

Fig. 5.16 Milne model. The cosmological time t and the trajectory of a cosmological observer O

and from (5.53) that

$$c_+ c_- = \frac{K}{4L} . \tag{5.55}$$

Equation (5.55) means that only one of the integration constants is arbitrary. The invariance of (5.52) and (5.53) with respect to time translations $t \to t + t_1$ enables us to use the remaining freedom to transform the constant to some standard value.

Assume further that $\Lambda > 0$ and introduce the shorthand $R = 1/\sqrt{|L|}$. The exponents in (5.54) are real, and (5.55) has the real solutions c_+ and c_- for all K. Let us begin with $K = 1$ and choose $c_+ = c_- = 1/(2\lambda)$. We obtain

$$a = R \cosh(t/R) .$$

The integration constant is chosen such that the minimal radius is attained at $t = 0$. In this case, the Robertson–Walker metric (5.11) has the form

$$ds^2 = dt^2 - R^2 \cosh^2(t/R) \left[\frac{dr^2}{1 - r^2} + r^2 \left(d\vartheta^2 + \sin^2 \vartheta \, d\varphi^2 \right) \right] . \tag{5.56}$$

The coordinates have the ranges $t \in (-\infty, \infty)$, $r \in [0, 1)$, and $(\vartheta, \varphi) \in S^2$ (and thus only cover half of the space-time). We can cover the whole space-time (except the poles) by replacing the coordinate r by $\chi := \arcsin r$ and let $\chi \in (0, \pi)$. We obtain the metric

$$ds^2 = dt^2 - R^2 \cosh^2(t/R) \left[d\chi^2 + \sin \chi \left(d\vartheta^2 + \sin^2 \vartheta \, d\varphi^2 \right) \right] .$$

The metric (5.56) is called *de Sitter metric*. Its exponential expansion again shows how unstable gravitation is. Arbitrarily large space with a constant energy density can be created.

The causal structure of de Sitter space-time can be exposed by a conformal deformation. We write the metric in the form

$$ds^2 = R^2 \cosh^2(t/R) \left(d\eta^2 - d\chi^2 - \sin^2 \chi \, d\Omega^2 \right) ,$$

where

$$\eta = \int \frac{dt}{R \cosh(t/R)} = \arctan \sinh(t/R) .$$

This implies that

$$\sinh(t/R) = \tan \eta \quad \cosh(t/R) = \frac{1}{\cos \eta}$$

and

$$t = R \, \text{arcsinh} \tan \eta .$$

For $t \in (-\infty, \infty)$, the values of η cover the interval $(-\pi/2, \pi/2)$. The light cones of the points with $r = 0$ and $\eta = \pm \pi/2$ (horizons) are given by (cf. (5.21)):

Fig. 5.17 Penrose diagram of de Sitter space-time. Every point is a sphere of radius $\sin\chi$. The *dashed lines* are the horizons

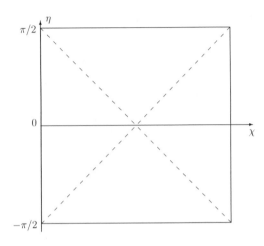

Fig. 5.17 Penrose diagram of de Sitter space-time. Every point is a sphere of radius $\sin\chi$. The *dashed lines* are the horizons

$$\chi = \pm(\eta \mp \pi/2) .$$

Figure 5.17, which illustrates the causal structure of de Sitter space-time, has been drawn with a frequently used method. This method only works for rotationally symmetric models, but nearly all models in these Notes are so. Every point represents a sphere (S^2) with radius $\sin\chi$. Light-like rotationally symmetric hypersurfaces (light cones of the points with $r = 0$) are represented by lines of slope 1, that is $\chi = \pm(\eta - \eta_0)$. Later we will see more such diagrams. They are called *Penrose diagrams*.

The horizons meet at $\eta = 0, r = 1$. Let us calculate the metric on such a horizon. We use the functions η, ϑ, and φ as coordinates along the horizons. The embedding functions are

$$t = \frac{1}{\lambda}\operatorname{arcsinh}\tan\eta ,$$

$$\chi = \pm\eta - \pi/2 ,$$

$$\theta = \vartheta ,$$

$$\phi = \varphi .$$

Then we have

$$ds^2 = -R^2 \left(d\vartheta^2 + \sin^2\vartheta\, d\varphi^2\right) .$$

This is a degenerate metric, since the coordinates on the hypersurface are η, ϑ, and φ, but the metric only contains two terms, one with $d\vartheta$ and another one with $d\varphi$. Such a hypersurface is called *light-like* (we will give a proper definition and discuss properties in the next chapter). The space-like sections $\eta =$const of the "cone" are spheres with constant radius $1/\lambda$.

de Sitter space-time can be visualized as a four-dimensional hypersurface in five-dimensional Minkowski space. The metric of the embedding space-time is

$$dS^2 = dT^2 - dU^2 - \left(dX^1\right)^2 - \left(dX^2\right)^2 - \left(dX^3\right)^2 , \tag{5.57}$$

and the equation of the hypersurface is

$$T^2 - U^2 - \left(X^1\right)^2 - \left(X^2\right)^2 - \left(X^3\right)^2 = -R^2 . \tag{5.58}$$

The hypersurface is described by the endpoints of space-like radial vectors of constant length λ^{-1} (Fig. 5.18). Hence it has a large symmetry group, the whole "Lorentz group" of five-dimensional Minkowski space (5.57). This space-time contains 10 mutually orthogonal 2-planes, thus the group has 10 generators (10 dimensions), 4 boosts, and 6 rotations. This group is denoted by SO(1, 4) and called *de Sitter group* [4].

We want to show that the induced metric on the hyperboloid (5.58) agrees with (5.56). To this end, we find out to which point (T, U, X^k) of the embedding space a point $(t, r, \vartheta, \varphi)$ is mapped. This yields the embedding formulas,

$$T = T(t, r, \vartheta, \varphi), \quad U = U(t, r, \vartheta, \varphi), \quad X^k = X^k(t, r, \vartheta, \varphi) .$$

The $(t = \text{const})$-surfaces of de Sitter space-time are 3-spheres. They result as sections of the hyperboloid with the $T = \text{const}$ planes. Thus we try $T = T(t)$, or better

$$T = Rf(t) \tag{5.59}$$

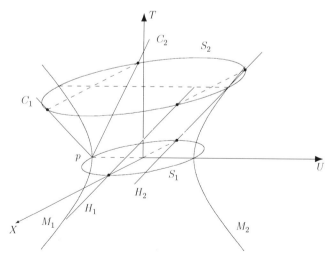

Fig. 5.18 Three-dimensional de Sitter space-time represented by the intersections M_1, M_2 with the (T, U)-plane and S_1 with the (X, U)-plane. The particle horizon of the trajectory M_1 is the event horizon of the trajectory M_2 and consists of two straight lines H_1 and H_2. The light cone of the point p is composed of the two straight lines C_1 and C_2

with a yet undetermined function $f(t)$. Let us determine the geometry of the sections. To this end, substitute (5.59) for T into the equation of the hyperboloid (5.58), and obtain

$$U^2 + \sum_k X^k X^k = R^2 \left(1 + f^2(t)\right) . \tag{5.60}$$

This is the equation of a 3-sphere of radius $\lambda^{-1}\sqrt{1 + f^2(t)}$. As we already know the embedding formulas for the 3-sphere, we obtain the spatial part of the metric (5.56). It is (5.4) with $U = X^4$, $U = \pm a(t)\sqrt{1 - r^2}$, and $X^k = a(t)rn^k$ where on the one hand

$$a(t) = R\cosh(t/R)$$

to get (5.56) and on the other hand

$$a(t) = R\sqrt{1 + f^2(t)} ,$$

to satisfy equation (5.60). Therefore we set

$$f(t) = \sinh(\lambda t).$$

Altogether, this yields

$$T = R\sinh(\lambda t) ,$$
$$U = \pm R\sqrt{1 - r^2}\cosh(\lambda t) ,$$
$$X^k = Rr\cosh(\lambda t)n^k .$$

The sign $+\ (-)$ gives the right (left) half of the hyperboloid (Fig. 5.18). From the derivation, it is clear that these functions satisfy (5.58) for all values of t, r, ϑ, and φ, and inserting these functions into (5.57) yields (5.56).

Inserting the horizon equations

$$t = R\,\text{,arcsin h}\tan\eta , \quad r = \cos\eta ,$$

into the embedding formulas, we obtain

$$T = R\tan\eta ,$$
$$U = \pm R\tan\eta ,$$
$$X^k = Rn^k .$$

Thus the horizons are the sections of the hyperboloid with the light-like planes $T \mp U = 0$ (Fig. 5.18).

Equation (5.55) is also valid for $K = 0$. This yields a foliation of de Sitter spacetime with Euclidean planes, which is advantageous for some applications. Let $c_- = 0$ and $c_+ = 1/\lambda$, so that

$$a = Re^{t/R} \, ,$$

and the Robertson–Walker metric becomes

$$ds^2 = dt^2 - R^2 e^{2t/R} \left(dr^2 + r^2 d\Omega^2 \right) \, . \tag{5.61}$$

For this metric, we obviously have

$$\frac{a'}{a} = \frac{1}{R} \, ,$$

and the Hubble constant is indeed constant.

We now turn to prove the following claim. The embedding formulas describing the metric on the hyperboloid are obtained by intersecting it with the family of light-like planes (Fig. 5.18).

$$T - U = \pm R e^{t/R} \, .$$

Each plane is defined by a constant t. As $t \in (-\infty, \infty)$, these planes range from $T - U = 0$ to $T - U = \pm\infty$. Inspired from (5.61), we set

$$X^k = R r e^{t/R} n^k$$

and obtain via (5.58) that

$$T + U = \pm R \left(r^2 e^{t/R} - e^{-t/R} \right) \, .$$

This yields the embedding formulas. The resulting metric indeed agrees with (5.61). We see that the sections are $\mathbb{E}^3$, and that they converge to the horizon $T - U = 0$ in the limit $t \to -\infty$. Furthermore, this horizon is the boundary of the region covered by these coordinates (one half of de Sitter space-time!).

Equation (5.55) is also valid for $K = -1$, which provides a foliation of de Sitter space-time with pseudo-spheres. To this end, let $c_+ = -c_-$ with $c_+ = R/2$, so that

$$a = R \sinh(t/R) \, ,$$

and

$$ds^2 = dt^2 - R^2 \sinh^2(t/R) \left(\frac{dr^2}{1 + r^2} + r^2 d\Omega^2 \right) \, . \tag{5.62}$$

The embedding formulas leading to this metric are obtained by intersecting the hyperboloid with the time-like planes (Fig. 5.18)

$$U = R \cosh(\lambda t) \, .$$

Again, each $t \in (0, \infty)$ yields a plane. We let

$$X^k = R r \sinh(t/R) n^k \, ,$$

obtain from (5.58) that

$$T = R\sqrt{1+r^2}\sinh(t/R) \,,$$

and, after some calculation, the metric (5.62). This metric covers the region of the hyperboloid bounded by the light cone

$$T^2 - \left(X^1\right)^2 - \left(X^2\right)^2 - \left(X^3\right)^2 = 0$$

of the point P with $T = X^k = 0$ in the plane $U = R$ in the five-dimensional space-time. This plane is the usual Minkowski space with coordinates T and X^k and the metric $dS^2 = dT^2 - (dX^1)^2 - (dX^2)^2 - (dX^3)^2$. The above light cone also lies on the hyperboloid and is a light cone there, namely that of P. Note that this light cone arises as the intersection of the hyperboloid with the plane tangential to the hyperboloid at P. As all points of the hyperboloid can be obtained from P by a symmetry, which leaves tangential planes tangential, we infer the following general statement. The light cone of a point Q in the hyperboloid is the intersection of the hyperboloid with its tangential plane in Q.

We found that there are parts of de Sitter space-time, which are Robertson–Walker space-times for each value of K. This bears the following significance. Each K-type of Robertson–Walker space-time can converge to de Sitter in the limit $t \to \infty$.

Yet another, but very important coordinate system in de Sitter space-time can be obtained as follows. Consider the planes

$$T \cosh\tau - U \sinh\tau = 0$$

for all constant τ. These are (T,U)-boosts of the plane $T = 0$ (Fig. 5.18). That is, the planes $\tau = $ const are mapped to each other by a sub-group of the symmetry group of the hyperboloid. The previous equation can be satisfied by

$$T = R\sqrt{1-\rho^2}\sinh\tau \,, \tag{5.63}$$

$$U = R\sqrt{1-\rho^2}\cosh\tau \,, \tag{5.64}$$

where we introduced another coordinate ρ. Equation (5.58) yields

$$\vec{X} \cdot \vec{X} = \rho^2 R^2 \,,$$

that is

$$X^k = R\rho n^k \,. \tag{5.65}$$

Substituting (5.63), (5.64) and (5.65) into (5.57) leads to

$$ds^2 = R^2\left[\left(1-\rho^2\right)d\tau^2 - \frac{d\rho^2}{1-\rho^2} - \rho^2 d\Omega^2\right] \,.$$

This is a static metric. It is singular at $\rho = 1$, and the coordinate regions $\tau \in (-\infty, \infty)$, $\rho \in (0,1)$, and $(\vartheta, \varphi) \in S^2$ cover quadrant I of de Sitter space-time (Fig. 5.17).

5.5.3 Anti-de Sitter Space-Time

We set $\Lambda < 0$ and define $R := 1/\sqrt{-\Lambda}$. Then the solution (5.54) is only real valued
if $c_- = c_+^*$. That is, (5.55) has only a solution for $K = -1$. It is

$$a = R\cos(t/R) ,$$

where we arranged $t = 0$ so that the scale factor has its maximum there. For $K = -1$,
(5.11) yields the metric

$$ds^2 = dt^2 - R^2 \cos^2(t/R) \left[\frac{dr^2}{1+r^2} + r^2 \left(d\vartheta^2 + \sin^2 \vartheta \, d\varphi^2 \right) \right] . \tag{5.66}$$

This is the metric of the so-called *anti-de Sitter space-time*. This metric is regular in
the finite interval $t/R \in \left(-\frac{\pi}{2}, \frac{\pi}{2} \right)$ and, at first, it looks like a recollapsing cosmology.
　　Again, this metric can be found on a highly symmetric hypersurface. The em-
bedding space is five-dimensional and has the flat metric

$$dS^2 = dT^2 + dU^2 - \left(dX^1 \right)^2 - \left(dX^2 \right)^2 - \left(dX^3 \right)^2 . \tag{5.67}$$

The equation of the hypersurface is (Fig. 5.19)

$$T^2 + U^2 - \left(X^1 \right)^2 - \left(X^2 \right)^2 - \left(X^3 \right)^2 = R^2 . \tag{5.68}$$

Again the symmetry group of this hypersurface is 10-dimensional. It is the group of
"rotations" around the origin in the embedding space with metric (5.67). It is called
anti-de Sitter group and is denoted by the symbol SO(2, 3) [4].

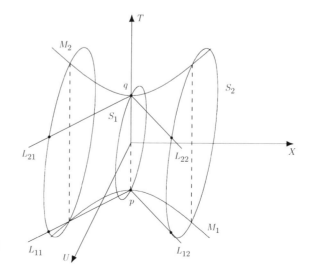

Fig. 5.19 Two-dimensional
anti-de Sitter space-
time represented by the
intersections M_1, M_2 with
the (T,X)-plane and S_1 with
the (T,U)-plane. The future
light cone of the point p is
composed of the two straight
lines L_{11} and L_{12}. Similarly,
the past light cone of the point
q is formed by L_{21} and L_{22}

We obtain the metric (5.66) by intersecting the hyperboloid with the planes $T = $ const. The corresponding embedding formulas are

$$T = R \sin(t/R) ,$$
$$U = \pm R\sqrt{1+r^2} \cos(t/R) , \tag{5.69}$$
$$X^k = Rr \cos(t/R) n^k . \tag{5.70}$$

The coordinates t, r, ϑ, and φ only cover the part of the hyperboloid with $T \in [-R,R]$. We thus assume these formulas to be valid only in the corresponding interval

$$-R\pi/2 < t < R\pi/2 , \tag{5.71}$$

where in addition $a \neq 0$. The hypersurfaces $t = \pm R\pi/2$ are given by $T = \pm R$ in the embedding space (Fig. 5.19). Inserting these values into (5.68), we find that the region (5.71) is bounded by the following two light cone halves in the planes $T = \pm R$:

$$U^2 - \left(X^1\right)^2 - \left(X^2\right)^2 - \left(X^3\right)^2 = 0 , \quad T = \pm R , \quad U > 0 . \tag{5.72}$$

Again, the planes $T = \pm R$ are Minkowski space-times with time coordinate U. The intersections with the hyperboloid with these spaces are the light cones in these space-times (the interior corresponds to $|U| > |X|$). These intersections also lie on the hyperboloid and play the role of the light cones of the points with $T = \pm R$ and $U = X^k = 0$ there. Furthermore, we have a characterization of the intersection of the light cone of an arbitrary point analogous to de Sitter space. Thus the coordinates t, r, ϑ, and φ only cover the region of the hyperboloid (5.68) bounded by the two light cones (5.72). Note, the hypersurfaces $t = $ const, where the constant varies in the range (5.71), are complete pseudo-spheres that do not intersect the light cones of the points $(\pm R, 0, 0, 0)$.

In the limit $t \rightarrow \frac{\pi}{2\lambda}$, all t-curves meet in the vertices $U = X^k = 0$ of these light cones. This is the reason why a vanishes (Fig. 5.19). This follows from (5.69) and (5.70).

However, even the hyperboloid is not the whole anti-de Sitter space-time. Note that the embedding implies the existence of closed time-like curves (acausality!). However, this is only a property of the embedding. To construct the actual anti-de Sitter metric, we introduce new coordinates on the hyperboloid covering it completely. Let us try the planes $T \cos \tau - U \sin \tau = 0$:

$$T = \frac{R \sin \tau}{\cos \chi}, \quad U = \frac{R \cos \tau}{\cos \chi} ,$$

where $\tau \in (0, 2\pi)$ and $\chi \in [0, \pi/2)$. Then

$$T^2 + U^2 = \frac{R^2}{\cos^2 \chi}$$

and (5.68) leads to

$$X^k = R \tan \chi n^k .$$

In these coordinates, the metric becomes

$$ds^2 = R^2 \left[\frac{d\tau^2}{\cos^2 \chi} - \frac{d\chi^2}{\cos^2 \chi} - \tan^2 \chi \, d\Omega^2 \right] . \tag{5.73}$$

This is a static space-time. The hyperboloid results from the assumption that τ be periodical, that is by identifying the points with $\tau = 0$ and $\tau = 2\pi$. We do not have to do this to the metric (5.73), and define instead: the whole anti-de Sitter space-time is given by the metric (5.73), where the coordinate ranges are $\tau \in (-\infty, \infty)$, $\chi \in [0, \pi/2)$, and $(\vartheta, \varphi) \in S^2$. This produces a causal space-time.

Introduce the following conformal deformation:

$$ds^2 = \frac{R^2}{\cos^2 \chi} \left[d\tau^2 - d\chi^2 - \sin^2 \chi \, d\Omega^2 \right] .$$

The last two terms in brackets equal the metric on a three-dimensional sphere (S^3) (cf. (5.6)), but $\chi \in [0, \pi/2)$ yields only a half 3-sphere from the pole to the equator. Thus the space-time has a time-like boundary, for each τ a 2-sphere, namely the equator of the 3-sphere $\tau = \text{const}$.

Consider the Penrose diagram of this space-time (Fig. 5.20) with the conformally deformed metric

$$ds^2 = d\tau^2 - d\chi^2 - \sin^2 \chi \, d\Omega^2 .$$

The hypersurface $\tau = 0$ lies between two light cones L_1 and L_2 asymptotic to it. L_1 is given by $\chi = \tau + \pi/2$ and L_2 by $\chi = -\tau + \pi/2$. For T this means

$$T = \frac{R \sin \tau}{\cos(\pm \tau + \pi/2)} = \pm R .$$

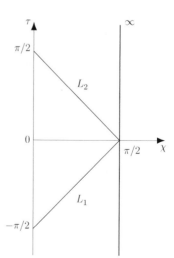

Fig. 5.20 Penrose diagram of anti-de Sitter space-time. The Cauchy horizons of the hypersurface $\tau = 0$ are L_1 and L_2. The diagram is endless in the future and past directions

These two light cones are called *Cauchy horizons* of the hypersurface $\tau = 0$ (for details on Cauchy horizons, see [11]).

5.6 The Early Universe

Sometimes modern cosmology is called "Big Bang cosmology", as the Big Bang is a dominant feature and a well-established assumption. Therefore, a frequent question is about the beginning of the universe and the reason for this huge explosion. These considerations are also motivated by the fact that the corresponding Friedmann–Lemaître model leads to very special initial conditions in several aspects. These are the so-called *problems of special initial conditions*: (1) horizons, (2) flatness, and (3) entropy. We will discuss these problems in turn.

5.6.1 Horizon Problem

Today, the cosmic microwave background has negligible interaction with the rest of matter (hydrogen gas is transparent, etc.), and thus forms a closed component of the whole system. For radiation in equilibrium, the Stefan–Boltzmann law implies $\rho_{\text{rad}} \sim T_{\text{rad}}^4$. Together with the equation $\rho_{\text{rad}} = (3M/8\pi G)a^{-4}$, this leads to the relation

$$T_{\text{rad}}a = \text{const}.$$

In combination with today's temperature, this implies that at the scale factor $a_R = 1500^{-1}$, the temperature of the radiation reached the so-called *recombination temperature* $T_R = 4500\,\text{K}$ at which hydrogen becomes ionized. Above T_R we have an opaque mixture of protons and electrons—plasma. The hypersurface $t = t_R$ is the so-called *surface of last scattering* (Fig. 5.21). The CMB observed today was emitted by part of the surface of last scattering. Denote the volume of this part of the surface by V_{R0} and its radius by R_{R0}. On the other hand, we have the particle horizons intersecting the surface. Define the *horizon radius* as the radius of the volume bounded by a particle horizon at $t = $const. A simple calculation yields that

$$R_H(t_R) \ll R_{R0}.$$

This calculation is based on the assumption that the radiation-dominated Friedmann model is a good approximation for the epoch before recombination. This means, if the Friedmann model is valid up to the beginning, then the particles in different regions of V_{R0} cannot interact. But then, how did they "come to know" that they all should have the same temperature? This problem can only be solved by a very special initial condition, which is objectionable.

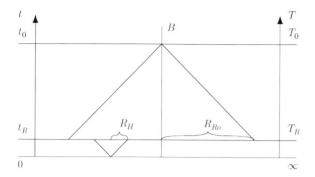

Fig. 5.21 Surface of the last scattering $t = t_R$. The particle horizon radius R_H is to be compared with the radius of the part of the surface visible to an observer B

5.6.2 Flatness Problem

The values that we obtain from direct measurements of the parameter Ω_M lie in the interval $(0.2, 4)$. This is very close to the value 1. The dynamics of the Friedmann–Lemaître model exhibit a strong instability of Ω_M. As soon as $K \neq 0$, Ω_M quickly diverges away from 1.

To illustrate this instability, we restrict ourselves to a very simple example, a radiation-dominated Friedmann model. For all other models, there are similar results, which are more difficult to obtain. The scale factor is given by (5.51). Equation (5.30) yields the definition of Ω_M (where we have to compute at an arbitrary time):

$$\Omega_M = \frac{\rho}{\rho_c} = \frac{M}{a^2 a'^2} .$$

A simple calculation gives

$$\Omega_M(t) = \left(1 - \frac{Kt}{\sqrt{M}}\right)^{-2} .$$

We see that Ω_M starts at 1 for all K, but diverges quickly if $K \neq 0$. Assume that $K = -1$, then the condition $\Omega_M > 1/5$ implies

$$1 + \frac{K}{\sqrt{M}} 10^{17} < \sqrt{5} ,$$

where we used that $t_0 = 10^{10}$ J or $t_0 = 10^{17}$ s $= 10^{25}$ m. This requires $\sqrt{M}/K > 10^{17}$ s. Thus at the beginning, it should be already determined that $\sqrt{M}/K$ is comparable to the age of the universe today. This is unnatural as at that time a natural time scale was given by the Planck time 10^{-33} s. Hence, right at the beginning, it would have to be known that the germ of the universe grows to a large cosmos. It is more desirable to find a natural process, which determines the value of $K/\sqrt{M}$.

5.6.3 Entropy Problem

The entropy S_{rad} of radiation in a volume V is

$$S_{rad} = \frac{32}{45} \frac{\pi^5 k}{h^3 c^3} V (kT)^3 \, ,$$

where we set the Boltzmann constant to 1. Estimating the number of baryons in the same volume results in 10^9 entropy per baryon. Hence, we live in a hot universe. These numbers stay the same, as all processes are adiabatic and preserve the number of baryons. This rises the question where that heat came from.

5.6.4 Cosmic Inflation: Orders of Magnitude

An important idea that is believed to solve these problems and to explain the driving force of the Big Bang is called *cosmic inflation* [14]. It is based on de Sitter model: the scale factor grows exponentially while the mass density remains constant. This enables formation of a macroscopic piece of matter from a microscopic one in a very short time interval. Still, energy is conserved, as the necessary work is done by the negative pressure and gravity (write energy equation (3.59) for ρ_Λ and p_Λ).

From where do we obtain the cosmological constant needed? It turns out that quantum fields can possess several ground states, each being stable only within a specific interval of temperatures similar to the well-known phases of ordinary matter. The temperature at which the stability properties of quantum fields change is again called the *critical temperature*, T_c. And again, there can be an energy-density jump between such ground states as there is one for ordinary phase transition of second kind, called latent heat. The jump is oriented so that it is negative if the temperature decreases through the critical value.

We can therefore imagine the following picture. If the temperature decreases and the higher energy ground state persists as a meta-stable state for some time interval Δt at temperatures lower than T_c, there is an effective vacuum energy density ρ_{vac} equal to the jump. Quantum field theory yields two numbers [14]

$$\rho_{vac} \approx (10^{15} \, \text{GeV})^4$$

and

$$T_c \approx 10^{15} \, \text{GeV} \, .$$

The density ρ_{vac} provides the desired cosmological constant.

During the time Δt—the inflationary epoch—the vacuum energy density ρ_{vac} of the false vacuum dominates the matter. In the beginning, since radiation has the same temperature, ρ_{rad} is comparable to ρ_{vac}. But the radiation is diluted quickly. It would be sufficient that $\Delta t \approx 10^{-33}$ s.

We use powers of GeV as our units ($\hbar$ and c are 1). Then the magnitudes of the transformation relations are useful $1 \, \text{GeV} \approx 10^{-27} \, \text{kg} \approx 10^{16} \, \text{m}^{-1} \approx 10^{24} \, \text{s}^{-1}$, $G \approx 10^{-38} \, \text{GeV}^{-2}$.

The energy density of vacuum yields an effective cosmological constant,

$$\Lambda = 8\pi G \rho_{vac} \approx 10^{23} \, (\text{GeV})^2 \approx 10^{55} \, \text{m}^{-2} \approx 10^{71} \text{s}^{-2} \,.$$

This is a huge value, compared to the current value of the cosmological constant of $\Lambda < 10^{-52} \, \text{m}^{-2}$. With this constant and the de Sitter scale factor $a \approx H^{-1} \exp(Ht)$, we obtain ($H = \sqrt{\Lambda/3}$ is roughly $10^{35} \, \text{s}^{-1}$) the scale factor for $\Delta t = 10^{-33}$

$$\frac{a(t_c + \Delta t)}{a(t_c)} \approx 10^{29} \,.$$

In this way, a tiny quantum fluctuation can indeed be blown up to a macroscopic piece. For example, a very small region with radius of a few Planck lengths $(10^{-19} \, \text{GeV}^{-1} \approx 10^{-35}$ m) is inflated to a macroscopic scale $\approx 10^{10} \, \text{GeV}^{-1} \approx 10^{-6}$ m. The total mass $\approx (10^{15} \, \text{GeV})^4 \times (10^{-19} \, \text{GeV}^{-1})^3 \approx 10^{-24}$ kg becomes considerable $\approx (10^{15} \, \text{GeV})^4 \times (10^{10} \, \text{GeV}^{-1})^3 \approx 10^{63}$ kg.

The horizon problem can then be solved by the trivial property that the volume elements of matter move along time-like trajectories and hence can not cross the horizon. If we thus begin with a small piece of matter, which lies inside the horizon radius before the inflationary epoch, then during inflation, this piece remains inside the particle horizon. The resulting volume is then larger than the part of the universe visible today.

The solution of the flatness problem is obtained from the nice property of de Sitter space that at any time $\Omega_M + \Omega_\Lambda = 1$ holds. We have $\Omega_M = 0$,

$$\rho_\Lambda = \frac{\Lambda}{8\pi G}$$

and

$$\rho_c = \frac{3}{8\pi G} H^2 = \frac{3}{8\pi G} \frac{\Lambda}{3} = \rho_\Lambda \,.$$

At the end of the inflationary epoch, the density of radiation is diluted by the factor $(10^{29})^4$, and the total density is very close to the vacuum energy density. Then after the inflation, the universe starts with Ω_M very close to 1!

Finally, the entropy problem can be solved by the latent heat. The large vacuum energy density decays completely to radiation, which yields enough heat.

The final state of the inflationary epoch is fixed, and practically independent of what happened before. To keep things simple, we assumed a radiation-dominated

Friedmann model (but with an arbitrary constant K). This assumption can be weakened, for example, we need not assume homogeneity and isotropy.

5.6.5 Quantum Cosmology

The three universal constants G, $\hbar$, and c can be combined to give units of all physical quantities: the so-called Planck units. For example, the *Planck length* $L_P = G^{1/2}\hbar^{1/2}c^{-1} \approx 10^{-34}$ m, the *Planck time* $T_P = L_P/c \approx 10^{-42}$ s, and the *Planck energy* $E_P = \hbar/T_P \approx 10^{19}$ GeV. When physical quantities surpass these horrific values, one speaks about *Planck regime*. It seems natural that gravity is to be quantized in the Planck regime [15]. This could be the case sometime before the inflationary epoch, and we are in the field of *quantum cosmology*. Today's quantum cosmology is based on the assumption of high symmetry. One utilizes the symmetry to reduce the degrees of freedom to a finite number. This leads to a form of quantum mechanics, not to a quantum field theory [16].

5.7 Exercises

1. Prove the following properties of the "vector" n^k (5.5):

   (a) The three vectors n^k, m^k, and l^k (in this order) form a positively oriented orthonormal 3-frame, where

   $$m^k := \frac{\partial n^k}{\partial \vartheta}, \quad l^k := \frac{1}{\sin \vartheta} \frac{\partial n^k}{\partial \varphi} .$$

   (b) Use this result to deduce the following identity

   $$\left(\vec{dn} \cdot \vec{dn} \right) = d\vartheta^2 + \sin^2 \vartheta \, d\varphi^2 .$$

   (c) Use this identity to compute the metric (5.6).

2. Calculate the curvature of S^3 and P^3 in the coordinates r, θ, and ϕ and show that

   $$R_{ijmn} = K(g_{im}g_{jn} - g_{in}g_{jm})$$

   everywhere on S^3 and P^3. Use the metric

   $$ds^2 = \frac{dr^2}{1 - Kr^2} + r^2 \left(d\theta^2 + \sin^2 \theta \, d\phi^2 \right) ,$$

   and the symmetry. (Consider only one point, the components of the curvature tensor with respect to a suitable *orthonormal basis*, and use the isotropy to determine the number of independent components of the curvature tensor.)

3. Prove that if $\bar{\xi}^k(r, \vartheta, \varphi)$ is a Killing vector field of the metric (5.9) then $\xi^\mu(t, r, \vartheta, \varphi)$ is a Killing vector field of the metric (5.11), where

$$\xi^0(t, r, \vartheta, \varphi) := 0, \quad \xi^k(t, r, \vartheta, \varphi) := \bar{\xi}^k(r, \vartheta, \varphi) .$$

4. Show that the Einstein tensor of the metric $ds^2 = dt^2 - a^2 \tilde{g}_{kl} dx^k dx^l$ has the form

$$G_{00} = 3a^{-2}\left(a'^2 + K\right) ,$$

$$G_{kl} = -\left(2aa'' + a'^2 + K\right)\tilde{g}_{kl} .$$

Hint: compute $\Gamma^\mu_{\nu\rho}$ from the geodesic equation. Then collect the terms with $\tilde{g}_{kl}$ and use the formula

$$\tilde{R}_{klmn} = K(\tilde{g}_{km}\tilde{g}_{ln} - \tilde{g}_{kn}\tilde{g}_{lm}) ,$$

where $\tilde{R}_{klmn}$ is the curvature tensor of the 3-metric $\tilde{g}_{kl}$.

5. On a manifold $\mathcal{M}$ let $g'_{\mu\nu}(x)$ and $g_{\mu\nu}(x)$ be two conformally related metrics:

$$g'_{\mu\nu}(x) = F(x)g_{\mu\nu}(x), \quad F(x) > 0 \quad \forall x .$$

What is the relation of the light-like geodesics of these two metrics, and what is that of the corresponding affine parameters?

6. A scale class contains all metrics that can be obtained from one of them by a *scale transformation*. These are the conformal transformations with constant factors. Show (1) all Robertson–Walker metrics can be divided into *scale classes* and (2) all space-times of a scale class have the same dynamical type.

7. Assume that the scale factor $a(t)$ is the solution of (5.31) with constants K, L, and M. How do we have to transform these constants in order to ensure that the scale factor $\bar{a}(t)$, obtained from a scale transformation, solves (5.31) with the transformed constants?

8. Compute the distance between two given cosmological observers at time t_0 in a Robertson–Walker model with $a(t) = a_1 + a_2 t + a_3 t^2$ with two different methods:

   (a) geodesic distance in the hypersurface $t = t_0$,
   (b) radar reflection time.

9. Consider the points with $\Omega_\Lambda = 0$ $(\Lambda = 0)$ in the Ω-diagram. Compute (a) the age function χ and (b) the scale-free luminosity distance $d_{\mathscr{L}}$.

10. A two-dimensional model of de Sitter or anti-de Sitter space-time results from the embedding formulas

$$dS^2 = dT^2 - \left(dX^1\right)^2 - \left(dX^2\right)^2 ,$$

$$T^2 - \left(X^1\right)^2 - \left(X^2\right)^2 = -H^{-2} ,$$

and

$$dS^2 = dT^2 + dU^2 - dX^2,$$
$$T^2 + U^2 - X^2 = H^{-2},$$

(a) Find the symmetry group and the Killing vector fields of the two-dimensional space-times defined this way.

(b) Show that the map $(T, X^1, X^2) \mapsto (-T, -X^1, -X^2)$ in the embedding space induces an isometry I of two-dimensional de Sitter space-time, and analogously for anti-de Sitter space-time.

11. Prove that each geodesic in the two-dimensional de Sitter space-time, respectively anti-de Sitter space-time, is an intersection of the two-dimensional de Sitter space-time, respectively anti-de Sitter space-time, with a 2-plane through the origin of the embedding space. (Show that this holds for all geodesics through one suitably chosen point, and then use symmetry.)

12. Use the result from Exercise 11 to show the following. Let p be a point in the two-dimensional de Sitter (anti-de Sitter) space-time and let q be a point inside (outside) of the light cone of $I(p)$ in the two-dimensional de Sitter (anti-de Sitter) space-time (the map I is from Exercise 10). Then there does not exist a geodesic that connects p to q.

References

1. A. A. Penzias and R. W. Wilson, Astrophys. J. **142** (1965) 419.
2. S. Weinberg, *Gravitation and Cosmology*. Wiley, New York, 1972.
3. L. Bergstöm and A. Goobar, *Cosmology and Particle Astrophysics*, Springer, Berlin, 2004.
4. A. O. Barut and R. Raczka, *Theory of Group Representation and Applications*, World Scientific, Singapore, 1985.
5. S. Kobayashi and K. Nomizu, *Foundations of Differential Geometry*, Wiley, New York, Vol. I 1963, Vol. II 1969.
6. J. Traschen and D. M. Eardley, Phys. Rev. D **34** (1996) 1665.
7. S. Perlmutter et al., Astrophys. J. **517** (1999) 565.
8. A. G. Riess et al., Astronomical J. **116** (1998) 1009.
9. Bičák, J.: *Einstein's Field Equations and Their Physical Implications*. Lect. Notes Phys. **540**. Springer, Berlin (2000).
10. B. Schutz, *Gravity from the Ground up*, Cambridge University Press, Cambridge, UK, 2003.
11. W. L. Freedman et al., Nature **371** (1994) 757.
12. M. J. Pierce et al., Nature **371** (1994) 371.
13. S. W. Hawking and G. F. R. Ellis, *The Large Scale Structure of Space-Time*, Cambridge University Press, Cambridge, UK, 1973.
14. A. H. Guth, *The Inflationary Universe: The Quest for a New Theory of Cosmic Origin*, Addison-Wesley, Reading, MA, 1997.
15. S. Weinberg, *The Quantum Theory of Fields* Vol. I, Cambridge University Press, Cambridge, UK, 1995.
16. A. Ashtekar, T. Pawlowski, P. Singh and K. Vandersloot, Phys. Rev. D **75** (2007) 024035.

Chapter 6
Rotationally Symmetric Models of Stars

In this chapter we shall construct models of stars, both static and dynamical ones, including a model of gravitational collapse. We will restrict ourselves to situations in which rotational symmetry is a reasonable approximation. Much of our attention will be focused on the so-called Schwarzschild solutions. These are rotationally symmetric solutions of the Einstein equations. Studying these solutions has led to significant insights into gravity which are still relevant today.

Similar to our treatment of cosmology, we will concentrate on the geometric aspects of stars, that is on their gravitational field. The properties of matter in the stars are represented by the simple equation of state $p = p(\rho)$. The rich details of the actual matter are of course important for serious star models (for the "Standard Model of the Sun", see [1]), but they do not interfere with the qualitative picture of the gravity part as described here.

In this section we set $\Lambda = 0$, as Λ is irrelevant for the astrophysics of stars. We begin with the treatment of the interior of rotationally symmetric stars.

6.1 Hydrostatic Equilibrium of Non-rotating Stars

6.1.1 Equations of the Hydrostatic Equilibrium

We consider models of static and rotationally symmetric stars consisting of an ideal fluid. Hence, the metric in the interior of the star has the form (2.91) and solves the Einstein equations (4.8) with $\Lambda = 0$. (It can be shown that a static, ideal fluid with reasonable equation of state which solves the Einstein equations is automatically rotationally symmetric. However, the proof is difficult and thus we simply assume rotational symmetry here [2].) The energy-momentum tensor must also have this symmetry, that is the 4-velocity of the fluid has to be parallel to the static Killing vector field. Otherwise static observers could measure energy currents. With respect to coordinates t, r, ϑ, and φ we then have

Hájíček, P.: *Rotationally Symmetric Models of Stars*. Lect. Notes Phys. **750**, 209–235 (2008)
DOI 10.1007/978-3-540-78659-7_6 © Springer-Verlag Berlin Heidelberg 2008

$$u^\mu = \frac{1}{\sqrt{B(r)}}(1,0,0,0) \ .$$

The pressure p and the matter density ρ must not depend on t, ϑ, or φ:

$$p = p(r) \ , \quad \rho = \rho(r) \ .$$

We do not specify the equation of state. To get an accurate model of the matter inside the star many other quantities have to be considered, e.g., temperature, conserved particle currents, etc. Here we only want to discuss the simplest model and represent the whole theory of matter by the equation of state.

The Einstein equations take a particularly simple form, if we introduce new functions $\Phi(r)$ and $m(r)$ instead of $A(r)$ and $B(r)$. We set

$$B(r) = e^{2\Phi(r)} , \tag{6.1}$$

$$A(r) = \frac{r}{r - 2m(r)} . \tag{6.2}$$

Then the tt- and rr-components of the Einstein equations become

$$m' = 4\pi G r^2 \rho \ , \tag{6.3}$$

$$\Phi' = \frac{m + 4\pi G r^3 p}{r(r - 2m)} \ . \tag{6.4}$$

The r-component of the Euler equation (3.60) is

$$p' = -(\rho + p)\Phi' \ ,$$

and by inserting (6.4), we obtain the so-called *Oppenheimer–Volkoff equation* [3]:

$$p' = -\frac{(\rho + p)(m + 4\pi G r^3 p)}{r(r - 2m)} \ . \tag{6.5}$$

Equations (6.3), (6.4), and (6.5) (in combination with one or multiple equations of state) form a complete system for the hydrostatic equilibrium of relativistic stars. (The Euler equation replaces the $\vartheta\vartheta$-component of the Einstein equations in a similar way as the energy (5.27) replaces the rr-component (5.25) in cosmology.) The resulting equations can be reduced to a system of two ordinary differential equations for two functions $p(r)$ and $m(r)$.

6.1.2 Conditions at the Center

Equations (6.3), (6.4), and (6.5) form a system of ordinary differential equations of first order with the independent variable r. Then one value of the desired solution for a given value of r determines the solution. We make a universal choice for this special value of r, namely $r = 0$. Which values can be assumed by the unknown functions at the center?

First, consider $m(0)$. The space–time shall be regular in all points with $r = 0$. Thus it must be locally flat there, and the surface $t = $ const must not have a conical singularity at the center. Then the area F_r of a small 2-sphere $r = $ const, $t = $ const with radius R_r must satisfy the Euclidean condition:

$$\lim_{r=0} \frac{F_r}{R_r^2} = 4\pi .$$

F_r and R_r have the form

$$F_r = 4\pi r^2 ,$$

and

$$R_r = \int_0^r dx \sqrt{A(x)} .$$

Using l'Hospital's rule, we obtain

$$\lim_{r=0} \frac{\sqrt{F_r}}{R_r} = \sqrt{4\pi} \lim_{r=0} \frac{1}{\sqrt{A(r)}} .$$

Hence, the center is only regular provided

$$\lim_{r=0} A(r) = 1 ,$$

which implies

$$m(0) = 0 . \tag{6.6}$$

The initial value $\Phi(0)$ does not bear any significance. Indeed, changing the function $\Phi(r)$ by a constant is equivalent to a coordinate transformation. To see this, we introduce a new time coordinate t' according to $t = t'e^C$. Then g_{00} transforms as follows:

$$g'_{00} = e^{2C} g_{00} = e^{2(\Phi+C)} .$$

Hence, $\Phi(r)$ is only determined up to an additive constant, as in Newton's theory. We fix the initial value in such a way that

$$\Phi(\infty) = 0 . \tag{6.7}$$

There is no condition (apart from $p > 0$) which restricts the initial value $p(0)$ of the function $p(r)$, that is the central pressure. For each value of the central pressure, we obtain a different model. Thus, for a fixed equation of state we find a one-dimensional family of models of stars. We only have to integrate (6.3), (6.4) and (6.5) with the initial values (6.6), (6.7), and a chosen central pressure. In this process, how do we know that we have arrived at the surface of the star?

6.1.3 Conditions at the Surface

Let $r = r_o$ at the surface of the star. At r_o the following conditions are obviously satisfied:

1. The density ρ can have a discontinuity at the surface, but the pressure $p(r)$ is continuous at r_o. Otherwise there are infinitely large forces at the surface (the force is proportional to p'). If we assume that there is vacuum outside of the star, then we must have $p(r) \equiv 0$ there. Thus the surface of the star can be identified with the first zero of the function $p(r)$.
2. The metric induced on each hypersurface $r = \text{const}$ is a continuous function of r at r_o. Otherwise one could measure different times or distances below or above the surface. This implies that the functions r and $\Phi(r)$ are continuous at r_o.

6.1.4 The Metric Outside the Star

Outside the star we set $\rho = p = 0$. For arbitrary r we obtain from (6.3) and (6.6) that

$$m(r) = 4\pi G \int_0^r \mathrm{d}x \, x^2 \rho(x) \, .$$

$m(r)$ is a continuous function of r, which takes a constant value m for $r > r_o$. The value of m is given by

$$m = 4\pi G \int_0^{r_o} \mathrm{d}r \, r^2 \rho(r) \, . \tag{6.8}$$

Then (6.4) implies for $r > r_o$

$$\Phi' = \frac{m}{r(r - 2m)} \, . \tag{6.9}$$

This has the solution

$$\Phi = \ln \sqrt{\frac{r - 2m}{r}} + \Phi_0 \, ,$$

where the constant Φ_0 has to be chosen according to (6.7). Thus the metric outside the star is of the following form

$$\mathrm{d}s^2 = \frac{r - 2m}{r} \mathrm{d}t^2 - \frac{r}{r - 2m} \mathrm{d}r^2 - r^2 (\mathrm{d}\vartheta^2 + \sin^2 \vartheta \, \mathrm{d}\varphi^2) \, . \tag{6.10}$$

This is the so-called *Schwarzschild solution* and the coordinates t, r, ϑ, and φ are called *Schwarzschild coordinates*.

The functions $A(r)$ and $B(r)$ are determined completely by the Einstein equations. Consider these functions in the limit $r \to \infty$ and expand them in the powers of $1/r$. We obtain

$$A(r) = \frac{1}{1 - \frac{2m}{r}} = 1 + \frac{2m}{r} + \cdots ,$$

$$B(r) = 1 - \frac{2m}{r} .$$

Comparison to the Eddington–Robertson expansion (2.98) and (2.99) yields

$$-\frac{2m}{r} = -2\alpha\frac{R_G}{r} ,$$

$$0 = 2(\beta - \alpha\gamma)\left(\frac{R_G}{r}\right)^2 ,$$

$$2\frac{m}{r} = 2\gamma\frac{R_G}{r} .$$

The first of these equations can be used to determine the mass of the star, as α must be 1 to result in the correct Kepler orbits. As $R_G = GM$ we thus find

$$M = 4\pi \int_0^{r_o} dr\, r^2 \rho . \tag{6.11}$$

This is the relativistic relation between the matter distribution ρ in the interior of the star and the mass of the star determined by the properties of the Kepler orbits of the satellites of the star.

The remaining equations represent the prediction from the Einstein equations for β and γ, namely,

$$\beta = \gamma = 1 , \tag{6.12}$$

which is in accordance with observations.

6.1.5 Comparison to Newtonian Gravity

The Newtonian theory of gravity leads to the following equations of hydrostatic equilibrium:

$$M' = 4\pi r^2 \rho , \tag{6.13}$$

$$\Phi'_N = G\frac{M}{r^2} , \tag{6.14}$$

$$p' = -G\frac{\rho M}{r^2} , \tag{6.15}$$

where $M(r)$ is the total mass inside the radius r, Φ_N the Newton potential, and ρ and p are the Newtonian matter density and pressure. The quantities ρ, p, and M_{tot} in the Newtonian theory can be considered equivalent to ρ, p, and M_{tot} in the Einstein theory, as they are defined and measured in the same way.

Comparing (6.3) and (6.13) shows that $m_{tot} = GM_{tot}$. In the relativistic theory M_{tot} is not simply an integral of ρ over a spatial volume. The solution of (6.13) can be written as such an integral:

$$M(r) = \int dV_N \rho ,$$

where V_N is the Newtonian volume element, as

$$dV_N = dr d\vartheta d\varphi r^2 \sin \vartheta ,$$

and

$$\int \int d\vartheta d\varphi \sin \vartheta = 4\pi .$$

The solution of (6.3) has an analogous form:

$$\frac{m(r)}{G} = \int \int \int dr d\vartheta d\varphi r^2 \sin \vartheta \rho , \qquad (6.16)$$

but the Einstein volume element of the hypersurface $t = $ const is given by the square root of the determinant of the induced metric,

$$ds^2 = A(r)dr^2 + r^2 d\vartheta^2 + r^2 \sin^2 \vartheta d\varphi^2 .$$

Hence,

$$dV_E = dr d\vartheta d\varphi r^2 \sin \vartheta \sqrt{A} ,$$

or

$$\frac{m(r)}{G} < \int dV_E \rho .$$

As $m(r) > 0$, (6.2) yields $A(r) > 1$, whence $dV_N < dV_E$. Nevertheless the quantity m_{tot}/G is the measurable total mass below the radius r_o. The reason of the difference is, roughly speaking, that in relativity all forms of energy contribute to the total mass and the contribution of gravity—the binding energy of the star—is negative. Then the total mass in relativistic theory must be smaller than the sum of the masses inside the radius r.

Comparison of (6.4) and (6.14) yields that the potential Φ grows faster in r in the relativistic theory than Φ_N in the Newtonian theory. This has two reasons. First, the pressure contributes to the source term, and second, the mass function $m(r)$ corrects the denominator. The force p' must be in equilibrium with gravity and thus must be larger in the Einstein theory. This can also be seen from the Oppenheimer–Volkoff equation where besides Φ' an extra term p occurs in the first factor. Hence, for relativistic stars collapse is more likely than for Newtonian ones.

6.1.6 Mass Limits

The equations of hydrostatic equilibrium with certain equations of state lead to the so-called *mass limits*. This is the maximal mass that can be reached by a star with reasonable equation of state. We now want to investigate the reasons for the existence of such mass. At the same time we will continue our comparison of the relativistic and the classical theories. An important difference between the Oppenheimer–Volkoff equation (6.5) and (6.15) is that the pressure p also appears on the right-hand side of (6.5). This leads to a positive feedback, a large value of the pressure p increases its own growth toward the center of the star.

These aspects can be illustrated by the simple example of the *incompressible fluid* with the following equation of state:

$$\rho = \text{const} . \tag{6.17}$$

We start by integrating the Newtonian equations (6.13), (6.14) and (6.15). From (6.13) we obtain

$$M(r) = \frac{\gamma \rho}{G} r^3$$

where we set $\gamma = 4\pi G/3$. Substituting into (6.15) yields

$$p(r) = P - \frac{1}{2}\gamma \rho^2 r^2 .$$

Here we expressed the constant of integration in terms of the central pressure $P = p(0)$.

The radius of the star is reached where $p(r_o) = 0$. This yields the relation

$$P = \frac{1}{2}\gamma \rho^2 r_o^2$$

between P and r_o for a given matter density ρ. Then $M_{\text{tot}} := (\gamma \rho /G) r_0^3$ satisfies

$$M_{\text{tot}} = \sqrt{\frac{6}{\pi G^3}} \frac{P^{3/2}}{\rho^2} .$$

Note that P remains finite for all values of r_o. Thus, in principle, r_o, and hence the total mass, can be arbitrarily large. If the pressure in the center is bounded by the respective realistic (finite) values, the radius r_o and thus the total mass is also bounded according to the above relations (Fig. 6.1). This yields a mass limit even for the Newtonian theory.

Integrating (6.3) implies that

$$m(r) = \gamma \rho r^3 .$$

Fig. 6.1 Dependence of the
total mass M_{tot} of the star on
its central pressure P, if the
matter is an incompressible
fluid. M_{totN} is the Newtonian
and M_{totE} the relativistic curve

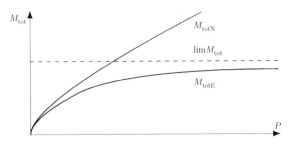

Then (6.5) becomes

$$p' = -\gamma\frac{(p+\rho)(3p+\rho)r}{1-2\gamma\rho r^2} .$$

Integrating this equation via separation of variables yields

$$\frac{3p+\rho}{p+\rho} = C\sqrt{1-2\gamma\rho r^2} ,$$

where C is a constant of integration, which can be expressed in terms of P:

$$\frac{3P+\rho}{P+\rho} = C .$$

This determines r_o to be

$$r_o = \sqrt{\frac{2}{\gamma\rho}\frac{P(2P+\rho)}{(3P+\rho)^2}} < \sqrt{\frac{1}{2\gamma\rho}}$$

and M_{tot} (Fig. 6.1),

$$M_{tot} = \sqrt{\frac{6}{\pi\rho G^3}}\left[\frac{P(2P+\rho)}{(3P+\rho)^2}\right]^{3/2} .$$

Consider the function $M_{tot}(P)$. For relatively small values of the central pressure

$$P \ll \rho ,$$

(which can actually be quite large), we have

$$M_{tot} = \sqrt{\frac{6}{\pi G^3}}\frac{P^{3/2}}{\rho^2}\left(1-6\frac{P}{\rho}+\cdots\right) .$$

This agrees with the Newtonian formula up to first order. For the derivative we obtain in the whole region $P \in (0,\infty)$ that

$$\frac{dM_{tot}}{dP} > 0 .$$

Hence M_{tot} is an increasing function of P. Finally,

$$\lim_{P \to \infty} M_{tot} = \frac{4}{9} \frac{1}{\sqrt{3\pi G^3}} \rho^{-1/2} .$$

This limit is finite. Hence, the pressure in the center diverges if the mass of the star is increased up to the limit. This is the effect of the relativistic corrections. However, it is clear that the chosen equation of state cannot hold for arbitrarily large values of the pressure and must break down way before $P = \infty$. If this happens in the non-relativistic region, then the Newtonian theory is sufficient to compute the mass limit, if this only happens in the relativistic region, then the Newtonian calculations have to be corrected.

In the so-called white dwarfs the pressure is dominated by the pressure of the degenerated electrons. For this equation of state one obtains the so-called Chandrasekhar mass limit or *Chandrasekhar mass* of 1.4 M_{Sun} [1]. The equation of state for baryonic matter (neutron stars) leads to a mass limit of 2–6 M_{Sun}, where the relativistic corrections have to be taken into account. White dwarfs and neutron stars are stars which can exist without a nuclear fuel burning in their interior [1].

6.1.7 Junction Conditions

When there are discontinuities in the matter, we need the so-called *junction conditions*. They complete the Einstein equations in the case of a jump surface in a similar way as the Poisson equation of electrostatics is completed by the condition that the potential be C^1 along the surface of the dielectric. We can formulate the junction conditions as follows:

> In the neighborhood of each point of a jump surface there exist coordinates so that the components of the metric in these coordinates are C^1 in that neighborhood.

We postulate that this condition holds in general.

Let us examine whether the resulting metric of the star satisfies these conditions. Recall that the metric has the form

$$ds^2 = B(r)dt^2 - A(r)dr^2 - r^2 d\vartheta^2 - r^2 \sin^2 \vartheta \, d\varphi^2 .$$

The surface of the star is at $r = r_o$. Let us examine the properties of the components in turn. For $B(r)$ we have (6.1) and (6.4). As $m(r)$ and $p(r)$ are continuous at $r = r_o$ also Φ' and thus B' is continuous. The components $g_{\vartheta\vartheta}$ and $g_{\varphi\varphi}$, that is the functions r^2 and $r^2 \sin^2 \vartheta$ have continuous derivatives of arbitrary order. Finally, from (6.2) we obtain that for $A(r)$ we have

$$\frac{dA}{dr} = 2 \frac{rm' - m}{(r - 2m)^2} .$$

But the function $m'(r)$ is *not* continuous,

$$\left.\frac{dm}{dr}\right|_{r<r_o} = 4\pi G r^2 \rho\,, \quad \left.\frac{dm}{dr}\right|_{r>r_o} = 0\,,$$

as ρ is allowed to jump. Hence, in the coordinates t, r, ϑ, and φ the metric is not C^1.

We try to introduce a new radial coordinate l, which is given near $r = r_o$ by the following equation:

$$l(r) := \int_{r_o}^r \sqrt{A(x)}\,dx\,.$$

It is continuous, equal to zero at $r = r_o$, and its derivatives

$$\frac{dl}{dr} = A(r)$$

is also continuous and non-zero. This implies that an arbitrary function $f(r)$ which is C^1 at $r = r_o$ defines a function $f(r(l))$ of l which is C^1 at $l = 0$. The transformed metric becomes

$$ds^2 = B(r(l))dt^2 - dl^2 - r^2(l)d\vartheta^2 - r^2(l)\sin^2\vartheta^2\,d\varphi^2\,.$$

All components of this metric are C^1. Hence the junction conditions are satisfied.

6.2 Properties of the Schwarzschild Solution

We shall now have a closer look at the special metric (6.10). We consider r in the range $(0, \infty)$, that is we set $r_o = 0$.

6.2.1 The Birkhoff Theorem

The first important property of the Schwarzschild metric which we shall describe (but not derive—that is not the scope of this book, a proof can be found in [4]) can be presented as follows.

Theorem 17 *Every rotationally symmetric solution to the Einstein equations with*

$$\Lambda = 0\,, \quad T_{\mu\nu} = 0$$

is identical to (a piece of) the Schwarzschild solution.

Hence the time independence need not be assumed.

This also means that a star moving such that it stays rotationally symmetric (only radial deformations) generates the Schwarzschild metric on its outside. This motivates to study this metric in the region up to $r = 0$.

6.2.2 Radial Light Rays

As shown in Sect. 6.1.4, the Schwarzschild metric yields the familiar geometry far outside the gravitational radius. But how does the solution look near that radius? The rr-component of the metric (6.10) diverges at $r = 2R_G$. This can either mean that the geometry is singular at these points or that we are looking at it in a singular coordinate system. To answer this question, we study the radial light-like geodesics of the metric (6.10). They not only help us find a better coordinate system but contain a lot of information about the geometry. For the time being we simply consider two space–times, one with $r > 2m$ and the other with $r < 2m$, the so-called *exterior* and *interior* Schwarzschild space–times.

The geodesics of the metric (6.10) are described by (2.95), (2.96), and (2.97). The proper choice of the constants for the case of radial light-like geodesics is $j = 0$ and $\mu = 0$, so that

$$\vartheta = \pi/2 , \quad \varphi = \text{const} , \quad \dot{r} = \pm e , \quad \dot{t} = \frac{e}{1 - \frac{2m}{r}} ,$$

where $e > 0$ is an arbitrary constant. The third equation implies that r itself is an affine parameter. Hence, we can choose the affine parameter $\lambda = \pm r$ in such a way that λ is always future pointing. This renders $e = 1$.

Thus we obtain two kinds of radial light-like geodesics:

1. $\dot{r} = +1$, so that $\frac{dr}{dt} > 0$ for $r > 2m$ and the geodesic is *outgoing* in the exterior Schwarzschild space–time. Then the solution (in both space–times) is

$$t - r - 2m \log \left| \frac{r}{2m} - 1 \right| = u , \tag{6.18}$$

where u is an arbitrary constant. Similarly
2. $\dot{r} = -1$, so that $\frac{dr}{dt} < 0$ for $r > 2m$ and the geodesic is *ingoing* in the exterior Schwarzschild space–time. The solution is

$$t + r + 2m \log \left| \frac{r}{2m} - 1 \right| = v , \tag{6.19}$$

where v is an arbitrary constant.

For constant radii r the constants u and v agree with the coordinate t up to an additive constant. Furthermore, in the region $r > 2m$, for very large r, t is the proper time of the asymptotic static observers. Hence u is the retarded and v the advanced time in the space–time with $r > 2m$.

6.2.3 Eddington–Finkelstein Coordinates

Equation (6.19) defines a function v on the space–time. We introduce the functions v, r, ϑ, and φ as new coordinates. These are called *Eddington–Finkelstein coordinates* [5]. To transform the metric (6.10) into these coordinates, we compute from (6.19) that

$$dt = -\frac{dr}{1 - \frac{2m}{r}} + dv \,,$$

and substitute this into (6.10) for both regular domains $r \in (0, 2m)$ and $r \in (2m, \infty)$. The result is

$$ds^2 = \frac{r - 2m}{r} dv^2 - 2 dv \, dr - r^2 (d\vartheta^2 + \sin^2 \vartheta \, d\varphi^2) \,. \tag{6.20}$$

The new components of the metric are all smooth at $r = 2m$, and the determinant

$$g = -r^4 \sin^2 \vartheta$$

only vanishes at $r = 0$ and $\vartheta = 0$, π. Hence the metric itself is regular at $r = 2m$. Thus the exterior and the interior Schwarzschild space–times can be identified as parts of the Eddington–Finkelstein space–time where $r \in (0, \infty)$.

We now consider the coordinates v, $r' = r$, $\vartheta' = \vartheta$, and $\varphi' = \varphi$ as regular (allowed) around the points with $r = 2m$. The old coordinates t, r, ϑ, and φ then have to be regarded as *singular* (not allowed) since, first

$$\begin{pmatrix} \dfrac{\partial v}{\partial t} & \dfrac{\partial v}{\partial r} \\[2mm] \dfrac{\partial r'}{\partial t} & \dfrac{\partial r'}{\partial r} \end{pmatrix} = \begin{pmatrix} 1 & \frac{r}{r-2m} \\ 0 & 1 \end{pmatrix} \,,$$

that is the matrix of the coordinate transformation is singular at $r = 2m$, and second, the metric is regular in the coordinates $(v, r, \vartheta, \varphi)$ and singular in $(t, r, \vartheta, \varphi)$. The class of regular (allowed) coordinates is determined *uniquely* by the requirement that the metric be regular, provided there exists at least *one* regular coordinate system. This holds in the general case and the proof is not difficult (exercise).

Why is the metric singular in Schwarzschild coordinates? As these coordinates are adapted to the symmetry, something must happen to the symmetry at $r = 2m$. We can easily find out what this is by computing the components of the Killing vector of the time translation. In Eddington–Finkelstein coordinates they are $\xi'^\mu = (1, 0, 0, 0)$. This vector is everywhere non-zero and smooth. But its square, $g_{\mu\nu} \xi^\mu \xi^\nu = (r - 2m)/r$ changes sign when passing through $r = 2m$. That is, the symmetry is time-like outside of $2m$, light-like at $2m$, and space-like below $2m$ (Fig. 6.2). This resembles the boost in the tx-plane where the corresponding generator also changes its signature. Hence the Schwarzschild space–time is not globally static.

The metric has a real singularity at $r = 0$, where also the curvature diverges and the space–time in no longer locally flat.

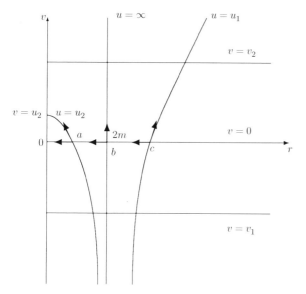

Fig. 6.2 The curves of constant u and v in Eddington–Finkelstein space–time. The form of the light cones is indicated by bold vectors

6.2.4 The Horizon

We can now draw the Eddington–Finkelstein space–time in the v–r diagram (Fig. 6.2). The ranges of the coordinates are $r \in (0, \infty)$ and $v \in (-\infty, \infty)$. The lines $v = \text{const}$ represent ingoing light-like rotationally symmetric hypersurfaces. The future direction along these hypersurfaces is the direction of decreasing r.

To understand what happens at $r = 2m$, we examine the outgoing radial light rays (6.18) near $r = 2m$ by rewriting them in Eddington–Finkelstein coordinates. Substituting t from (6.19) into (6.18) yields

$$v - 2r - 4m \log \left| \frac{r}{2m} - 1 \right| = u . \tag{6.21}$$

Thus, along the curves where $u = \text{const}$, we have that

$$\frac{dr}{dv} = \frac{r - 2m}{2r} . \tag{6.22}$$

This results in three different kinds of outgoing radial light rays (Fig. 6.2):

1. $r > 2m$. Then

$$\frac{dr}{dv} > 0$$

and the radial coordinate r increases in the future direction along the ray. For the "ends" of the ray we find from (6.21) that

$$v \to \infty \qquad r = \infty \qquad r' = \tfrac{1}{2}$$
$$v \to -\infty \qquad r = 2m \qquad r' = 0 .$$

Each ray beginning at a point with $r > 2m$ reaches arbitrarily distant points if v is large enough. Furthermore, it also stays in this region for all values of v in the past.

2. $r = 2m$. It can be checked that this equation, together with $\vartheta = \text{const}$, $\varphi = \text{const}$ defines a light-like geodesic in the Eddington–Finkelstein space–time. It corresponds to the coordinate value $u = \infty$. Thus light is trapped at constant radius $r = 2m$.

3. $r < 2m$. Then

$$\frac{dr}{dv} < 0 ,$$

and the radial coordinate r decreases along the ray in future direction. For the "ends" of the ray we obtain

$$v \to u \qquad r = 0 \qquad r' = -\infty$$
$$v \to -\infty \qquad r = 2m \qquad r' = 0 .$$

Thus, each outgoing ray $u = \text{const}$ which starts at a point with $r < 2m$ falls into the center at $v = u$. It also remains in this region for all values of v in the past.

We can easily show that for no (including non-radial) causal signals the r coordinate can increase faster with v than for the outgoing radial ones. Let such a signal move along a curve given by the functions $v(\lambda)$ $r(\lambda)$, $\vartheta(\lambda)$, and $\varphi(\lambda)$. As it is causal and future directed, its tangential vector $(\dot{v}, \dot{r}, \dot{\vartheta}, \dot{\varphi})$ satisfies

$$\dot{v} \geq 0$$

with $\dot{r} < 0$ if $\dot{v} = 0$. Furthermore,

$$\left(1 - \frac{2m}{r}\right) \dot{v}^2 - 2\dot{v}\dot{r} \geq 0 .$$

In this inequality we only have equality for $\dot{\vartheta} = \dot{\varphi} = 0$. Distinguish the following cases:

1. $\dot{v} = 0$. This is only possible for $\dot{\vartheta} = \dot{\varphi} = 0$, that is on a radial ray.
2. $\dot{v} > 0$. Then we have in the interior, $r < 2m$, and outside, $r > 2m$, from the equation above that

$$\frac{\dot{r}}{\dot{v}} \leq \frac{r - 2m}{2r} ,$$

and the largest increase in r is given by the equality case, which corresponds to a radial ray.

Hence we find that light originating from the region $r \leq 2m$ can never reach points outside this region. The hypersurface $r = 2m$ is an event horizon for the observers outside. They can observe events with $r > 2m$ but cannot see beyond the hypersurface $r = 2m$.

The fact that the r coordinate decreases in the region $r \in (0, 2m)$ along any causal curve means that r is a time coordinate there. This is in agreement with the form of the metric as the coefficient of dr^2 is positive.

The situation near the horizon can be illustrated by an "acoustic analogy" (Fig. 6.3). Imagine two horizontal parallel glass plates. There is a hole in the center of the lower glass plate. From all directions there is a laminar flow of fluid under pressure between the plates. Assume that the speed of the fluid reaches the sound velocity at a certain radius R_s (the so-called *sonic point*) and exceeds sound velocity at smaller radii. Now an acoustic source can be introduced at different points between the plates. If it lies outside of R_s, sound can travel outside. At R_s the sound can only reach the interior of the circle R_s and at all smaller radii almost all the sound directly flows into the hole.

A way of thinking that is widely established today is to consider the event horizon as the surface of an independent physical object with the name *black hole* [6]. This is not a matter of course, as the horizon is only an imaginary surface. However, it proved useful in astrophysics as well as in the theory in general. Numerical simulations of the Einstein equations show that black holes move like ordinary objects in the gravitational field of other objects, for example in binary stars.

Today the existence of black holes is practically undisputed in astrophysics. Very strong candidates are constantly observed. These candidates are often very powerful sources of radiation and energy (galactic nuclei, binary x-ray sources, etc. [1]). This is because the vicinity of the event horizon is a deep potential well. The energy which can be gained by letting things fall into the well is the source of the activity of these objects in the sky. A good candidate for a black hole is an object which can be shown to have a large enough mass with a small enough volume. Different mass limits can be used in this respect.

Observations of the center of our galaxy [7, 8] show that there is a strong point-like source of x-ray radiation, Sagittarius A*, in the center. The region at distance

Fig. 6.3 Acoustic model of a black hole. Spherically symmetric fluid current with velocity field $\vec{V}(r)$ represented by vectors $\vec{V}_1$, $\vec{V}_2$, and $\vec{V}_3$ reaches the sonic point at the cylinder S where $V = V_2 = v_s$, v_s being the sound velocity. The sound is carried along by the fluid. The three swimmers a, b, and c each send out a sound signal with effective velocity v_+ in the outward, and v_- in the inward directions

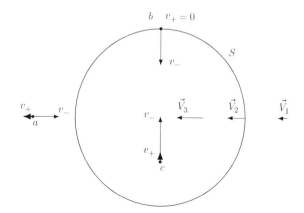

of 3LY is observed since 1992. It was possible to identify single stars there. In particular, the orbital segments of the six closest stars were extrapolated to complete Kepler orbits. These orbits were used to compute the mass of the central object $M \approx 3 \times 10^6 \, M_{Sun}$. For example, the star S2 with mass 15–20 M_{Sun} has an orbital period of ≈ 17 years. Four of these stars have passed their pericenter during the time of observation. These observations lead to an estimate of the object's radius of about $R \leq 500 R_S$ where R_S is the Schwarzschild radius for the mass M. It is difficult to construct an object with that much mass and so small radius which is not a black hole.

6.3 Oppenheimer–Snyder Collapse Model

In this section we study the formation of a horizon during stellar collapse. To simplify the discussion, we consider a very simple model of the star [9]. The high symmetry of the model and the trivial equation of state lead to equations of motion which can be solved explicitly. The main characteristics of this solution hold for more general situations. To study a more realistic model we would need a very powerful computer, but we would find the same qualitative properties.

The space–time of the model consists of three parts: the interior, the exterior, and the surface of the star.

6.3.1 The Interior

We assume that the matter of the star is distributed in a locally homogeneous and isotropic way. The junction condition $p = 0$ at the surface then yields $p = 0$ everywhere. Thus the metric in the interior is part of a Friedmann solution for dust (Fig. 6.3); recall that we assumed $\Lambda = 0$ at the beginning of this chapter. We write the corresponding metric (5.11) in the form

$$ds^2 = dt^2 - a^2(t) \frac{dx^2}{1 - Kx^2} - a^2(t)x^2 \, d\Omega^2 \, .$$

Instead of r we used the letter x to denote the radial coordinate so that r can be reserved for the radius of the 2-sphere $t = \text{const}, x = \text{const}$. Thus

$$r(t) = a(t)x \, . \tag{6.23}$$

The function $a(t)$ satisfies (5.31) with $L = 0$ and $\gamma = 1$:

$$\left(\frac{da}{dt} \right)^2 - \frac{M}{a} + K = 0 \, . \tag{6.24}$$

As we are studying collapse we want that $a' < 0$. In the closed model ($K > 0$ in (5.49) and (5.50)) this corresponds to the second half of the cosmic cycle, $\pi < \sqrt{K}\eta < 2\pi$, and the time reversal ($t \rightarrow -t$) of the $K \leq 0$ solutions ((6.24) is invariant with respect to time reversal). We further assume that the collapse starts at a time where $a' = 0$. Such times only exist when $K > 0$ ($\sqrt{K}\eta = \pi$) or $K = 0$ ($t = -\infty$). Let us focus on the case $K > 0$. The maximal value of $a(t)$ is finite for $K > 0$, we denote it by a_M. From (6.24) we obtain that

$$a_M = \frac{M}{K}. \tag{6.25}$$

We can now re-scale according to (5.12) to achieve $a_M = 1$, that is $M = K$.

We transform the solution (5.49) and (5.50) so that the time t as well as the parameter $\tilde{\eta} = \sqrt{K}\eta - \pi$ vanish at the time when the collapse starts:

$$a = \frac{1 + \cos\tilde{\eta}}{2}, \tag{6.26}$$

$$t = \frac{\tilde{\eta} + \sin\tilde{\eta}}{2\sqrt{K}}. \tag{6.27}$$

6.3.2 The Outside

For the metric outside the star we assume (due to Birkhoff's theorem) the form (6.20). For $r > 2m$ this metric is static with the corresponding Killing vector field ξ^μ.

6.3.3 The Surface

At a star surface the Einstein equations have to be completed by the junction conditions. They are needed to uniquely determine the solution.

If the star is rotationally symmetric, so must be its surface. It forms an interface between the Friedmann and the Schwarzschild parts. In the interior it is generated by the dust trajectories:

$$t = \tau, \quad x = x_o, \quad \vartheta = \vartheta_0, \quad \varphi = \varphi_0,$$

where x_o is a fixed constant, which determines the part of the Friedmann space–time that plays the role of the star, and ϑ_0 and φ_0 assume values in the intervals

$$0 \leq \vartheta_0 < \pi, \quad 0 \leq \varphi_0 < 2\pi.$$

These dust trajectories are time-like geodesics, as $p = 0$ and no force (besides gravity) acts on the dust particles. According to (6.23), the radius $r(t)$ of the sphere $t = \text{const}$ is given by

$$r(t) = a(t)x_o .$$ (6.28)

Equation (6.24) yields the following equation for the function $r(t)$:

$$\frac{\dot{r}^2}{Kx_o^2} - \frac{x_o}{r} + 1 = 0 .$$ (6.29)

Its solution is given by (6.26) and (6.27):

$$r = \frac{x_o}{2}(1 + \cos \tilde{\eta}) ,$$ (6.30)

$$t = \frac{\tilde{\eta} + \sin \tilde{\eta}}{2\sqrt{K}} .$$ (6.31)

The function $r(t)$ determines the geometry of the surface:

$$d_3 s^2 = dt^2 - r^2(t)d\Omega^2 .$$

The junction conditions imply the following. From the outside the surface is also generated by time-like geodesics. Physically, this means that dust particles slightly outside the surface of the star move parallel to the surface. More generally, the trajectories of free-falling particles have to vary smoothly if the limit is taken from above and from below (if the 4-velocity jumps, we have infinite acceleration and need infinite force). Furthermore, the two hypersurfaces have to "fit together", that is the 3-metric on the interior equals the 3-metric on the exterior, otherwise the two surfaces cannot be glued together (Fig. 6.4). Finally, the area of the $x = \text{const}$ sphere has to grow with the same rate in the interior and the exterior when we move perpendicularly to the surface. Let us calculate the form of the surface from the outside.

Let the time-like geodesics generating the surface be given by

$$v = v(t) , \quad r = r(t) , \quad \vartheta = \vartheta_0 , \quad \varphi = \varphi_0 ,$$

where t is the proper time along the geodesic. The functions $v(t)$ and $r(t)$ satisfy the geodesic equations of the metric (6.20). These can be reduced to the following first-order integrals:

$$\left(1 - \frac{2m}{r}\right)\dot{v}^2 - 2\dot{v}\dot{r} = 1 ,$$ (6.32)

$$\left(1 - \frac{2m}{r}\right)\dot{v} - \dot{r} = e .$$ (6.33)

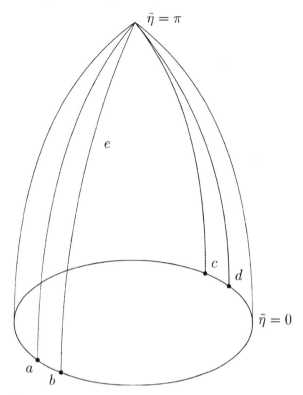

$\tilde{\eta} = \pi$

e

c

d

$\tilde{\eta} = 0$

a

b

Fig. 6.4 Two-dimensional Friedmann model. The part *abe* represents the inside of the star. The pairs *ab* and *cd* of points represent 2-spheres of the same radius

Eliminating $\dot{v}$ from these equations, we obtain the radial equation

$$\dot{r}^2 + 1 - \frac{2m}{r} - e^2 = 0 \, . \tag{6.34}$$

Assume that the radius of the star has a finite value r_M when the collapse begins. At that time also $\dot{r} = 0$. Inserting into (6.34) yields

$$e^2 = 1 - \frac{2m}{r_M} \, ,$$

and thus (6.34) becomes

$$\frac{r_M}{2m} \dot{r}^2 - \frac{r_M}{r} + 1 = 0 \, . \tag{6.35}$$

Equations (6.29) and (6.35) must lead to the same solutions. This implies that

$$K = \frac{2m}{r_M^3} \, , \quad x_o = r_M \, . \tag{6.36}$$

In view of (6.36) we can compute the parameters K and x_0 of the solution in the interior from the parameters m and r_M of the outside. The solution of (6.35) follows from (6.30) and (6.31) by inserting (6.36):

$$r = \frac{1}{2} r_M (1 + \cos \tilde{\eta}),$$ (6.37)

$$t = \frac{1}{2} r_M \sqrt{\frac{r_M}{2m}} (\tilde{\eta} + \sin \tilde{\eta}).$$ (6.38)

Thus the total mass m/G of the star and its initial radius r_M determine the Friedmann metric in the interior (K) and the x-coordinate of the surface in a unique way. However, the same x-coordinate can mean two very different parts of the Friedmann space–time as on the S^3 there are two S^2-surfaces with the same x-coordinate. We have to take the closer one, as the area of the spheres $x = $ const increases at this surface and decreases at the other one when approaching it from the inside. It can be shown that then all junction conditions are satisfied.

This completes our construction. For every pair (m, r_M) we found exactly one model. The metric and the matter in the interior part are uniquely determined by the junction conditions (6.36).

Equations (6.37) and (6.38) lead to the following conclusions about the final state of the collapse. The collapse begins at the value of the parameter $\tilde{\eta} = 0$, where $t = 0$ and $r = r_M$. It ends at $\tilde{\eta} = \pi$, $t_s = \frac{1}{2} \pi r_M \sqrt{r_M/2m}$, and $r = 0$. The end of the collapse is a singularity as the matter density ρ and the curvature become infinite. The surface crosses the horizon when $r = 2m$. The corresponding value $\tilde{\eta}_H$ of the parameter $\tilde{\eta}$ is given by

$$\cos \frac{1}{2} \tilde{\eta}_H = \sqrt{\frac{2m}{r_M}}.$$

In the interior, the metric is regular at that time. In combination with $\tilde{\eta}_H \in (\pi/2, \pi)$ this yields that

$$\tilde{\eta}_H = \pi - 2 \arcsin \sqrt{\frac{2m}{r_M}}.$$

Inserting this into (6.38), we obtain the proper time t_H of an observer at the surface of the star as it crosses the horizon:

$$t_H = r_M \sqrt{\frac{r_M}{2m}} \left(\frac{\pi}{2} - \arcsin \sqrt{\frac{2m}{r_M}} + \sqrt{\frac{2m}{r_M}} \sqrt{1 - \frac{2m}{r_M}} \right).$$

We assume that the star was "normal" before the collapse, that is

$$\frac{2m}{r_M} \ll 1,$$

and expand the arcsin:

$$\arcsin x = x + \frac{1}{6} x^3 + \cdots.$$

Then $t_H \approx t_s - 4m/3$. Here m is in our units the time that light needs to travel the distance m. The final state of the collapse, from $r_M \approx 2m$ to $r = 0$, appears to run at almost double the speed of light. Of course, a difference in the coordinate r is not equal to the corresponding radial distance.

6.3.4 Radial Light-Like Geodesics

Let us first consider radial light-like geodesics in the whole Friedmann space–time. We can later restrict our considerations to the interior of the star. As $x = 0$ consists of the north and south poles of the 3-sphere $t = $ const, the radial light-like geodesics generate the light cones of these points (future or past). It is favorable to introduce other coordinates in place of t and x. Define

$$t = \frac{1}{2\sqrt{K}}(\tilde{\eta} + \sin\tilde{\eta}) , \quad x = \frac{\sin\chi}{\sqrt{K}} .$$

Then the metric becomes

$$ds^2 = \left[\frac{1}{2\sqrt{K}}(1 + \cos\tilde{\eta})\right]^2 (d\tilde{\eta}^2 - d\chi^2 - \sin^2\chi \, d\Omega^2) . \tag{6.39}$$

The whole space–time is covered by the following coordinate ranges for η and χ (Fig. 6.5):

$$-\pi < \tilde{\eta} < \pi , \quad 0 \le \chi \le \pi , \tag{6.40}$$

since the complete 3-sphere $\tilde{\eta} = $ const contains the two hemispheres with $0 \le x \le 1$. Each point $(\tilde{\eta}, \chi)$ in the region (6.40) represents a 2-sphere in space–time. The radius of this sphere can be read off the metric (6.39):

$$r = \frac{1}{2\sqrt{K}}(1 + \cos\tilde{\eta})\sin\chi . \tag{6.41}$$

The right-hand side is positive provided (6.40) holds. On the boundary of the rectangle (6.40) we have $r = 0$. At $\chi = 0, \pi$ this holds, since these points are the poles of the 3-sphere. For $\eta = -\pi, +\pi$ we are in the singularity where the radius of the complete 3-sphere vanishes.

With respect to the new coordinates, the radial light-like geodesics are given by the four functions $\tilde{\eta}(\lambda)$, $\chi(\lambda)$, $\vartheta(\lambda)$, and $\varphi(\lambda)$, which must satisfy the relations

$$\left[\frac{1}{2\sqrt{K}}(1 + \cos\tilde{\eta})\right]^2 (\dot{\tilde{\eta}}^2 - \dot{\chi}^2) = 0 , \quad \vartheta(\lambda) = \vartheta_0 , \quad \varphi(\lambda) = \varphi_0 .$$

It follows that

$$\chi = \pm(\tilde{\eta} - \tilde{\eta}_0) . \tag{6.42}$$

Fig. 6.5 Penrose diagram of
a closed Friedmann model.
The inside of the star is given
by $\tilde{\eta} \in (0, \pi)$ and $\chi \in (0, \chi_0)$.
The points of turn around of
outgoing null hypersurfaces
fill the *dashed line*. The
surfaces H and S separate
types 1, 3, and 4

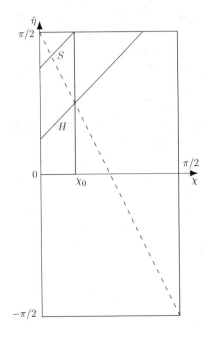

We do not need the affine parameter, and thus this result is sufficient for our purposes. In the rectangle (6.40) the (+)-solutions (6.42) determine radial outgoing
(with respect to the north pole) light-like hypersurfaces. When they reach a point
with $\chi = 0$, they can be interpreted as the future light cone of that point, and similarly at a point with $\chi = \pi$ as its past light cone. All geodesics which start at the north
pole meet at the south pole (as soon as $\tilde{\eta}_0 < 0$). Analogously for the (−)-solutions.
We see that the light needs the $\tilde{\eta}$-interval 2π to travel around the 3-sphere—this is
the complete cosmic cycle.

A hypersurface (6.42) starts at $r = 0$, then expands, that is the radius $r(\tilde{\eta})$ of its
intersections with the hypersurfaces $\tilde{\eta} = $ const increases, reaches a maximum—the
turning point—and then contracts back to $r = 0$. Let us determine the turning points.
To this end we substitute (6.42) for χ into (6.41):

$$r(\tilde{\eta}) = \pm \frac{1}{2\sqrt{K}}(1 + \cos\tilde{\eta})\sin(\tilde{\eta} - \tilde{\eta}_0) \, .$$

The turning points are determined by the equation

$$r'(\tilde{\eta}_u) = 0 \, ,$$

or

$$\cos(2\tilde{\eta} - \tilde{\eta}_0) = -\cos(\tilde{\eta} - \tilde{\eta}_0) \, .$$

The solutions in the interior of (6.40) are

$$\tilde{\eta}_u = \frac{2}{3}\tilde{\eta}_0 + \frac{1}{3}(2n+1)\pi , \tag{6.43}$$

where n must be an integer. Equation (6.42) then implies that

$$\chi_u = \pm(\tilde{\eta}_u - \tilde{\eta}_0) . \tag{6.44}$$

Eliminating η_0 from (6.43) and (6.44), we obtain the equation for the turning points. The turning points for outgoing hypersurfaces are

$$\tilde{\eta}_u^+ = -2(\chi_u - \frac{1}{2}\pi) , \tag{6.45}$$

and for ingoing ones

$$\tilde{\eta}_u^- = 2(\chi_u - \frac{1}{2}\pi) . \tag{6.46}$$

The interior of the star is given by the following rectangle in (6.40)

$$0 \le \tilde{\eta} \le \pi , \quad 0 \le \chi \le \chi_o , \tag{6.47}$$

where

$$\chi_o = \arcsin(\sqrt{K}x_o) .$$

Obviously, we have to choose

$$\chi_o < \frac{\pi}{2} ,$$

as otherwise the derivatives of the metric are not continuous at the surface. Then only the curve (6.45) intersects the rectangle (6.47). This means that the ingoing light-like hypersurfaces contract everywhere in the interior of the star. On the other hand the behavior of the outgoing light-like hypersurfaces it not so simple. There are four cases, depending on the time at which the hypersurface starts at the north pole $\chi = 0$ (Fig. 6.5).

1. $\tilde{\eta}_0 < \pi - 3\chi_o$. These hypersurfaces expand and reach the surface of the star with $r' > 0$. They connect smoothly to the $u = \text{const}$ (outgoing) hypersurfaces outside of the horizon.
2. $\tilde{\eta}_0 = \pi - 3\chi_o$. This hypersurface expands in the interior of the star and reaches its surface with $r' = 0$ and $r = 2m$. It connects smoothly to the horizon $r = 2m$ outside of the star.
3. $\pi - 3\chi_o < \tilde{\eta}_0 < \pi - \chi_o$. These hypersurfaces expand, reach their points of turn around inside the star and subsequently contract. Still contracting, they reach the surface with $r < 2m$. They connect smoothly to the contracting $u = \text{const}$ hypersurfaces on the outside.
4. $\tilde{\eta}_0 \ge \pi - \chi_o$. These hypersurfaces are inside of the star as in 3., but they do not reach the surface of the star. Instead, they fall into the singularity $\eta = \pi$.

We now see how the horizon forms during the collapse. It is the hypersurface of type 2 (Fig. 6.5). It starts as an ordinary light cone in the center of the star at time $\tilde{\eta}_0 = \pi - 3\chi_o$, emerges divergence free (that is with $r' = 0$) at the surface of the star, and connects smoothly to the hypersurface $r = 2m$ in the Eddington–Finkelstein space–time. The decrease in the divergence of the light rays which form the light cone is caused by the matter which crosses the light cone (this is shown by the so-called Raychaudhuri equation [4]). No event beyond this hypersurface can be observed from the outside. Thus this hypersurface is the *event horizon*. Note in particular that the singularity lies beyond the event horizon and thus cannot influence any events outside the horizon and cannot be observed from the outside either.

We shall now examine how the collapse looks like for an outside observer. Consider a light source at the surface of the star. Its trajectory is given by the functions $v(t)$ and $r(t)$ which are determined by (6.32) and (6.33). In particular, t is the proper time of the source.

The proper time of the asymptotic observer agrees with the retarded time u up to an additive constant. The light which is emitted by the source travels along the hypersurfaces $u = $ const until it reaches the asymptotic observer. The relation between the time of arrival at the asymptotic observer and the time of emission at the source is thus given by the function $u(t)$. It results by inserting the functions $v(t)$ and $r(t)$ into (6.21) for the variables v and r.

Along the complete trajectory we have

$$\dot{r} < 0, \quad \dot{v} > 0, \quad e > 0.$$

Thus, (6.34) implies that

$$\dot{r} = -\sqrt{e^2 - \left(1 - \frac{2m}{r}\right)}, \tag{6.48}$$

$$\dot{v} = \frac{e + \dot{r}}{1 - \frac{2m}{r}}. \tag{6.49}$$

From (6.48) and (6.49) we calculate the values of the derivatives at the horizon $r = 2m$:

$$\dot{r}_H = -e, \quad \dot{v}_H = \frac{1}{2e},$$

since

$$\lim_{x \to 0} \frac{e - \sqrt{e^2 - x}}{x} = \frac{1}{2e}.$$

Thus we have

$$r \approx 2m - e(t - t_H) + \cdots,$$

and

$$v \approx v_H + \frac{1}{2e}(t - t_H) + \cdots.$$

If we insert this in u,

$$u = v - 2r - 4m \ln \left| \frac{r}{2m} - 1 \right| ,$$

it results in

$$u \approx -4m \ln |t - t_H| + \left(v_H - 4m - 4m \ln \frac{e}{2m} \right) + \left(2e + \frac{1}{2e} \right) (t - t_H) .$$

The leading term on the right-hand side is *independent* of the trajectory of the ingoing light source, only the correction terms depend on it (via the constant e). The redshift is given by $z = \dot{u} - 1$, and thus diverges when the surface of the star approaches the horizon.

We can draw the following conclusions. The asymptotic observer will never see the fall of the star through the horizon. The light signal which carries this information will not reach him in finite time! Even before that he will no longer be able to receive light signals from the surface of the star due to the increasing redshift, which dims the signal. In summary, we obtained the following important properties of our model:

1. The final state of the collapse is a singularity.
2. A regular horizon forms. It hides the singularity from an outside observer and limits the influence the singularity has on the rest of the space–time.
3. The redshift diverges and the movement of the star freezes for the asymptotic observer when the surface of the star approaches the horizon.
4. The fall through the horizon is not observable from the outside.

We elaborate on the first point. The singularity represents infinite density and curvature. Furthermore, it formed from regular initial data. This yields an intrinsic contradiction in general relativity: regular initial data, together with the Einstein equations, that is certain assumptions of the theory, imply a state, which violates other assumptions of the theory—the singular points do not have a nearly flat neighborhood.

It is important to emphasize that we mean a complete collapse here (complete means that it reaches the horizon). A more realistic equation of state can stop the collapse in one of its phases in certain special situations. However, if the collapse reaches the horizon, there is nothing that can prevent further collapse.

We should also understand the main difference to, say, classical electrodynamics. It could be objected that point charges in electrodynamics also form singular fields. However, in classical electrodynamics the charges are modeled as continuous distributions. To form a point charge thus requires an infinite amount of energy since charges of the same sign repel each other. Consequently, there are no point charges in classical electrodynamics. In the theory of gravity, the same "charges", that is mass, attract each other. They can spontaneously form a point mass, that is a singularity. It is believed that quantum theory might be able remove this problem.

These properties remain valid for far more general models of gravitational collapse [1] and are considered universal. A generalization of point (2) to an arbitrary complete gravitational collapse is called the *cosmic censorship hypothesis* ("naked" singularities are not allowed). In [4], this is called "future asymptotic predictability".

6.4 Exercises

1. Let $(\mathcal{M}, g)$ be a static, rotationally symmetric space–time with the metric

$$ds^2 = B(r)dt^2 - A(r)dr^2 - r^2 d\vartheta^2 - r^2 \sin^2 \vartheta \, d\phi^2 .$$

   Furthermore, let $W_{\mu\nu}(x)$ be an arbitrary tensor of type $(0,2)$ on $\mathcal{M}$.
   Prove that if $W_{\mu\nu}(x)$ has the same symmetries as $g_{\mu\nu}$, then it must have the form
   $W_{\mu\nu}(x) = 0$ for all $\mu \neq \nu$ and all x, and $W_{\vartheta\vartheta} \sin^2 \vartheta = W_{\phi\phi}$. Hint: work with
   components with respect to a suitable orthonormal basis. Rotate or reflect the
   basis and examine how the components change under these operations.
2. For the spherically symmetric metric, compute the Christoffel symbols $\{^{\mu}_{\rho\sigma}\}$ and
   then use the above result to calculate all components of the Ricci tensor $R_{\mu\nu}$.
3. Use the results from Exercise 1 to write down the Einstein equations for the
   functions $A(r)$, $B(r)$, $\rho(r)$, and $p(r)$.
4. Show that the following two systems of equations are equivalent:

   (1) the tt-, rr-, and $\vartheta\vartheta$-components of the Einstein equations and
   (2) the tt-, rr-components of the Einstein equations and the Euler equation.

5. Transform the metric of the exterior Schwarzschild space–time to the retarded
   Eddington–Finkelstein coordinates u, r, ϑ, and φ, and construct the corresponding
   extension (retarded Eddington–Finkelstein space–time) of exterior Schwarzschild
   space–time in analogy to our extension in the advanced Eddington–Finkelstein
   coordinates v, r, ϑ, and φ. Which kind of horizon can be found in the retarded
   Eddington–Finkelstein space–time?
6. Compute the density ρ_H of the star in the Oppenheimer–Snyder collapse model
   at the instant when the surface of the star crosses the horizon. The total mass m
   and the initial radius r_M are given. Then insert $m = m_{Galaxy}$ and $m = m_{Sun}$ for the
   mass and transform the resulting values to kg m^{-3}.
7. Compute the tidal forces (in N/kgm) which act on a co-moving observer on the
   surface of the star. Insert the values from Exercise 6 for the mass. When does it
   start to get uncomfortable?
   Hint: assume that the observer is in the Schwarzschild part of the space–time.
   First compute for an observer at rest ($r = $ const), and show that in the special
   case of the Schwarzschild space–time the tidal forces are independent of the
   radial velocity of the observer.

References

1. B. Schutz, *Gravity from the Ground up*, Cambridge University Press, Cambridge, UK, 2003.
2. Beig, R., Schmidt B. G.: *Einstein's Field Equations and their Physical Implications*. Lect. Notes Phys. **540**, Springer, Berlin (2000).
3. J. R. Oppenheimer and G. M. Volkov, Phys. Rev. **55** (1939) 374.

4. S. W. Hawking and G. F. R. Ellis, *The Large Scale Structure of Space-Time*, Cambridge University Press, Cambridge, UK, 1973.
5. A. S. Eddington, Nature **113** (1924) 192; D. Finkelstein, Phys. Rev. **110** (1958) 965.
6. C. W. Misner, K. S. Thorne and J. A. Wheeler, em Gravitation, Freeman, San Francisco, CA, 1973.
7. R. Schödel et al., Astrophys. J. **596** (2003) 1015.
8. A. M. Ghez et al., arXiv astro-ph/ 0306130.
9. J. R. Oppenheimer and H. Snyder, Phys. Rev. **56** (1939) 455.

Chapter 7
Stationary Black Holes

In the previous chapter we studied two space–times containing black holes. First we inspected the advanced Eddington–Finkelstein space–time and second the Oppenheimer–Snyder space–time. This background now allows us to define a black hole in a more rigorous way than before (cf. p. 223):

Definition 18 Let $(\mathcal{M}, g)$ be an asymptotically flat space–time and let $\mathcal{M}_\infty$ be an asymptotically flat region in $\mathcal{M}$. Let H be a light-like hypersurface in $\mathcal{M}$ which separates $\mathcal{M}$ into two regions $\mathcal{M}_+$ and $\mathcal{M}_-$ such that $\mathcal{M}_\infty \subset \mathcal{M}_+$. If every event $p \in \mathcal{M}_+$ is visible for an observer in $\mathcal{M}_\infty$, but no event $q \in \mathcal{M}_-$, then H is called *absolute event horizon* with respect to $\mathcal{M}_\infty$ and $\mathcal{M}_-$ is called *isolated black hole*.

We shall now study black holes satisfying the additional condition of stationarity (for a completely rigorous theory, see [1]). The black hole which emerges from the collapsing star has begun as the light cone of an event in the center of the collapsing star and is therefore not stationary. One example for a stationary black hole is the one in the advanced Eddington–Finkelstein space–time, as one of the Killing vector fields is light-like at the horizon. We will elaborate on this notion and shall need some more knowledge about hypersurfaces.

7.1 Hypersurfaces

Sometimes it is advantageous to define hypersurfaces in a different way than by an equation $u = 0$. For instance, the latter definition produced only oriented hypersurfaces and is therefore less general. We investigate an alternative way of defining a hypersurface and describe some important properties of hypersurfaces in space–time.

Hájíček, P.: *Stationary Black Holes*. Lect. Notes Phys. **750**, 237–274 (2008)
DOI 10.1007/978-3-540-78659-7_7   © Springer-Verlag Berlin Heidelberg 2008

7.1.1 Definition

Let $(\mathcal{M}, g)$ be a space–time and $\bar{S}$ a three-dimensional manifold. A smooth hyper-surface $S = \iota(\bar{S})$ in $\mathcal{M}$ is defined by a map $\iota : \bar{S} \mapsto \mathcal{M}$ with the following properties: (a) ι has an inverse ι^{-1} on S and (b) if $\{y^k\}$ are coordinates in a neighborhood of $p \in \bar{S}$ in $\mathcal{M}$, then the map ι can be represented by the embedding functions $x^\mu(y^1, y^2, y^3)$ and we require that the (3×4)-matrix $\partial x^\mu / \partial y^k$ be of rank 3. We can regard y^1, y^2, and y^3 as functions on S and hence as coordinates there (Fig. 7.1).

7.1.2 Tangential Vectors

A vector tangent to S can be defined as a tangent vector of $\mathcal{M}$ which is the tangent of a curve contained in S. The four functions $x^\mu(\lambda, y^2, y^3)$ of a variable λ, where y^2 and y^3 remain fixed, define such a curve in S. Similar reasoning for y^2 and y^3 yields that the three vectors with components

$$\left(\frac{\partial x^0}{\partial y^k}, \ldots, \frac{\partial x^3}{\partial y^k} \right), \quad k = 1, 2, 3,$$

are three tangent vectors of S in each point. Hence condition (b) means that the three vectors $\partial x^\mu / \partial y^k$, $k = 1, 2, 3$, are linearly independent everywhere on S.

Each curve C in S determines a curve $\bar{C}$ in $\bar{S}$ by $\bar{C} := \iota^{-1} \circ C$. Then

$$x^\mu(\lambda) = x^\mu(y(\lambda)),$$

where $x^\mu(\lambda) = h \circ C$ and $y^k(\lambda) = \bar{h} \circ \iota^{-1} \circ C$. Hence each tangent vector of S can be written as a linear combination of $\partial x^\mu / \partial y^k$, $k = 1, 2, 3$, since

$$\dot{x}^\mu(\lambda) = \frac{\partial x^\mu}{\partial y^k} \dot{y}^k(\lambda).$$

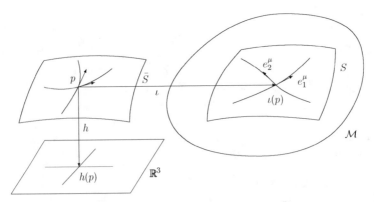

Fig. 7.1 The embedding of the three-dimensional manifold $\bar{S}$ in the four-dimensional manifold $\mathcal{M}$, the coordinate curves, and their tangential vectors $e_k^\mu = \partial x^\mu / \partial y^k$

Thus the vectors $\partial x^\mu / \partial y^k$, $k = 1, 2, 3$, form a basis for the space of tangential vectors to the hypersurface in the corresponding point (Fig. 7.1). Consequently, each tangent vector u^μ of S in $p \in S$ can be written as

$$u^\mu = \bar{u}^k \frac{\partial x^\mu}{\partial y^k} ,$$

where $\bar{u}^k$ can be considered as components of u with respect to the coordinates $\{y^k\}$.

7.1.3 Induced Metric

The metric of $\mathscr{M}$ can be projected to the following 2-tensor on S:

$$\gamma_{kl} := g_{\mu\nu} \frac{\partial x^\mu}{\partial y^k} \frac{\partial x^\nu}{\partial y^l} .$$

γ_{kl} is called the *metric induced by $g_{\mu\nu}$ on S*. We observe that the components of γ_{kl} are scalar products of the basis vectors $\frac{\partial x^\mu}{\partial y^k}$. In general, if e_k^μ is an arbitrary basis of the tangential space, then the induced metric has the components $\gamma_{kl} = g_{\mu\nu} e_k^\mu e_l^\nu$ with respect to this basis. As we will see later, γ_{kl} need not be a metric in our sense (non-degenerate). We thus want to extend our notion of a metric slightly to also allow *degenerate* metrics.

7.1.4 Normal

A normal N^μ of S is every non-vanishing vector which satisfies the equation $g_{\mu\nu} N^\mu \frac{\partial x^\nu}{\partial y^k} = 0$ for all k. That is, it is orthogonal to all tangent vectors of S. There are many such vectors in each point $p \in S$, but they are all contained in a one-dimensional sub-space of $T_p\mathscr{M}$. The set $N_p S$ of all possible normals in this point is this sub-space without the origin, that is a set with two components.

Let N^μ be a normal and let t^μ be a vector at $p \in S$ which satisfies the equation $g_{\mu\nu} N^\mu t^\nu = 0$. Then t^μ is tangential to S.

Proof Such a vector satisfies the equation $N_\mu t^\mu = 0$. This is a non-trivial linear equation and the solutions of this equation form a three-dimensional sub-space of $T_p\mathscr{M}$. The tangent vectors also satisfy this equation and they span a three-dimensional sub-space. Hence these two sub-spaces must agree.

7.1.5 Classification of Hypersurfaces

Depending on the signature of its normal, we classify a hypersurface to be either *time-like*, *space-like*, or *light-like*. This corresponds to the condition that N^μ is, in that order, space-like, time-like, or light-like everywhere on S. We will also use the notion of non-time-like, which corresponds to a normal which is non-space-like everywhere.

What are the consequences of this definition for the signature of the induced metric and the relative position of the hypersurface with respect to the local light cones? To answer these questions, we choose a local inertial frame x^μ at p. The metric at p then has the form $g_{\mu\nu}(p) = \eta_{\mu\nu}$. We denote the basis at p which is associated to the coordinates by (e_0, e_1, e_2, e_3). These vectors are given by their components with respect to x^μ by $e_\alpha^\mu := \delta_\alpha^\mu$. The tangent space $T_p\mathcal{M}$ equipped with the metric $g_{\mu\nu}(p)$ on $T_p\mathcal{M}$ is Minkowski space $(T_p\mathcal{M}, g_{\mu\nu}(p))$.

The light cone $L_p\mathcal{M}$ of the point p in this Minkowski space–time plays an important role. The tangent vectors of all causal curves through p must lie in this light cone, or more precisely in the future light cone $L_p^+\mathcal{M}$. The light cone is bounded by vectors of the form

$$l\left(e_0^\mu + n^k(\vartheta, \varphi)e_k^\mu\right),$$

where $l \in \mathbb{R}$ and $n^k(\vartheta, \varphi)$ is given by (5.5). The boundary of the future light cone is given by $l \in (0, \infty)$.

Assume first that the hypersurface is time-like. Then the local inertial frame can be chosen so that $N^\mu = (0, 0, 0, n)$ and $N_\mu = (0, 0, 0, -n)$, where n is a number. Hence, the equation $N_\mu t^\mu = 0$ becomes $t^3 = 0$, and all solutions thereof have the form $t^0 e_0^\mu + t^1 e_1^\mu + t^2 e_2^\mu$, with $(t^0, t^1, t^2) \in \mathbb{R}^3$. They form the tangent space T_pS for S at p. Thus T_pS is a time-like plane in $T_p\mathcal{M}$, and e_0^μ, e_1^μ and e_2^μ are three orthonormal vectors tangential to S at p. The induced metric can then be transformed to $\mathrm{diag}(1, -1, -1)$ and hence has signature -1.

What is the relative position of T_pS with respect to $L_p\mathcal{M}$ or $L_p^+\mathcal{M}$, respectively? The plane T_pS intersects $L_p\mathcal{M}$ so that the common vectors satisfy

$$l e_0^\mu + l n^k(\vartheta, \varphi)e_k^\mu = t^0 e_0^\mu + t^1 e_1^\mu + t^2 e_2^\mu,$$

that is $n^3(\vartheta, \varphi) = 0$ or $\vartheta = \pi/2$, $t^0 = l$, $t^1 = l\cos\varphi$, and $t^2 = l\sin\varphi$. The component of the vectors in the future light cone with respect to the normal $N^\mu = e_3^\mu$ is $l\cos\vartheta e_3^\mu$. As $l > 0$, this component points into the direction of N^μ if $\vartheta < \pi/2$. Hence, there are curves which cross S at p in the direction of the normal N^μ, and also in the opposite direction. That is, a time-like hypersurface can be crossed by future-oriented causal curves in both directions.

For a space-like hypersurface we proceed analogously. This leads to three space-like orthonormal tangent vectors and to the signature -3. We thus have that T_pS is spanned by the vectors of the form $t^1 e_1^\mu + t^2 e_2^\mu + t^3 e_3^\mu$ and the normal is of the form $N^\mu = e_0^\mu$. The equation

$$l e_0^\mu + l n^k(\vartheta, \varphi)e_k^\mu = t^1 e_1^\mu + t^2 e_2^\mu + t^3 e_3^\mu$$

now has the only solution $l = 0$ and $t^1 = t^2 = t^3 = 0$. A space-like hypersurface separates the light cone $L_p\mathcal{M}$ into its future and past components. The normal $N^\mu = e_0^\mu$ is chosen such that it lies in the future half of the light cone. Thus, particles and light can cross a space-like hypersurface only in one direction, namely in the direction of the future-oriented normal.

Finally, for a light-like hypersurface we can choose the local inertial frame so that $N^\mu = n(1,0,0,1)$ and $N_\mu = n(1,0,0,-1)$. Then the equation $N_\mu t^\mu = 0$ yields that $t^0 - t^3 = 0$. All solutions to this equations are given by $t^0(e_0^\mu + e_3^\mu) + t^1 e_1^\mu + t^2 e_2^\mu$ with $(t^0, t^1, t^2) \in \mathbb{R}^3$. These define a light-like hyperplane in $T_p\mathcal{M}$ such that $e_0^\mu + e_3^\mu$, e_1^μ, and e_2^μ are three independent tangent vectors. They are mutually orthogonal and the square of their norm is, in that order, $0, -1, -1$. Hence the induced metric can be written in the form $\mathrm{diag}(0,-1,-1)$. It has no well-defined signature and is degenerate. The tangent vector $e_0^\mu + e_3^\mu$ is orthogonal to each other tangent vector, including itself: it is the normal! Thus the normal is simultaneously tangential to S and annihilates the induced metric $\gamma_{kl}N^k = 0$. It is furthermore the only non-space-like direction tangential to a light-like hypersurface.

Again, we consider the equation

$$l e_0^\mu + l n^k(\vartheta, \varphi) e_k^\mu = t^0(e_0^\mu + e_3^\mu) + t^1 e_1^\mu + t^2 e_2^\mu.$$

It follows that $l = t^0$ and $n^3 = 1$, that is $\vartheta = 0$. Hence there is a single common direction in T_pS and $L_p\mathcal{M}$, namely $t^0(e_0^\mu + e_3^\mu) = t^0 N^\mu$, that is, the normal. The plane T_pS touches $L_p\mathcal{M}$ in this direction. Thus the future half of the light cone lies on one side of T_pS, and again future-oriented causal curves can cross S at p only in one direction.

The type of the hypersurface can also be recognized by looking at the form of the induced metric, as there are no other possibilities besides the three discussed above. In particular $\det(\gamma_{kl}) > 0$ if the hypersurface is time-like, $\det(\gamma_{kl}) < 0$ if it is space-like, and $\det(\gamma_{kl}) = 0$ if it is light-like.

We have proved the following important theorem:

Theorem 18 *Causal, future-directed curves can cross a time-like hypersurface in both directions, but a space-like or light-like hypersurface only in one direction at each point.*

Now we can define a stationary black hole.

Definition 19 Let $(\mathcal{M}, g)$ be an asymptotically flat space–time with an absolute event horizon H. The isolated black hole is called *stationary* if there exists a Killing vector field in $\mathcal{M}$ which is everywhere normal to the horizon H.

Examples: 1. advanced Eddington–Finkelstein space–times contain a stationary black hole according to our definition. 2. Cosmological horizons in de Sitter space–time are not horizons of isolated stationary black holes, although there exist Killing vector fields which are orthogonal to these horizons. The problem is that the space–time is not asymptotically flat.

7.2 Rotating Charged Black Holes

Consider an object in an asymptotically flat space–time which experienced a complete gravitational collapse. The hypothesis of cosmic censorship implies that in the last phase of the collapse a black hole forms. In the rotationally symmetric case the space–time outside the black hole will not necessarily be described by the Schwarzschild solution, as we expect strong, but rotationally symmetric radiation around the star and the horizon (in this case there is no gravitational radiation). After some time, the radiation disperses as it partly falls into the black hole or is radiated to infinity. Hence, a rotationally symmetric vacuum solution will be a very good approximation for the last stage of collapse. By the Birkhoff theorem, this is the Schwarzschild metric. Thus, the final state of the collapse will be the Schwarzschild black hole.

In the non-symmetric case the geometry outside the object should be time dependent and sensitive to the details of the collapse. We thus expect that large quantities of energy can be radiated away. In particular there may be gravitational radiation. Physical intuition tells us that after a long enough period of time the storm will fade and metric and matter approach a stationary state. The matter and radiation present will either be swallowed by the black hole or escape to infinity. We thus expect that a *stationary* vacuum solution or a *stationary* electro-vacuum solution, in the case that the collapsing object carries a strong charge, describes the space–time long after the collapse in a satisfactory way. These expectations are supported by model simulations. This leads us to the following basic assumption for the gravitational collapse [1]:

> In the final phase of the gravitational collapse of an isolated object the space–time is a part of an electro-vacuum solution up to an arbitrarily small error. This solution is stationary, asymptotically flat, contains a smooth event horizon, and is smooth between the horizon and the asymptotic regime.

An electro-vacuum solution is a solution of the equations:

$$G_{\mu\nu} = 8\pi G T_{\mu\nu} \,,$$
$$\nabla_\rho F^{\mu\rho} = 0 \,,$$
$$F_{\mu\nu} = \partial_\mu A_\nu - \partial_\nu A_\mu \,,$$
$$T_{\mu\nu} = -\frac{1}{4\pi} \left(F_{\mu\rho} F_\nu^\rho - \frac{1}{4} g_{\mu\nu} F_{\rho\sigma} F^{\rho\sigma} \right) \,.$$

Here, we will not discuss how such solutions are found, and how the solutions described in this chapter have been discovered, as this is much more complicated as our search for the cosmological solutions or the Schwarzschild solution. For us it will be sufficient to know that all solutions of the Einstein–Maxwell equations with the properties stated above have been identified. They form the so-called *Kerr–Newman family*. To show that these are really all such solutions is the content of the so-called *uniqueness theorems* [2]. These theorems are difficult to prove and we will thus assume their assertions here.

The uniqueness theorems are remarkable in two different ways. First, in the above sense, we know everything about the stationary black holes. Second, there are a lot (an infinite dimensional family) of stationary electro-vacuum solutions which are asymptotically flat. The Kerr–Newman family forms a subset of measure zero (that is a very small set). Similarly, all solutions that contain completely collapsing stars form an infinite dimensional family. However, each of these converges to a Kerr–Newman solution. If that is true (a mathematically rigorous proof is not yet completed, but there is evidence from model simulations) and if we assume that the result of the collapse is a black hole whose state is determined by the region outside the horizon, then in the collapse a lot of information is lost.

7.2.1 First Look at Kerr–Newman Space–Time

7.2.1.1 Metric and Symmetry

The metric of a rotating, charged black hole in equilibrium is

$$ds^2 = \frac{\Delta}{\Sigma} \left(dt - a \sin^2 \vartheta \, d\varphi \right)^2$$
$$- \frac{\sin^2 \vartheta}{\Sigma} \left[a \, dt - \left(r^2 + a^2 \right) d\varphi \right]^2 - \frac{\Sigma}{\Delta} dr^2 - \Sigma d\vartheta^2 , \tag{7.1}$$

and the electromagnetic potential is

$$A_\mu dx^\mu = \frac{Qr}{\Sigma} \left(dt - a \sin^2 \vartheta \, d\varphi \right) \tag{7.2}$$

[3]. This so-called quasi-diagonal form is favorable for many calculations (exercise). The coordinates t, φ, r, and ϑ used here are called *generalized Boyer–Lindquist coordinates* (the order of the coordinates is different from the conventions used previously: $x^0 = t$, $x^1 = \varphi$, $x^2 = r$, and $x^3 = \vartheta$). The following abbreviations are frequently used:

$$\Sigma = r^2 + a^2 \cos^2 \vartheta , \tag{7.3}$$

$$\Delta = r^2 - 2Mr + a^2 + Q^2 . \tag{7.4}$$

The symbols M, a, and Q denote constants with dimensions of length.

The Kerr–Newman family is only three dimensional, it only has the three independent parameters M, a, and Q. This is the remarkable simplicity of the structure of black holes in equilibrium ("black holes have no hair", see [4], p. 876). We immediately see that setting $a = Q = 0$ leads to $A_\mu = 0$ and the metric (7.1) becomes the Schwarzschild metric. The Schwarzschild horizon is at the solution $r = 2M$ of the equation $\Delta = 0$. There are further interesting sub-families: $Q = 0$ yields the so-called Kerr metric and $a = 0$ the Reissner–Nordström metric. The Reissner–Nordström family ($a = 0$) is rotationally symmetric and static, the orbits of the rotation group are given by the equations $t = $ const and $r = $ const. The coordinate r then has the known relation to the area of the orbits which only allows $r \geq 0$.

The general Kerr–Newman metric and the potential are independent of the coordinates t and φ. Thus the vector fields $(1,0,0,0)$ and $(0,1,0,0)$ are Killing vector fields. We denote them by ξ^μ and φ^μ, respectively. There are two additional discrete symmetries: the reflection at the equatorial plane $\vartheta \mapsto -\vartheta + \pi$, and the inversion

$(t, \varphi) \mapsto (-t, -\varphi)$. If $a \neq 0$, then there are no other Killing vector fields and the metric is only axi-symmetric. (The two symmetry axes are given by $\vartheta = 0, \pi$.) The coordinate r does not have a simple geometric meaning in this case and can also be negative without creating a contradiction. The hypersurfaces $r = \text{const}$ are topologically cylinders $\mathbb{R} \times S^2$.

7.2.1.2 Asymptotic Properties

An important property of the metric (7.1) is that it is asymptotically flat both at $r \to \infty$ and (if $a \neq 0$) at $r \to -\infty$. Indeed, if we expand the components of the metric with respect to powers of r^{-1}, we obtain

$$g_{tt} = 1 - 2M \frac{1}{r} + O\left(r^{-2}\right) ,$$

$$g_{t\varphi} = \left(2Ma \sin^2 \vartheta\right) \frac{1}{r} + O\left(r^{-2}\right) ,$$

$$g_{\varphi\varphi} = -r^2 \sin^2 \vartheta \left(1 + O\left(r^{-2}\right)\right) ,$$

$$g_{rr} = -\left[1 + 2M\frac{1}{r} + O\left(r^{-2}\right)\right] ,$$

$$g_{\vartheta\vartheta} = -r^2 \left(1 + O\left(r^{-2}\right)\right) .$$

The last three terms show that r is approximately the usual radial coordinate in the asymptotic regime. The potential (7.2) has the expansion:

$$A_\mu = \left(Q\frac{1}{r}, -aQ \sin^2 \vartheta \frac{1}{r}, 0, 0\right) + O\left(r^{-2}\right).$$

Observe that the coefficient of the $1/r$-term in the electrostatic potential A_0 is Q. From electrodynamics in flat space–time we know that a similar expansion would correspond to a source with charge $Q_{\text{elstat}} = Q \times c^2 \times G^{-1/2}$. Thus the same holds for the asymptotic region of asymptotically flat space–times. Hence, our source has also this charge and that is the meaning of the parameter Q. An analogous relation for the gravitation is given by (4.67).

Transforming the coordinates x^1, x^2, and x^3 into spherical coordinates and setting $J^k = J\delta_3^k$ at the same time, we find that in the region $r \to +\infty$ the metric corresponds to a source of mass M, with angular momentum $J = aM$ and charge Q. The asymptotic observers on the other side, $r \to -\infty$, see an object of mass $-M$, angular momentum $J = -aM$, and charge $-Q$ (exercise). Recall that we chose units so that $G = c = 1$. Then all quantities can be measured in dimensions of length and its powers.

7.2.1.3 Singularities of the Metric

We infer that we are dealing with a rotating, charged source. What kind of source? The singularities of the metric (7.1) can shed light on this. Note that the metric becomes singular when $\Sigma = 0$ or $\Delta = 0$. Is this a true singularity or the effect of a bad choice of coordinates (similar to the Schwarzschild metric at the horizon)? This can be seen from the components of the curvature tensor and its corresponding invariants (which are preferably calculated using a computer). They only diverge at $\Sigma = 0$. We postpone the examination of the points with $\Delta = 0$ to the next section. Thus consider $\Sigma = 0$. If $a = 0$ then $r = 0$ is similar to the corresponding location in the Schwarzschild metric. For $a \neq 0$ we have an interesting situation: the singularity is at $(r = 0, \vartheta = \pi/2)$ but for $r = 0$ and $\vartheta \neq \pi/2$ both the metric and the potential are regular.

Consider the surface $r = t = 0$; its metric is

$$ds^2 = -a^2 \cos^2 \vartheta \, d\vartheta^2 - \tan^2 \vartheta (a^2 \cos^2 \vartheta - Q^2 \sin^2 \vartheta) d\varphi^2.$$

This is the metric of two isometric discs, the first given by $\varphi \in [0, 2\pi], \vartheta \in [0, \pi/2)$ and the second by $\varphi \in [0, 2\pi], \vartheta \in (\pi/2, \pi]$. Both are singular at the boundary $\vartheta = \pi/2$. The singularity has the form of a ring. Hence, the value 0 of the r-coordinate not necessarily describes a point ("the origin"), and along the curves of constant ϑ one can analytically reach negative values of r (Fig. 7.2). This implies in particular that the two axes $\vartheta = 0$ and $\vartheta = \pi$ never meet, as it is the case at the $(r = 0)$-point in flat space.

7.2.1.4 Kerr–Newman Coordinates

Consider the signature of the hypersurfaces $r = $ const. It is determined by the determinant D of the induced metric, the hypersurface is time-like if $D > 0$, space-like if $D < 0$, and light-like if $D = 0$. A simple calculation yields $D = \Delta \Sigma \sin^2 \vartheta$ (exercise). Assume that

$$a^2 + Q^2 > M^2. \tag{7.5}$$

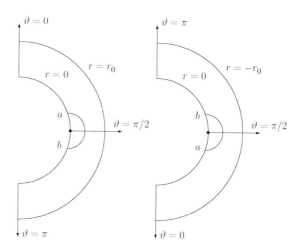

Fig. 7.2 The structure of the Kerr–Newman space–time around the singularity $r = 0, \vartheta = \pi/2$. The left picture is the part $r > 0$, the right one $r < 0$. The axes $\vartheta = 0$ and $\vartheta = \pi$ do not intersect. A closed path through a and b around the singularity must cross the line $\vartheta = \pi/2$ twice

Then the equation $\Delta = 0$ has no solution. In this case $D > 0$, and the hypersurfaces are all time-like, that is causal signals can cross in both directions. This makes it plausible that the singular ring is visible and can be influenced from $r = \pm\infty$. Later we shall identify the corresponding light-like geodesics. Here, the singularity is called "naked", and we can consider the singular ring as the source of the solution.

In the other case the equation has two roots:

$$r_\pm = M \pm \sqrt{M^2 - a^2 - Q^2} \,. \tag{7.6}$$

In the sequel we will restrict to the cases with $M^2 > a^2 + Q^2$. The so-called *extreme* solutions with $M^2 = a^2 + Q^2$ are very interesting, but their significance in astrophysics is marginal. In this case the metric (7.1) is well defined in the following three regions: the external Kerr–Newman space–time with negative radial coordinate, where $-\infty < r < r_-$ and the hypersurfaces $r = $ const are time-like, the internal Kerr–Newman space–time, where $r_- < r < r_+$ and the hypersurfaces $r = $ const are space-like, and the external Kerr–Newman space–time with positive radial coordinate, where $r_+ < r < \infty$ and the hypersurfaces are again time-like.

We suspect that these regions are subsets of a larger space–time, where they are separated by regular hypersurfaces $r = r_\pm$, similar to the Eddington–Finkelstein space–time. To investigate this, we try to find coordinates (v, η, r, ϑ), which are analogous to Eddington–Finkelstein coordinates (and in fact become the Eddington–Finkelstein coordinates when setting $a = Q = 0$). They are called advanced *Kerr–Newman coordinates*. We make the following ansatz for the transformation: we fix r and ϑ and define the advanced coordinates by

$$t = v - X(r), \quad \varphi = \eta - Y(r), \tag{7.7}$$

where $X(r)$ and $Y(r)$ satisfy

$$X'(r) = \frac{r^2 + a^2}{\Delta},$$

$$Y'(r) = \frac{a}{\Delta}.$$

The coordinate η obtained in this way is again an angular coordinate. In particular the values (v, η, r, ϑ) and $(v, \eta + 2\pi, r, \vartheta)$ represent the same points. The corresponding transformation of the differential forms in the metric (7.1) are

$$dt - a\sin^2\vartheta \, d\varphi = dv - a\sin^2\vartheta \, d\eta - \frac{\Sigma}{\Delta}dr,$$

$$a \, dt - (r^2 + a^2) \, d\varphi = a \, dv - (r^2 + a^2) \, d\eta.$$

Substituting into the expression for the metric yields

$$ds^2 = \frac{\Delta}{\Sigma}\left(dv - a\sin^2\vartheta \, d\eta\right)^2 - 2\left(dv - a\sin^2\vartheta \, d\eta\right)dr$$

$$- \frac{\sin^2\vartheta}{\Sigma}\left[a \, dv - (r^2 + a^2) \, d\eta\right]^2 - \Sigma d\vartheta^2, \tag{7.8}$$

and for the potential we get

$$A_\mu \, dx^\mu = \frac{Qr}{\Sigma} \left(dv - a\sin^2 \vartheta \, d\eta \right) - \frac{Qr}{\Delta} dr \,. \tag{7.9}$$

We see that the metric is regular everywhere, except at the points with $\Sigma = 0$ and $\vartheta = 0, \pi$ as $\det g_{\mu\nu} = -\Sigma^2 \sin^2 \vartheta$ (exercise) and all terms which had Δ in the denominator vanished. The "singularity" at $\vartheta = 0, \pi$ is the well-known coordinate singularity of spherical coordinates. In the potential the last term on the right-hand side is singular, but can be removed by a gauge transformation. The transformation of the Killing vector fields into Kerr–Newman coordinates yields $\xi^\mu = (1,0,0,0)$ and $\varphi^\mu = (0,1,0,0)$.

The meaning of the coordinate v in the asymptotic region $r \to \infty$ can be found by considering only the leading order terms in the metric (7.8):

$$ds^2 \to dv^2 - 2 \, dv \, dr - r^2 d\vartheta^2 - r^2 \sin^2 \vartheta \, d\eta^2 \,.$$

Hence, v is asymptotically the advanced time. In the interior of the space–time the surfaces $v = \text{const}$ are *not* light-like but *time-like* as the induced metric in coordinates η, r, and ϑ is

$$dS^2 = F \, d\eta^2 - \Sigma d\vartheta^2 + 2a\sin^2 \vartheta \, dr \, d\eta \,,$$

where

$$F = \frac{\Delta}{\Sigma} a^2 \sin^4 \vartheta - \frac{\sin^2 \vartheta}{\Sigma} (r^2 + a^2)^2 \,.$$

The determinant $a^2 \Sigma \sin^4 \vartheta$ is positive for $a \neq 0$. The curve which is given by

$$v = v_0, \quad \eta = \eta_0, \quad r = -\lambda, \quad \vartheta = \vartheta_0 \,, \tag{7.10}$$

with parameter λ, is a light-like geodesic and becomes an analogy to the radial light-like geodesic in the Kerr–Newman space–times with $a = 0$ (exercise). This geodesic is future oriented for $r \to \infty$ and thus everywhere future pointing.

For many calculations we need to know the contravariant metric. Its components with respect to the coordinates (v, η, r, ϑ) are (exercise):

$$g^{\mu\nu} = \begin{pmatrix} -\dfrac{a^2 \sin^2 \vartheta}{\Sigma}, & -\dfrac{a}{\Sigma}, & -\dfrac{r^2 + a^2}{\Sigma}, & 0 \\[2mm] -\dfrac{a}{\Sigma}, & -\dfrac{1}{\Sigma \sin^2 \vartheta}, & -\dfrac{a}{\Sigma}, & 0 \\[2mm] -\dfrac{r^2 + a^2}{\Sigma}, & -\dfrac{a}{\Sigma}, & -\dfrac{\Delta}{\Sigma}, & 0 \\[2mm] 0, & 0, & 0, & -\dfrac{1}{\Sigma} \end{pmatrix} \,. \tag{7.11}$$

7.2.1.5 The Horizons

Consider the points with $r = r_\pm$ in the advanced Kerr–Newman space–time. They form regular three-dimensional hypersurfaces. The signature of these hypersurfaces can be found by computing the metric induced by the space–time metric. To do so, we simply set $r = r_\pm$ in the formula (7.8)

$$ds^2 = -\frac{\sin^2\vartheta}{\Sigma}\left[a\,dv - \left(r_\pm^2 + a^2\right)d\eta\right]^2 - \Sigma d\vartheta^2 . \tag{7.12}$$

These are metrics with signature $(-1,-1,0)$ (exercise), that is light-like hypersurfaces.

As we have seen, light-like hypersurfaces are semi-permeable membranes, that is signals can only pass in one direction. In the advanced Kerr–Newman space–time both hypersurfaces can only be crossed from the exterior to the interior. We can see this from the fact that the future-pointing light-like geodesic (7.10) crosses to the inside. An observer sitting at large values of r cannot see behind either of the two hypersurfaces. On the other hand, events arbitrarily close to the outer hypersurface with radial coordinate r_+ can be observed. This can be seen similarly as in the case of the advanced Eddington–Finkelstein space–time. Hence, the hypersurface $r = r_+$ is an absolute event horizon, and the interior is a black hole.

Some geometric properties of the horizon play an important role in the theory of black holes. A particularly interesting quantity is the area of a space-like section of the horizon. We show the following important property. Let S be an arbitrary space-like section of the horizon (Fig. 7.3). Let $A(S)$ be its area. Then

$$A(S) = 4\pi\left(r_+^2 + a^2\right) . \tag{7.13}$$

Note that this is independent of S. Let $v = v(\vartheta, \eta)$ be the equation of the section, that is $v(\vartheta, \eta)$ is a smooth function on the sphere. The induced metric is

$$ds^2 = -\frac{\sin^2\vartheta}{\Sigma}\left[av_{,\vartheta}\,d\vartheta + \left(av_{,\eta} - r_+^2 - a^2\right)d\eta\right]^2 - \Sigma d\vartheta^2.$$

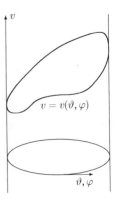

Fig. 7.3 The section
$v = v(\vartheta, \varphi)$ of a horizon

Then a new coordinate ψ can be introduced:

$$\psi := \eta - \frac{a}{r_+^2 + a^2} v(\vartheta, \eta) \, .$$

It follows that ψ is an angular coordinate like η, in particular $\psi(0) = \psi(2\pi)$. The transformed metric is

$$ds^2 = -\left(r_+^2 + a^2\right)^2 \frac{\sin^2 \vartheta}{\Sigma} d\psi^2 - \Sigma d\vartheta^2 \tag{7.14}$$

This metric does not depend on $v = v(\vartheta, \eta)$. We thus have shown that all sections of the horizon are isometric. In particular, they all have the same area. For the metric (7.14) we obtain

$$\det g_{kl} = \left(r_+^2 + a^2\right)^2 \sin^2 \vartheta \, .$$

The area of the section is an integral over the determinant:

$$\begin{aligned} A(S) &= \int_S d\vartheta \, d\psi \sqrt{\det(g_{kl})} \\ &= \left(r_+^2 + a^2\right) \int_0^\pi \sin \vartheta \, d\vartheta \int_0^{2\pi} d\psi \\ &= 4\pi \left(r_+^2 + a^2\right) \, . \end{aligned}$$

This proves the claim.

7.3 Dynamics of Charged Particles

All further considerations depend on understanding the motion of test particles in Kerr–Newman space–time. This will be investigated in this section.

7.3.1 Integrals of Motion

We shall now study the motion of a test particle with rest mass μ and charge q in a space–time with metric $g_{\mu\nu}$ and electromagnetic potential A_μ. This will not be a geodesic, as the Lorentz force generates a 4-acceleration. First, consider the non-trivial case $\mu \neq 0$. To "deduce" the equations of motion we use the equivalence principle. To this end, we make the initial assumption that space–time is flat, choose an inertial frame, and express the known Lorentz force with respect to this inertial frame in its relativistic form:

$$F^\kappa = q F^\kappa{}_\nu \frac{dx^\nu}{ds} \, ,$$

where $F^{\mu\nu}$ is the tensor of the electromagnetic field, $x^\mu(s)$ denotes the trajectory of the particle, and s is the proper time along the trajectory. The equation of motion becomes

$$\mu \frac{d^2 x^\kappa}{ds^2} = qF^\kappa{}_\nu \frac{dx^\nu}{ds} .$$

We have to rewrite this in the generally covariant form. Besides the expression for the 4-acceleration everything is already covariant. We must only write the acceleration with the covariant derivative, and thus obtain

$$\mu \frac{D^2 x^\kappa}{ds^2} = qF^\kappa{}_\nu \frac{dx^\nu}{ds} , \tag{7.15}$$

where we introduced the useful shorthand

$$\frac{D^2 x^\kappa}{ds^2} := \frac{d^2 x^\kappa}{ds^2} + \Gamma^\kappa_{\rho\sigma} \frac{dx^\rho}{ds} \frac{dx^\sigma}{ds} .$$

Now, we simply postulate that the equation of motion (7.15) be valid in arbitrary space–times. We see that for $q = 0$ this becomes the geodesic equation. Also note the role of the parameter: it *must* be the proper time. We can also show that the acceleration is perpendicular to the tangent vector:

$$\frac{dx^\rho}{ds} \left(\frac{D}{ds} \frac{dx^\sigma}{ds} \right) g_{\rho\sigma} = \frac{q}{\mu} F_{\rho\sigma} \frac{dx^\rho}{ds} \frac{dx^\sigma}{ds} = 0 ,$$

that is that the norm of the tangent vector is preserved under these equations.

Equation (7.15) follows from the variation principle for the action:

$$S = \int d\lambda \left(\mu \sqrt{g_{\rho\sigma} \dot{x}^\rho \dot{x}^\sigma} + qA_\rho \dot{x}^\rho \right) .$$

Here $\dot{x}^\kappa = dx^\kappa/d\lambda$, with λ an arbitrary parameter and A_κ an electromagnetic potential yielding the field $F_{\rho\sigma}$. This can be shown as follows. Denote the expression $\sqrt{g_{\rho\sigma} \dot{x}^\rho \dot{x}^\sigma}$ by l. Note that $\dot{s} = l$. Then the Lagrangian is $L = \mu l + qA_\kappa \dot{x}^\kappa$ and the left-hand side of the Euler–Lagrange equation is

$$\frac{\partial L}{\partial x^\kappa} - \frac{d}{d\lambda} \frac{\partial L}{\partial \dot{x}^\kappa}$$

$$= \frac{\mu}{2l} g_{\rho\sigma,\kappa} \dot{x}^\rho \dot{x}^\sigma + q\partial_\kappa A_\rho \dot{x}^\rho - \frac{d}{d\lambda} (\mu l^{-1} g_{\kappa\rho} \dot{x}^\rho + qA_\kappa)$$

$$= \frac{\mu}{2l} g_{\rho\sigma,\kappa} \dot{x}^\rho \dot{x}^\sigma + q\partial_\kappa A_\rho \dot{x}^\rho - \mu g_{\kappa\rho} \left(\frac{\dot{x}^\rho}{l} \right)^\circ - \frac{\mu}{l} g_{\kappa\rho,\sigma} \dot{x}^\rho \dot{x}^\sigma - q\partial_\rho A_\kappa \dot{x}^\rho$$

$$= -\mu g_{\kappa\rho} \left(\frac{\dot{x}^\rho}{l} \right)^\circ - \frac{\mu}{2l} (g_{\kappa\rho,\sigma} + g_{\kappa\sigma,\rho} - g_{\rho\sigma,\kappa}) \dot{x}^\rho \dot{x}^\sigma + qF_{\kappa\rho} \dot{x}^\rho .$$

This implies that

$$\frac{\mu}{l}\left(\frac{\dot{x}^{\kappa}}{l}\right)^{\circ} + \frac{\mu}{l^2}\Gamma^{\kappa}_{\rho\sigma}\dot{x}^{\rho}\dot{x}^{\sigma} = \frac{q}{l}F^{\kappa}_{\rho}\dot{x}^{\rho} .$$

Equation (7.15) follows by substituting $\dot{s}$ for l, qed.

The practical use of the variation principle is that we can use it to quickly find integrals of motion. If the Lagrangian is independent of one coordinate, then the corresponding momentum is conserved. Thus let $\partial_{\kappa}L = 0$ for some κ. Then

$$\frac{\partial}{\partial \dot{x}^{\kappa}}L = \text{const} .$$

From the expression for the Lagrangian, we find that

$$\frac{\partial}{\partial \dot{x}^{\kappa}}L = \mu g_{\kappa\rho}\frac{dx^{\rho}}{ds} + qA_{\kappa},$$

since $\dot{s} = \sqrt{g_{\mu\nu}\dot{x}^{\mu}\dot{x}^{\nu}}$. We can write the right-hand side as a scalar function by introducing the vector $\chi^{\mu} = \delta^{\mu}_{\kappa}$. Then

$$P_{\chi} := \chi^{\mu}\left(\mu g_{\mu\rho}\frac{dx^{\rho}}{ds} + qA_{\mu}\right)$$

is conserved along the trajectory of the particle. The covariant 4-momentum of the test particle is

$$p_{\mu} = \mu g_{\mu\nu}\frac{dx^{\nu}}{ds} .$$

Thus we can write

$$P_{\chi} = \chi^{\mu}(p_{\mu} + qA_{\mu}) .$$

We can now mix the particles with photons. Since then $q = 0$ and χ^{μ} is a Killing vector field, P_{χ} is also conserved for photons.

Example: From the form of the metric and the potential with respect to Kerr–Newman coordinates we see that ξ^{μ} and φ^{μ} satisfy the conditions on χ. The corresponding conserved quantities can be expressed in components with respect to the advanced Kerr–Newman coordinates as follows:

$$e = p_{\nu} + qA_{\nu}, \quad -j = p_{\eta} + qA_{\eta} . \tag{7.16}$$

The quantity in question is also conserved by scattering processes. Assume that n charged particles with 4-momenta $p^{\mu}_1, \ldots, p^{\mu}_n$ and charges $q_1, \ldots, q_n$ collide at the point p and that n' particles with momenta $p'^{\mu}_1, \ldots, p'^{\mu}_{n'}$ and $q'_1, \ldots, q'_{n'}$ are generated (Fig. 2.1). Then we must have

$$p^{\mu}_1 + \ldots + p^{\mu}_n = p'^{\mu}_1 + \ldots + p'^{\mu}_{n'},$$

and

$$q_1 + \ldots + q_n = q'_1 + \ldots + q'_{n'} .$$

Multiply the first equation by $g_{\mu\nu}\chi^\nu$ at p and add the second one multiplied by $A_\mu \chi^\mu$ at p. Then we obtain

$$P_{1\chi} + \ldots + P_{n\chi} = P'_{1\chi} + \ldots + P'_{n'\chi},$$

which is the desired conservation law.

We want to express the condition for χ^μ in arbitrary coordinates. In adapted coordinates $\{\bar{x}^\mu\}$ it is equivalent to the following three equations:

$$\bar{\chi}^\mu = \delta^\mu_\kappa\,,$$
$$\bar{\chi}^\mu \partial_{\bar\mu} \bar{g}_{\rho\sigma} = 0\,,$$
$$\bar{\chi}^\mu \partial_{\bar\mu} \bar{A}_\rho = 0.$$

We already transformed the first two equations into general coordinates and obtained the Killing equation:

$$g_{\mu\nu,\rho}\chi^\rho + g_{\rho\nu}\chi^\rho_{,\mu} + g_{\mu\rho}\chi^\rho_{,\nu} = 0.$$

In an analogous way we obtain the symmetry equation for the potential (exercise):

$$A_{\mu,\rho}\chi^\rho + A_\rho \chi^\rho_{,\mu} = 0.$$

A vector field satisfying this equation is called symmetry vector field of the potential. If the vector field χ^μ is a Killing vector field and a symmetry vector field for the potential at the same time—in this case we say that χ^μ is a symmetry vector—then the corresponding quantity P_χ is conserved along the trajectory of the particle. This can also be proven directly from the Killing equation and the symmetry equation for the potential (exercise).

7.3.2 The Equatorial Plane and the Axes of Symmetry

In general there exist only two independent symmetry vector fields in the Kerr–Newman space–time. The conservation laws corresponding to these fields are not sufficient to reduce the general problem to simple quadratures. There is also an integral of motion quadratic in the components of the momentum—the so-called Carter integral [5]. However, in the special case of motion in the equatorial plane $\vartheta = \pi/2$ or along the axes of symmetry $\vartheta = 0$ or $\vartheta = \pi$, we can do with only the former two quantities.

We have to start by showing that the respective motion remains in these submanifolds if they start there with an initial velocity tangent to these sub-manifolds. Something similar can be shown in general. We assume that

1. the Lagrangian of the particle is invariant with respect to the transformation $x^\mu \mapsto x'^\mu$ in the configuration space $\mathcal{M}$,

2. this transformation is the identity on the sub-manifold $\mathcal{N}$ of the configuration space $\mathcal{M}$, and
3. there is a neighborhood U of $\mathcal{N}$ in $\mathcal{M}$ so each point $r \in U \setminus \mathcal{N}$ the transformation is non-trivial.

Let the trajectory $x^\mu(\lambda)$ be a solution to the Euler–Lagrange equation for the initial data $x^\mu(0) = r \in \mathcal{N}$ and $\dot{x}^\mu \in T_r\mathcal{N}$ with $x^\mu(\lambda_0) \notin \mathcal{N}$ for some λ_0. Without loss of generality we can assume that $x^\mu(\lambda_0) \in U$. Then $x'^\mu(\lambda_0) \neq x^\mu(\lambda_0)$. That is, the solution $x'^\mu(\lambda)$, which can be obtained from $x^\mu(\lambda)$ by applying the transformation, is different from the original one. On the other hand, it has the same initial data which leads to a contradiction.

This general fact can be applied to the equatorial plane by using the reflection at the equatorial plane $\vartheta \mapsto \pi - \vartheta$ and to the axes of symmetry by using the rotation $\varphi \mapsto \varphi + c$.

We are only interested in the region outside the horizon and we can thus work with the Boyer–Lindquist metric (7.1) and the corresponding potential (7.2).

7.3.2.1 The Equatorial Plane

The integrals of motion with $\vartheta = \pi/2$ are

$$g_{tt}\dot{t} + g_{t\varphi}\dot{\varphi} + \frac{Qq}{r} = e, \tag{7.17}$$

$$g_{t\varphi}\dot{t} + g_{\varphi\varphi}\dot{\varphi} - \frac{aQq}{r} = -j, \tag{7.18}$$

$$g_{tt}\dot{t}^2 + 2g_{t\varphi}\dot{t}\dot{\varphi} + g_{\varphi\varphi}\dot{\varphi}^2 - \frac{r^2}{\Delta}\dot{r}^2 = \mu^2, \tag{7.19}$$

where we chose the parameter λ in the following way: for massive particles $\lambda := s/\mu$, and for photons λ is simply the physical parameter. Hence we avoid the difficulty that massive particles can be charged and hence do not move along geodesics. For convenience, we rewrite the (7.17), (7.18), and (7.19) in matrix form. We define the following matrices:

$$\mathbf{g} = \begin{pmatrix} g_{tt} & g_{t\varphi} \\ g_{\varphi t} & g_{\varphi\varphi} \end{pmatrix}, \quad \mathbf{h} = \begin{pmatrix} 1 \\ -a \end{pmatrix}, \quad \mathbf{p} = \begin{pmatrix} e \\ -j \end{pmatrix}, \quad \mathbf{u} = \begin{pmatrix} \dot{t} \\ \dot{\varphi} \end{pmatrix},$$

where

$$g_{tt}\,dt^2 + 2g_{t\varphi}\,dt\,d\varphi + g_{\varphi\varphi}\,d\varphi^2 := \frac{\Delta}{r^2}(dt - a\,d\varphi)^2 - \frac{1}{r^2}\left[a\,dt - (r^2 + a^2)\,d\varphi\right]^2.$$

Then (7.17), (7.18), and (7.19) can be written as follows:

$$\mathbf{g}\mathbf{u} + \mathbf{h}\frac{Qq}{r} = \mathbf{p},$$

$$\mathbf{u}^\top \mathbf{g} \mathbf{u} - \frac{r^2}{\Delta}\dot{r}^2 = \mu^2 \, .$$

Outside of $\Delta = 0$ we can solve these equations for the derivatives:

$$\dot{r}^2 + V_{\text{eff}}(r) = 0 \, , \tag{7.20}$$

$$\mathbf{u} = \mathbf{g}^{-1}\left(\mathbf{p} - \frac{Qq}{r}\mathbf{h}\right) , \tag{7.21}$$

where

$$V_{\text{eff}}(r) = -\frac{\Delta}{r^2}\left(\mathbf{p}^\top - \frac{Qq}{r}\mathbf{h}^\top\right)\mathbf{g}^{-1}\left(\mathbf{p} - \frac{Qq}{r}\mathbf{h}\right) + \frac{\Delta}{r^2}\mu^2 \, . \tag{7.22}$$

Substituting for $\mathbf{g}, \mathbf{p}, \mathbf{h}$, and Δ into (7.22) yields

$$V_{\text{eff}}(r) = Q^2(ae - j)^2 r^{-4} + 2\left[Qaq(ae - j) - M(ae - j)^2\right]r^{-3}$$
$$+ \left[j^2 - a^2 e^2 + \mu^2(a^2 + Q^2) - Q^2 q^2\right]r^{-2} + 2(Qqe - M\mu^2)r^{-1} - e^2 + \mu^2 \, . \tag{7.23}$$

$V_{\text{eff}}(r)$ is an effective potential in the sense that it is not valid for all particles with given mass μ and charge q but depends on the values for e and j. It equals $\mu^2 - e^2$ at $r = \pm\infty$. The motion is only possible where $V_{\text{eff}}(r)$ is negative. For example, all particles with $|e| > \mu$ can reach the asymptotic region $r = \infty$.

7.3.2.2 The Axes of Symmetry

The motion along the axes of symmetry ($\vartheta = 0, \pi$) can be studied in an analogous way. The coordinates in this two-dimensional sub-manifold are t and r. The integrals of motion yield:

$$\dot{r}^2 + V_{\text{eff}}^a(r) = 0 \, ,$$

$$\dot{t} = \frac{r^2 + a^2}{\Delta}\left(e - \frac{Qqr}{r^2 + a^2}\right) ,$$

with

$$V_{\text{eff}}^a(r) = \mu^2\frac{\Delta}{r^2} - \left(e - \frac{Qq}{r^2}\right)^2 \, .$$

7.4 Energetics of Black Holes

In astrophysics, black holes play the role of a rich reservoir of energy. Whenever a very efficient source is needed to explain the huge energy output in an astrophysical object, a black hole is postulated (x-ray binaries, active galactic nuclei, etc.). We will now study why this is the case. In general, there are two kinds of energy that can be extracted from a region containing a black hole:

1. the black hole's own energy,
2. the energy of the test particles in the field of the black hole.

7.4.1 Available Energy of a Black Hole

(following the diploma thesis of Ch. Farrugia, Bern 1978)

It turns out that rotation and electric energy of black holes can be extracted. Only the rotation energy seems to have astrophysical significance because no viable process of charging black holes could be thought of. The gravitomagnetic field associated with the rotation seems to play a crucial role in the origin of the so-called jets [6, 7]. These are huge geysers (of power in a broad interval around $10^{38} \, Js^{-1}$) of relativistic particles gushing in two opposite directions from young active galactic nuclei, sometimes hundreds of kiloparsec into the space.

For the rest of the theory we can restrict ourselves to the external Kerr–Newman solution as it contains all events outside a black hole which emerged from a collapse.

7.4.1.1 The Ergosphere

The symmetry vectors of the Kerr–Newman solution are $\chi^\mu = \alpha \xi^\mu + \beta \varphi^\mu$. Let us examine the integral curves of these vector fields. These are the curves $x^\mu = x^\mu(\lambda)$ whose tangential vectors agree with the vector field. Hence, an integral curve corresponding to the vector field with two arbitrary constants α and β is given by $t = \alpha \lambda$, $\varphi = \beta \lambda$, $r = $ const, and $\vartheta = $ const. This curve is only closed for $\alpha = 0$. If in addition $\beta = 1$, then $\chi^\mu = \varphi^\mu$ and φ^μ infinitesimally generates the rotation around the axis of symmetry. That is, the transformation $\lambda \mapsto \lambda + \psi$ is such a rotation by angle ψ. Hence the vector φ^μ is uniquely characterized, including its normalization. Similarly, the vector ξ^μ is characterized by the fact that it generates the translation of the proper time of asymptotic observers.

Let us examine the significance of the corresponding conserved quantities, as they will play an important role. We denoted P_χ for $\chi^\mu = \xi^\mu$ by e and for $\chi^\mu = \varphi^\mu$ by $-j$. We see that the electromagnetic term in P_χ vanishes in the limit $r = \infty$ as the potential decays like r^{-1}. The rest is identical to the conserved quantities of uncharged particles. We already know the meaning of this (cf. Sect. 2.8.2): e is the energy of the particle and j the angular momentum with respect to the symmetry axis, both with respect to the asymptotic observer.

Consider an uncharged, massive particle with 4-velocity u^μ at the point p of space–time. The particle has energy $e = \mu g_{\mu\nu} \xi^\mu u^\nu$. If we assume that ξ^μ is timelike at p then e has a positive minimum with value $\mu \xi$, where $\xi = \sqrt{g_{\mu\nu} \xi^\mu \xi^\nu}$ is the norm of the Killing vector. This minimum is assumed for particles with $u^\mu = \xi^{-1} \xi^\mu$

(exercise), that is a particle "at rest" with respect to the asymptotic observer. We thus want to say that ξ is the *gravitational potential* with respect to the asymptotic observer. This is in analogy to the Newtonian theory, where the minimal energy of a particle in the gravitational field in a fixed point is equal to the mass of the particle times the potential.

However, if the vector ξ^μ is space-like at p, then the energy of the particle at p has no minimum, and we can have negative energy with arbitrarily large absolute value (exercise). Consequently, in the region where ξ^μ is space-like, there is no analogue to the gravitational potential (no particle there can be at rest with respect to the asymptotic observer). The part of this region outside of the horizon is called *ergosphere*. It is defined by the inequality $M + \sqrt{M^2 - a^2 - Q^2} < r < M + \sqrt{M^2 - a^2 \cos^2 \vartheta - Q^2}$ (Fig. 7.4). The boundary of the ergosphere is thus given by $r^2 + a^2 \cos^2 \vartheta - 2Mr + Q^2 = 0$, it touches the horizon at the poles $\vartheta = 0, \pi$, and lies outside of the horizon for all other values of ϑ, provided $a \neq 0$. There is no ergosphere for $a = 0$.

Now consider a charged particle. Its minimal energy at a point p with time-like ξ^μ is $\mu\xi + qA_\mu\xi^\mu$. This is analogous to the energy of a similar particle in the non-relativistic theory which is $\mu\times$ gravitational potential $+ q\times$ electrostatic potential. We can therefore say that the term $\Phi = A_\mu\xi^\mu$ is the *electrostatic potential* with respect to the asymptotic observer.

The region outside the horizon, where $\mu g_{\mu\nu}\xi^\mu u^\nu + qA_\mu\xi^\mu$—the energy of a particle with charge q with respect to the observer at infinity—can be negative, is called *generalized ergosphere*. It is clear that the generalized ergosphere always contains the ergosphere. The boundary of the generalized ergosphere is thus given by either the equation $\mu\xi + q\Phi = 0$ when the norm of $q\Phi$ is negative at points with $\xi = 0$, or by $\xi = 0$ otherwise.

7.4.1.2 Penrose Processes

The significance of the ergosphere is that it can be used to extract energy from the black holes. This is based on the following assumption: if a particle with small

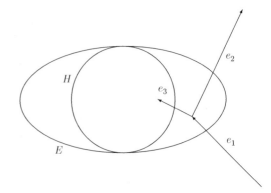

Fig. 7.4 Penrose process. A test particle with energy $e_1 > 0$ comes from infinity and decays at point p into two particles with energies $e_2 > 0$ and $e_3 < 0$

values of e, j, and q falls through the horizon, then the parameters M, J, and Q of the black hole change like

$$\delta M = e, \quad \delta J = j, \quad \delta Q = q.$$

This assumption is physically natural, it means that the total energy, the total angular momentum, and the total charge is conserved. If a particle vanishes in the black hole, the parameters of the black hole change accordingly. However, the proof is too difficult for us, as the small perturbations in the metric and the potential caused by the particle can no longer be neglected (the test particle approximation is based on such neglecting). We will thus make this assumption without proof (for a proof, see e.g., [4, 8]). Then our claim immediately follows as e can only be negative in the presence of an ergosphere.

Such particles can only exist inside the ergosphere, since e is conserved and must be non-negative outside the ergosphere. Thus, no such particle can be thrown into the black hole from the asymptotic region! If the energy gain were only possible in close vicinity of the black hole, there would not be much practical use of it. Thus we have to construct a process that can transfer this energy to the asymptotic region, at least in principle.

This is what the Penrose processes achieve [9]. The idea is to utilize the conservation of e, j, and q in scattering processes. We throw one particle with e_1, j_1, and q_1 from the asymptotic region into the ergosphere. There it decays into two particles, one with e_2, j_2, and q_2 and another one with e_3, j_3, and q_3. We have $e_1 = e_2 + e_3$, $j_1 = j_2 + j_3$, and $q_1 = q_2 + q_3$. If $e_3 < 0$ then $e_2 > e_1$. Provided the third particle falls into the black hole, it does not need to leave the ergosphere. If the second particle again reaches the asymptotic region, the observers there have gained energy. Such processes are actually possible, but we do not want to go into details here.

7.4.1.3 Second Law of Black Hole Energetics

We shall investigate, which particles can be captured by the black hole, and how the parameters of the black hole are affected. We are working at the horizon and thus need Kerr–Newman coordinates. If the trajectory of the particle is given by $v(\lambda)$, $\eta(\lambda)$, $r(\lambda)$, and $\varphi(\lambda)$ then it crosses the horizon at $r = r_+$ if $dr/d\lambda < 0$ at that point. That is,

$$p^r|_{r=r_+} < 0 \,.$$

We want to express this condition in terms of e, j, and q of the particle's trajectory. To this end, we rewrite p^r in terms of the covariant components of the momentum and the contravariant metric (7.11),

$$p^r = -\frac{r^2 + a^2}{\Sigma} p_v - \frac{\Delta}{\Sigma} p_r - \frac{a}{\Sigma} p_\eta \,, \tag{7.24}$$

and use (7.16), to express p_v and p_η in terms of e, j, and q:

$$p_v = e - \frac{qQr}{\Sigma},$$

$$p_\eta = -j + \frac{qQra\sin^2\vartheta}{\Sigma}.$$

Evaluating this at the horizon yields the inequality

$$e > \frac{aj + Qr_+q}{a^2 + r_+^2}. \tag{7.25}$$

Equation (7.25) is called *entry condition*.

   Equation (7.25) holds for all, in particular space-like, curves. To show that there are particles that enter this way, we need some more work. We have to show that there is a time-like trajectory for each value of the quantities e, j, q, M, a, Q, v, η, and ϑ, which satisfies the entry condition. The coordinates of the point of entry, the parameters of the black hole, (M,a,Q), and the parameters of the particle, (e,j,q), determine the components p_v and p_η. The components p_ϑ and p_r can then be chosen arbitrarily.

   The condition that the curve be non-space-like is $g^{\mu\nu}p_\mu p_\nu \geq 0$. Evaluated at the horizon it yields

$$(ap_v + p_\eta)^2 + 2p_r\left[(r_+^2 + a^2)p_v + ap_\eta\right] + p_\vartheta^2 \leq 0.$$

From (7.24) we infer that

$$(r_+^2 + a^2)p_v + ap_\eta = -(\Sigma p^r)_{r=r_+}.$$

The right-hand side is positive, thus

$$p_r \leq -\frac{(ap_v + p_\eta)^2 + p_\vartheta^2}{2\left[(r_+^2 + a^2)p_v + ap_\eta\right]}.$$

Hence we can always choose such a p_r without affecting inequality (7.25). Thus there always exists at least one particle which satisfies (7.25), qed.

   There is a surprising interpretation of the entry condition (7.25). Calculate the growth of the area of the black hole (7.13),

$$dA = 8\pi(r_+\,dr_+ + a\,da).$$

If we substitute $\sqrt{M^2 - a^2 - Q^2}$ by $r_+ - M$ everywhere after differentiation, we obtain for dr_+ that

$$dr_+ = \frac{r_+}{r_+ - M}dM - \frac{a}{r_+ - M}da - \frac{Q}{r_+ - M}dQ.$$

Then

$$r_+\,dr_+ + a\,da = \frac{r_+^2 + a^2}{r_+ - M}dM - \frac{a}{r_+ - M}dJ - \frac{Qr_+}{r_+ - M}dQ,$$

where we substituted $da = d(J/M)$. Altogether this yields

$$dA = \frac{8\pi}{r_+ - M} \left[\left(r_+^2 + a^2 \right) dM - a\, dJ - Qr_+\, dQ \right] . \tag{7.26}$$

If we further use the assumption that $dM = e$, $dJ = j$, and $dQ = q$ we obtain that

$$dA = \frac{8\pi(r_+^2 + a^2)}{r_+ - M} \left(e - \frac{aj + Qr_+ q}{r_+^2 + a^2} \right) .$$

It follows that inequality (7.25) is equivalent to the simple condition

$$dA > 0 . \tag{7.27}$$

The area of the black hole must increase in every process with particles entering the black hole. This is the so-called *second law of black hole energetics* [10]. In the form of (7.27) it can be proven in a much more general setting, for example for colliding black holes [11].

7.4.1.4 First Law of Black Hole Energetics

We can solve (7.26) for the differential of the mass:

$$dM = \frac{\kappa}{8\pi} dA + \omega\, dJ + \phi\, dQ , \tag{7.28}$$

where the coefficients κ, ω, and ϕ are defined as follows:

$$\kappa = \frac{r_+ - M}{r_+^2 + a^2}, \quad \omega = \frac{a}{r_+^2 + a^2}, \quad \phi = \frac{Qr_+}{r_+^2 + a^2} .$$

Equation (7.28) is the so-called *differential mass formula*. It describes the change of the total mass of the black hole depending on the changes of area, the angular momentum, and the charge. It is sometimes also called the first law of black hole energetics [10]. Let us examine the meaning of the coefficients κ, ω, and ϕ in the differential mass formula.

From the metric of the event horizon, given by (7.12), we can see that a light-like vector L^μ, which is tangential to the horizon, has the following components with respect to the coordinates (v, η, r, ϑ):

$$L^\mu = \left(1, \frac{a}{r_+^2 + a^2}, 0, 0 \right) .$$

The normalization is such that

$$L^\mu = \xi^\mu + \frac{a}{r_+^2 + a^2} \varphi^\mu , \tag{7.29}$$

where $\xi^{\mu} = (1,0,0,0)$ is the Killing vector field of stationarity and $\varphi^{\mu} = (0,1,0,0)$ the one of axi-symmetry. The vector L^{μ} represents the only future-pointing causal direction along the horizon. Intuitively, it can be interpreted as the direction into which the horizon is moving. The relation (7.29) then means that the horizon is rotating with respect to the observer at infinity. The angular velocity is given by ω. This yields the interpretation of ω. Note that the rotation is *rigid*, that is $\omega = $ const along the horizon.

Another important property of the vector field L^{μ} is that it is a symmetry of the whole space–time if we regard it as a linear combination of two vector fields of symmetry with constant coefficients. It is the only *causal* vector field of symmetry which has the above normalization. The square of the norm of this vector is

$$
L = g_{\mu\nu} N^{\mu} N^{\nu}
$$
$$
= \frac{\Delta}{\Sigma} \left(\frac{r_+^2 + a^2 \cos^2 \vartheta}{r_+^2 + a^2} \right)^2 - \frac{a^2 \sin^2 \vartheta}{\Sigma} \left(\frac{r^2 - r_+^2}{r_+^2 + a^2} \right)^2
$$
$$
= b(\vartheta)(r - r_+) + O(r - r_+)^2,
$$

where
$$
b(\vartheta) = \frac{r_+ - r_-}{(r_+^2 + a^2)^2} (r_+^2 + a^2 \cos^2 \vartheta) > 0,
$$

and $O(x^n)$ is a term of order x^n. It follows that $L > 0$ above the horizon. On the horizon we have $L = 0$. Hence L^{μ} is the only vector field of symmetry which is non-space-like in a whole neighborhood of the horizon in external Kerr–Newman space–time (where $r \geq r_+$).

The Killing vector field L^{μ} is thus time-like for observers which are arbitrarily close and above the horizon. For such observers, which are at rest relative to the horizon, $\sqrt{L}$ plays the role of the gravitational potential and $A_{\mu} N^{\mu}$ the role of the electric potential. Furthermore, we have that

$$
\lim_{r=r_+} g^{\mu\nu} \sqrt{L}_{,\mu} \sqrt{L}_{,\nu} = -\kappa^2 ,
$$
$$
\lim_{r=r_+} A_{\mu} N^{\mu} = \phi
$$

(exercise). This yields the interpretation of the coefficients κ and ϕ: κ is the surface gravity and ϕ the electric potential of the black hole (for all local observers near the horizon). It is important to remark that all three quantities ω, κ, and ϕ need not be constant at the horizon according to their definition. The fact that they are constant is sometimes referred to by the term "0th law" of black hole dynamics. Formally, the constantness resembles an equilibrium condition, similar to the condition that temperature and pressure be constant in thermodynamics.

7.4.1.5 Irreducible Mass

The entry condition can be used to answer the question about the maximal energy gain. Inequality (7.25) implies that the energy of a particle falling into the black hole cannot be arbitrarily small. The energy gained for given values of j and q (and a given black hole) cannot exceed the value

$$e_{\sup} = -\frac{aj + Qr_+q}{a^2 + r_+^2} \, .$$

This value cannot be achieved, it is only possible to get arbitrarily close. The reason is that for this value we have $(p^r)_{r=r_+} = 0$ and thus the particle moves tangential to the horizon. This is only possible in a light-like direction, which is not achievable for massive particles. Also for light-like particles this state cannot be created from the outside.

Ideal processes, where equality holds in the above equation, are called *reversible* for the following reason. Consider a process in which a particle with e_1, j_1 and q_1 falls into a black hole with parameters M, J, and Q. There, the parameters of the hole become $M' = M + e_1$, $J' = J + j_1$, and $Q' = Q + q_1$. Does there exists another process in which a particle with e_2, j_2, and q_2 falls into the black hole, so that the final parameters of the black hole, $M'' = M + e_1 + e_2$, $J'' = J + j_1 + j_2$, and $Q'' = Q + q_1 + q_2$, agree with the initial ones? Obviously, this implies $e_1 = -e_2$, $j_1 = -j_2$, and $q_1 = -q_2$. The last two equations, together with inequality (7.25) imply:

$$e_1 > \frac{aj_1 + Qr_+q_1}{a^2 + r_+^2}, \quad e_2 > -\frac{a'j_1 + Q'r_+'q_1}{a'^2 + r_+'^2},$$

where $a' = J'/M'$ and $r_+' = r_+(M', J', Q')$. To first order the primes in the second equation can be omitted. It then follows that $e_1 + e_2 > 0$, and equality can only hold if both processes are ideal.

To proceed, we need the expression of the energy of the black hole in terms of the parameters A, J, and Q. To this end, we rewrite relation (7.13) in terms of M, J, and Q:

$$A = 4\pi \left[\left(M + \sqrt{M^2 - J^2/M^2 - Q^2} \right)^2 + \frac{J^2}{M^2} \right]$$

$$= 4\pi \left(2M^2 - Q^2 + 2\sqrt{M^4 - J^2 - M^2 Q^2} \right),$$

and solve this equation for M:

$$M = \frac{1}{2} \sqrt{\frac{4\pi}{A} \left[\left(\frac{A}{4\pi} + Q^2 \right)^2 + 4J^2 \right]}. \tag{7.30}$$

This is the so-called *mass formula*. The differential mass formula is simply its differential.

The mass formula can be used to determine the total available energy of a black hole. Consider a black hole with parameters A, J, and Q. Its total energy is given by the mass formula. Now energy can be extracted via Penrose processes, but not more than the limit for reversible processes. This decreases the rotational energy and the electric energy of the black hole so that A remains constant until the state with minimal mass for the constant A is reached. The minimum can be computed from (7.30). The first derivatives of M are

$$\frac{\partial M}{\partial J} = \frac{4\pi}{A} \frac{J}{M}, \quad \frac{\partial M}{\partial Q} = \frac{4\pi}{A} \left(\frac{A}{4\pi} + Q^2 \right) \frac{Q}{2M}.$$

It follows that the only extremum of this function is at $J = Q = 0$. It is an absolute minimum, as one easily proves the inequality

$$M(A,J,Q) > M(A,0,0)$$

for all $J \neq 0$ and $Q \neq 0$ by taking the square of the expression for M.

This is a Schwarzschild black hole with area A, from which no more energy can be extracted. According to the mass formula, its energy is

$$M_{\text{irr}}^2 = \frac{A}{16\pi}.$$

This energy is called the *irreducible energy* of the black hole with parameters A, J, and Q. The available energy of such a black hole thus equals to the difference of the total mass, given by the mass formula, and the irreducible mass.

The similarity of our formulas to the fundamental laws of thermodynamics is striking. If somehow an entropy of a black hole could be introduced so that

$$S = bA , \tag{7.31}$$

where b is a constant and the quantity T,

$$T = \frac{\kappa}{8\pi b} , \tag{7.32}$$

could be understood as a temperature, then there would be more than formal similarity. So far the nature of temperature could be justified to some extent. As it turned out (theoretically), in the presence of a black hole, a quantum gas can only be in equilibrium at the temperature (7.32) with $b = \alpha k l_P^{-2}$ ($k =$ Boltzmann constant, $\alpha =$ a numerical coefficient close to one, $L_P =$ Planck length) [12]. A similar understanding of formula (7.31) has not been reached up to now.

7.4.2 Energy of Particles in the Field of a Black Hole

We have seen that the part of the energy of a black hole which is stored in its charge and rotation can be extracted. However, a large quantity of energy can be gained if only a black hole is present, without reducing its energy (in fact it will grow in most cases). The source of the energy is the rest mass of the particles which approach the black hole. The minimal energy of an uncharged particle with respect to an observer at infinity is given by the gravitational potential ξ (we will stay out of the ergosphere). In the asymptotic region we have $\xi = 1$, thus the energy equals the rest mass μ. At the ergosphere $\xi = 0$ and thus the energy is zero. If the rest of the energy, $\mu(1 - \xi)$, can be transferred to the asymptotic region, then we gain up to 100% of the rest energy of the particle. In principle, we can get arbitrarily close to this efficiency.

A real system has always less efficiency. The problem is to decelerate a particle to energy $\mu\xi$. We shall now construct a model of such a system, namely the accretion disc. As we are interested in the basic principle, we shall simplify the model severely. (The research on accretion discs is nowadays a large and complicated field of its own [13].) We assume that around a Kerr black hole ($Q = 0$) there are uncharged dust particles moving along practically circular orbits in the equatorial plane. There is friction among the particles which heats them up. They radiate the thermal energy away and sink to lower circular orbits. As they do not exactly follow the circular orbits, these orbits need to be *stable*. This works before they reach the so-called *last stable orbit*, then they quickly fall into the black hole without much radiation. This will increase or decrease the angular velocity of the black hole and its energy will grow in general. A part of the rest mass of the particles will, however, be radiated to the asymptotic region. The subsequent calculations are based on [8].

7.4.2.1 Energy and Angular Momentum of the Circular Orbits

Consider the circular orbits in the equatorial plane of the Kerr space–time. These satisfy (7.20), (7.21), and (7.23) with $Q = q = 0$. In particular

$$V_{\text{eff}} = -\frac{2M(j - ae)^2}{r^3} + \frac{j^2 - a^2e^2 + \mu^2a^2}{r^2} - \frac{2M\mu^2}{r} - e^2 + \mu^2 . \tag{7.33}$$

The effective potential and its derivative must be equal to zero along such an orbit, and the orbit is stable if the second derivative is positive there. We use the scale invariance of the problem to get rid of two parameters. Introducing the dimensionless quantities

$$\bar{e} = \frac{e}{\mu} , \quad \bar{j} = \frac{j}{\mu r} , \tag{7.34}$$

and

$$\alpha = \frac{a}{M} , \quad p = \sqrt{\frac{M}{r}} , \tag{7.35}$$

we can rewrite the potential and its derivatives in the form

$$\frac{1}{\mu^2}V_{\text{eff}} = -2p^2\left(\bar{j} - \alpha p^2\bar{e}\right)^2 + \bar{j}^2 - \alpha^2 p^4\bar{e}^2 - \bar{e}^2 + \alpha^2 p^4 - 2p^2 + 1 , \tag{7.36}$$

$$\frac{r}{2\mu^2}\frac{\partial V_{\text{eff}}}{\partial r} = 3p^2\left(\bar{j} - \alpha p^2\bar{e}\right)^2 - \bar{j}^2 + \alpha^2 p^4\bar{e}^2 - \alpha^2 p^4 + p^2 , \tag{7.37}$$

$$\frac{r^2}{2\mu^2}\frac{\partial^2 V_{\text{eff}}}{\partial r^2} = -12p^2\left(\bar{j}-\alpha p^2\bar{e}\right)^2 + 3(\bar{j}^2 - \alpha^2 p^4\bar{e}^2) + 3\alpha^2 p^4 - 2p^2. \qquad (7.38)$$

For a black hole we need $a \leq M$, that is $\alpha \leq 1$. Furthermore, it does not make sense to consider circular orbits with a radius smaller than that of the horizon. That is $r \geq M + \sqrt{M^2 - a^2}$, which leads to $p^2 \leq 1/(1 + \sqrt{1-\alpha^2})$. In summary the ranges of α and p are

$$0 \leq \alpha \leq 1, \quad 0 \leq p^2 \leq \frac{1}{1+\sqrt{1-\alpha^2}} \leq 1. \qquad (7.39)$$

We consider the system $V_{\text{eff}} = 0$ and $V'_{\text{eff}} = 0$ as equations for e and j, and assume M, a, and r to be given. To simplify this task, we introduce new variables:

$$x = \bar{j} - \alpha p^2\bar{e}, \quad y = \bar{j} + \alpha p^2\bar{e},$$

so that

$$\bar{j} = \frac{1}{2}(y+x), \quad \bar{e} = \frac{1}{2\alpha p^2}(y-x). \qquad (7.40)$$

Equation (7.37) then yields

$$y = 3p^2 x + \frac{p^2 - \alpha^2 p^4}{x}. \qquad (7.41)$$

Inserting into (7.36) leads to the following quadratic equation for x^2:

$$\left[4\alpha^2 p^6 - (3p^2 - 1)^2\right]x^4 + 2\left(\alpha^2 p^6 - 3p^4 + \alpha^2 p^4 + p^2\right)x^2 - \left(-\alpha^2 p^4 + p^2\right)^2 = 0.$$

The first miracle is that the discriminant is a square:

$$\Delta = [2\alpha p^3(-\alpha^2 p^4 + 2p^2 - 1)]^2.$$

The solution therefore does not contain a square root:

$$x^2 = \frac{-\alpha^2 p^6 - \alpha^2 p^4 + 3p^4 - p^2 - 2\iota\alpha p^3\left(-\alpha^2 p^4 + 2p^2 - 1\right)}{4\alpha^2 p^6 - (3p^2 - 1)^2},$$

here ι is an arbitrary sign yielding both solutions of the quadratic equation.

The second miracle is that the numerator can be factorized and one can cancel as follows:

$$-\alpha^2 p^6 - \alpha^2 p^4 + 3p^4 - p^2 - 2\iota\alpha p^3\left(-\alpha^2 p^4 + 2p^2 - 1\right)$$
$$= \left(2\iota\alpha p^3 + 3p^2 - 1\right)\left(-\iota\alpha p^2 + p\right)^2,$$

or

$$x = \iota'\frac{p - \iota\alpha p^2}{\sqrt{1 - 3p^2 + 2\iota\alpha p^3}}, \qquad (7.42)$$

where ι' is another arbitrary sign.

A simple calculation yields that (7.34), (7.40), (7.41), and (7.42) give the following results for e and j:

$$e = \iota'\iota\mu \frac{1 - 2p^2 + \iota\alpha p^3}{\sqrt{1 - 3p^2 + 2\iota\alpha p^3}}, \tag{7.43}$$

$$j = -\iota'\mu M \frac{1 - 2\iota\alpha p^3 - \alpha^2 p^4}{p\sqrt{1 - 3p^2 + 2\iota\alpha p^3}}. \tag{7.44}$$

Equations (7.43) and (7.44) yield four solutions (e, j), depending on two signs ι and ι'. Two of them correspond to a sign change of both e and j, which changes the orientation of the tangent vector of the orbit. We can choose this sign so that the resulting orbit is future oriented. We make the usual assumption that the time orientation is given by the Killing vector ξ. Thus, where ξ is time-like, we must have that $e > 0$ to give a future-oriented orbit. This is the case outside the ergosphere $r > 2M$ or $p^2 < 1/2$. The numerator on the right-hand side of (7.43) is then positive provided α is small enough. Thus for future-oriented orbits we must have

$$\iota'\iota = 1.$$

This yields $\iota' = \iota$ and we obtain

$$e = \mu \frac{1 - 2p^2 + \iota\alpha p^3}{\sqrt{1 - 3p^2 + 2\iota\alpha p^3}}, \tag{7.45}$$

$$j = \iota\mu M \frac{1 - 2\iota\alpha p^3 + \alpha^2 p^4}{p\sqrt{1 - 3p^2 + 2\iota\alpha p^3}}. \tag{7.46}$$

As it will turn out, this formula describes the future-oriented orbits in all cases. Equations (7.45) and (7.46) are invariant with respect to the transformation

$$\alpha \mapsto -\alpha, \quad j \mapsto -j, \quad \iota \mapsto -\iota. \tag{7.47}$$

Hence, it is sufficient to consider the cases with $\alpha \geq 0$ as the others can be obtained by the transformation (7.47). We thus assume subsequently that $\alpha \geq 0$. From (7.46) we then obtain that $j > 0$ if $\iota = +1$, and $j < 0$ if $\iota = -1$, since

$$1 - 2\alpha p^3 + \alpha^2 p^2 = (1 - \alpha p^3)^2 + \alpha^2 p^4 (1 - p^2) \geq 0.$$

The sign ι therefore determines one of the two possible directions of the rotation, the orbits with $\iota = +1$ are co-rotating, the ones with $\iota = -1$ are counter-rotating. Equation (7.45) shows that the energy of the co-rotating orbits is different from the counter-rotating ones if $\alpha \neq 0$. This is an important gravitomagnetic effect. Obviously the counter-rotating orbits are attracted more than the co-rotating ones. The gravitomagnetic force thus has the opposite sign than the magnetic force in Maxwell theory.

Equations (7.45) and (7.46) simplify a lot in the Schwarzschild case, $\alpha = 0$:

$$e = \mu \, \frac{1-2p^2}{\sqrt{1-3p^2}} \, , \quad j = \frac{\imath \mu m}{p\sqrt{1-3p^2}} \, , \tag{7.48}$$

and for the extreme black hole, $\alpha = 1$. In this case we have the identities

$$1 - 3p^2 + 2\imath p^3 = (1 - \imath p)^2 (1 + 2\imath p) \, , \tag{7.49}$$

$$1 - 2p^2 + \imath p^3 = (1 - \imath p)(1 + \imath p - p^2) \, , \tag{7.50}$$

$$1 - 2\imath p^3 - p^4 = (1 - \imath p)(1 + \imath p + p^2 + -\imath p^3) \, , \tag{7.51}$$

and we obtain

$$e = \mu \, \frac{1 + \imath p - p^2}{\sqrt{1 + 2\imath p}} \, , \tag{7.52}$$

$$j = \imath \mu M \, \frac{1 + \imath p + p^2 - \imath p^3}{p\sqrt{1 + 2\imath p}} \, . \tag{7.53}$$

7.4.2.2 Signature of the Circular Orbits

We now examine the conditions for the circular orbits to be time-like, light-like, or space-like. Thus, we have to study the sign of the expression under the square root. The square root always appears in the combination

$$\frac{\mu}{\sqrt{1 - 3p^2 + 2\imath \alpha p^3}} \, .$$

Here μ^2 is the norm of the vector tangent to the circular orbit, and thus the time-like orbits have $\mu > 0$, the light-like ones have $\mu = 0$ and for the space-like ones μ is purely imaginary. We can thus interpret the sign of the expression $1 - 3p^2 + 2\imath \alpha p^3$ as the signature of the orbit. The space-like orbits can be considered as orbits on which the particles have to travel faster than light to escape the drag of gravity. Thus, we are only interested in

$$1 - 3p^2 + 2\imath \alpha p^3 \geq 0 \, . \tag{7.54}$$

Let us distinguish two cases:

1. $\imath = +1$: In this case (7.54) is equivalent to

$$\alpha \geq \frac{3p^2 - 1}{2p^3} \, .$$

If $p^2 \in [0, 1/3)$ then $(3p^2 - 1)/(2p) < 0$ and there are only time-like circular orbits for all $\alpha \in [0, 1]$. If $p^2 \in [1/3, 1]$ then we have

$$0 \le \frac{3p^2 - 1}{2p^3} \le 1$$

and there are non-space-like circular orbits for all

$$\alpha \in \left[\frac{3p^2 - 1}{2p^3}, 1 \right] .$$

For p^2 in the above interval, $\alpha = (3p^2 - 1)/(2p^3)$ yields a light-like orbit.

2. $\iota = -1$: This time we have

$$\alpha \le \frac{1 - 3p^2}{2p^3} ,$$

which can only be satisfied if $p^2 \le 1/3$, that is $r \ge 3M$. The inequality

$$\frac{1 - 3p^2}{2p^3} \le 1$$

is equivalent to $2p^3 + 3p^2 - 1 \le 0$. Identity (7.49) with $\iota = -1$ implies that this inequality only holds for $p^2 \ge 1/4$. In the interval $p^2 \in [0, 1/4)$ we therefore have time-like circular orbits for all $\alpha \in [0, 1]$. In the interval $p^2 \in [1/4, 1/3]$ we have time-like as well as light-like orbits, the former for $\alpha \in [0, (1 - 3p^2)/(2p^3))$, and the latter at $\alpha = (1 - 3p^2)/(2p^3)$.

In summary we find (in the original parameters M and a):

	Co-rotating circular orbits	
	$r \in (M, 3M)$	$r \in (3M, \infty)$
Time-like	$a > \sqrt{r/M}(3M - r)/2$	$\forall a$
Light-like	$a = \sqrt{r/M}(3M - r)/2$	None
Space-like	$a < \sqrt{r/M}(3M - r)/2$	None

At $r = 3M$ and $a = 0$ there is the light-like circular orbit in Schwarzschild space–time and at $r = M$ and $a = M$ there is the co-rotating light-like circular orbit of the extremal black hole.

	Counter-rotating circular orbits		
	$r \in (M, 3M)$	$r \in (3M, 4M)$	$r \in (4M, \infty)$
Time-like	None	$a > \sqrt{r/M}(3M - r)/2$	$\forall a$
Light-like	None	$a = \sqrt{r/M}(3M - r)/2$	None
Space-like	$\forall a$	$a < \sqrt{r/M}(3M - r)/2$	None

At $r = 3M$ and $a = 0$ there is again the light-like circular orbit of Schwarzschild space–time and at $r = 4M$ and $a = M$ the counter-rotating light-like orbit around an extremal black hole.

Note that the denominator in (7.45) can be rewritten as follows:

$$1 - 2p^2 + \iota\alpha p^3 = (1 - 3p^2 + 2\iota\alpha p^3) + p(p - \iota\alpha p^2) .$$

As $\alpha \leq 1$ and $p^2 \leq p$, both terms on the right-hand side have to be non-negative. They can only vanish for $\iota = +1$ and $p = \alpha = 1$. This is the extremal black hole and the circular orbit is light-like at the horizon. In any case we see that all circular orbits have positive energy e.

It is interesting that the orbit at the horizon can be considered as a limit of co-rotating time-like circular orbits in the space–time of an extremal black hole. We thus let $\alpha = 1$, that is $a = M$, and $\iota = +1$. The limit $r \to M$ then is equivalent to $p \to 1$ with $p < 1$. Equations (7.52) and (7.53) yield

$$\lim_{p \to 1} e = \frac{\mu}{\sqrt{3}} , \tag{7.55}$$

$$\lim_{p \to 1} j = \frac{2M\mu}{\sqrt{3}} . \tag{7.56}$$

This is only possible since there is an infinite redshift between the horizon and the asymptotic region. As the orbits become light-like, their energy diverges for every local observer. However, the infinite redshift obviously can render the corresponding energy finite for an observer at infinity.

7.4.2.3 Stability of Circular Orbits

Another question we have to discuss is the stability of the circular orbits. Let us examine the potential (7.33). It is a polynomial of order three in $1/r$ that increases from $-\infty$ near $r = 0$ to $-e^2 + \mu^2$ near $r = \infty$. Its first extremum is therefore a maximum, which corresponds to an unstable circular orbit. The second extremum is a minimum and thus implies a stable orbit. This requires $V''_{\text{eff}} > 0$, where we have to insert the values for j and e which annihilate V_{eff} and V'_{eff}. We can also study the equivalent inequality

$$\frac{r^2}{2p^2} V''_{\text{eff}} + \frac{3r}{2p^2} V'_{\text{eff}} > 0 .$$

Equations (7.37) and (7.38) yield

$$-3(\bar{j} - \alpha p^2 \bar{e})^2 + 1 > 0 .$$

Substituting $x = \bar{j} - \alpha p^2 \bar{e}$ from (7.42) with $\iota' = \iota$ leads to

$$\frac{3 \left(p - \iota\alpha p^2\right)^2}{1 - 3p^2 + 2\iota\alpha p^3} < 1 . \tag{7.57}$$

If $1 - 3p^2 + 2\iota\alpha p^3 > 0$, that is the circular orbit is time-like, then (7.57) can be written in the form

$$3\alpha^2 p^4 - 8\iota\alpha p^3 + 6p^2 - 1 < 0 \tag{7.58}$$

and vice versa, if (7.58) holds, the circular orbit is non-space-like. Inequality (7.58) is therefore the condition for stability for non-space-like orbits with parameters ι and p around a black hole described by α. Let us examine this inequality.

First, note that the left-hand side,

$$L(\iota, \alpha, p) = 3\alpha^2 p^4 - 8\iota\alpha p^3 + 6p^2 - 1 \,,$$

has boundary values

$$L(\iota, \alpha, 0) = -1$$

and

$$L(\iota, \alpha, 1) = 3\alpha^2 - 8\iota\alpha + 5 \,.$$

But

$$3\alpha^2 - 8\iota\alpha + 5 = (5 - 3\iota\alpha)(1 - \iota\alpha) \,,$$

whence $L(\iota, \alpha, 1) \geq 0$. Equality only holds for $\iota\alpha = 1$, that is $\iota = 1$ and $\alpha = 1$.

For given ι and α, L is an increasing function of p in the interval $(0, 1)$. Indeed

$$\frac{\mathrm{d}L}{\mathrm{d}p} = 12p(1 - \iota\alpha p)^2 \,. \tag{7.59}$$

Thus there exist exactly one solution $p = p_0(\iota, \alpha)$ of the equation $L = 0$ in the interval $[0, 1]$ and thus inequality (7.58) is equivalent to

$$p < p_0(\iota, \alpha) \,. \tag{7.60}$$

Hence, the circular orbits are stable if and only if their radius satisfies

$$r \in \left(\frac{M}{p_0^2(\iota, \alpha)}, \infty \right) \,,$$

and they are unstable otherwise. The orbit with radius

$$r_0 = \frac{M}{p_0^2(\iota, \alpha)}$$

is called the *last stable orbit*.

Let us consider the behavior of the last stable orbit. First we note that

$$p_0(\iota, 0) = 1/\sqrt{6} \,,$$

that is the last stable orbit of the Schwarzschild space–time lies at $r_0 = 6M$ independent of ι. We furthermore have that

$$L(\iota, p) = -(1 - \iota p)^3 (1 + 3\iota p) \,,$$

whence

$$p_0(1, 1) = 1 \,, \quad p_0(-1, 1) = 1/3 \,.$$

That is, the last stable co-rotating orbit lies at $r_0 = M$ and the counter-rotating one at $r_0 = 9M$.

The derivative of p_0 with respect to α is given by the formula

$$\frac{\partial p_0}{\partial \alpha} = -\frac{\partial L}{\partial \alpha}\left(\frac{\partial L}{\partial p_0}\right)^{-1}. \tag{7.61}$$

We have that $\partial L/\partial p \geq 0$ in view of (7.59) and

$$-\frac{\partial L}{\partial \alpha} = 2\iota p^3 (4 - 3\iota \alpha p). \tag{7.62}$$

The term in parenthesis on the right-hand side is always positive and thus p_0 increases with α (from $1/\sqrt{6}$ to 1) for $\iota = +1$ and decreases (from $1/\sqrt{6}$ to $1/3$) for $\iota = -1$.

Finally, consider the behavior of the energy e_0 and the angular momentum j_0 on the last stable orbit. There, by (7.57), we have $3x^2 = 1$. Equation (7.42) with $\iota' = \iota$ implies that $|x| = \iota x$, that is $x = \iota/\sqrt{3}$. Equations (7.40) and (7.41) yield:

$$e = \frac{\mu}{2\alpha p^2}\left[(3p^2 - 1)x + \frac{p^2 - \alpha^2 p^4}{x}\right],$$

$$j = \frac{\mu M}{2p^2}\left[(3p^2 + 1)x + \frac{p^2 - \alpha^2 p^4}{x}\right],$$

whence

$$e = \frac{\iota\mu}{2\sqrt{3}}\left(\frac{6p_0^2 - 1}{\alpha p_0^2} - 3\alpha p_0^2\right),$$

$$j = \frac{\iota\mu M}{2\sqrt{3}}\left(\frac{6p_0^2 + 1}{p_0^2} - 3\alpha^2 p_0^2\right).$$

These expressions can be simplified further by using the identity $L(\iota, \alpha, p_0) = 0$:

$$e = \frac{\mu p_0}{\sqrt{3}}(4 - 3\iota \alpha p_0), \tag{7.63}$$

$$j = \frac{2\iota\mu M}{\sqrt{3}}(3 - 2\iota \alpha p_0). \tag{7.64}$$

A simple calculation using (7.61) yields

$$\frac{de_0}{d\alpha} = -\frac{\iota\mu}{3\sqrt{3}}\frac{p_0^2}{(1 - \iota\alpha p_0)^2}. \tag{7.65}$$

Hence the energy of the last stable orbits decreases in the co-rotating case and increases in the counter-rotating case.

In summary: for co-rotating circular orbits and increasing a

1. r_0 decreases from $6M$ at $a = 0$ to M at $a = M$.
2. e_0 decreases from $(2\sqrt{2}/3)\mu$ at $a = 0$ to $(1/\sqrt{3})\mu$ at $a = M$.

For counter-rotating circular orbits and increasing a

1. r_0 increases from $6M$ at $a = 0$ to $9M$ at $a = M$.
2. e_0 increases with from $(2\sqrt{2}/3)\mu$ at $a = 0$ to $(5\sqrt{3}/9)\mu$ at $a = M$.

In particular we have that $2\sqrt{2}/3 \approx 94.3\%$, $1/\sqrt{3} \approx 58\%$, and $5\sqrt{3}/9 \approx 95.9\%$. The energy efficiency by decelerating the particles thus lies between 5.7 and 42% for the co-rotating and between 5.7 and 4.2% for the counter-rotating circular orbits.

7.4.2.4 Time Variability

In a realistic accretion disc the temperature is not distributed homogeneously, but there are hotter and colder regions. The hotspots create most of the radiation. This radiation then shows an approximate periodic structure, which agrees with the period of rotation of the spot. Let us compute the corresponding frequency.

A circular orbit has the form

$$t = \dot{t}\lambda , \quad \varphi = \dot{\varphi}\lambda , \quad r = r_0 , \quad \vartheta = \pi/2 ,$$

where $\dot{t}$ and $\dot{\varphi}$ result from (7.21) with $Q = 0$ and are constant. A period corresponds to the interval $\Delta\lambda = 2\pi/\dot{\varphi}$ in λ. The two points of the orbit which differ by $\Delta\lambda$ thus have the same coordinates φ, r, and ϑ. The difference in the t-coordinate is $\Delta t = 2\pi\dot{t}/\dot{\varphi}$. Imagine that at each of these points a light signal is emitted to an observer at infinity. These signals travel along two light-like autoparallels. The second one arises from the first one by a translation of t by Δt, as the symmetry does not change the properties of autoparallels and the trajectory of the observer. But t is the proper time of the observer. We thus arrive at the important result that the frequency measured by an asymptotic observer equals $\dot{\varphi}/\dot{t}$.

From equation (7.21) we obtain

$$\begin{pmatrix} \dot{t} \\ \dot{\varphi} \end{pmatrix} = \mathbf{g}^{-1} \begin{pmatrix} e \\ -j \end{pmatrix} .$$

A simple calculation implies

$$\mathbf{g}^{-1} = \frac{1}{\Delta} \begin{pmatrix} \dfrac{2M}{r}a^2 + r^2 + a^2 & \dfrac{2M}{r}a \\[2mm] \dfrac{2M}{r}a & \dfrac{2M}{r} - 1 \end{pmatrix} ,$$

whence

$$\frac{\dot{\varphi}}{\dot{t}} = \frac{2\alpha p^4 re + \left(1 - 2p^2\right)j}{r^2\left(1 + \alpha^2 p^4 + 2\alpha p^6\right)e - 2\alpha p^4 rj} .$$

Substituting for e and j from (7.45) and (7.46) yields the following relation for the frequency

$$\nu = \frac{p^3}{1 + \iota \alpha p^3} \frac{1}{2\pi M} .$$

This holds for arbitrary circular orbits. The highest frequencies arise from the edge of the disc, that is from the last stable orbits where $p = p_0$. In Schwarzschild space–time ($\alpha = 0$) this is

$$\nu = \frac{1}{6\sqrt{6}} \frac{1}{\pi M} = \frac{1}{\sqrt{6}} \frac{1}{2\pi r_0} ,$$

for an extremal black hole ($\alpha = 1$) with $\iota = +1$

$$\nu = \frac{1}{4\pi M} = \frac{1}{2} \frac{1}{2\pi r_0} ,$$

and with $\iota = -1$

$$\nu = \frac{1}{52} \frac{1}{\pi M} = \frac{9}{52} \frac{1}{2\pi r_0} .$$

Note that the expression $2\pi r_0$ gives the length of the orbit only in the case $\alpha = 0$. In the general case the length of the orbit with radius r is given by

$$2\pi R = 2\pi r \sqrt{1 + 2\alpha p^3 + \alpha^2 p^4} .$$

For $\alpha = 1$ and $\iota = +1$ this yields

$$\nu = \frac{1}{2\pi R} .$$

This is a rather astonishing formula, as this frequency already includes the infinite redshift between the circular orbit and infinity. Nevertheless, this frequency looks like the orbit has been traversed with light speed, as it indeed has been.

More about the orbits of test particles can be found in [8].

7.5 Exercises

1. Show that the metric (7.1) is asymptotically flat for $r \to -\infty$ and that the asymptotic observers there see the field of an object with mass $-M$, angular momentum $J = -aM$, and charge $-Q$.
   Hint: introduce a new radial coordinate $r' = -r$.
2. Consider a metric of the form

$$ds^2 = A_{\alpha\beta}(x) \left(e^{\alpha}_{\mu} dx^{\mu} \right) \left(e^{\beta}_{\nu} dx^{\nu} \right) ,$$

where $A_{\alpha\beta}(x)$ is a matrix with scalar elements and $\{e^{\alpha}_{\mu}(x)\}$ is a basis for the covectors.

Show:

(a) The signature of the metric is the signature of $A_{\alpha\beta}$,

(b) $\det g_{\mu\nu} = \det A_{\alpha\beta} (\det e^{\alpha}_{\mu})^2$, and

(c) $g^{\mu\nu} = (A^{-1})^{\alpha\beta} e^{\mu}_{\alpha} e^{\nu}_{\beta}$, where $\{e^{\mu}_{\alpha}\}$ is a basis dual to the basis $\{e^{\alpha}_{\mu}\}$, that is $e^{\mu}_{\alpha} e^{\beta}_{\mu} = \delta^{\beta}_{\alpha}$.

3. Show that the curve $v = v_0$, $\eta = \eta_0$, $r = -\lambda$, and $\vartheta = \vartheta_0$ is a light-like geodesic of the Kerr–Newman metric.

4. Show that the vector field $\xi^{\mu}(x)$ is a symmetry of the potential $A_{\mu}(x)$ if and only if

$$A_{\mu,\rho} \xi^{\rho} + A_{\rho} \xi^{\rho}_{,\mu} = 0.$$

5. Deduce the conservation law $\dot{P}_{\xi} = 0$ directly from the Killing equation for the metric and the symmetry equation for the potential.

6. Deduce the equations of motion (7.20) and (7.21) and show that along a future pointing trajectory with decreasing (or increasing) coordinate r only one of the pairs (u, ξ) and (v, η) can be regular at the horizon.

7. Show that the energy $e = p_{\mu} \xi^{\mu}$ of a particle with respect to asymptotic observers has a minimum at each point of the space–time where the Killing vector field is non-space-like and that there is no minimum if ξ^{μ} is space-like.

8. Compare the rotational energy of black holes to the rotational energy of a simple Newtonian model for a star: a rigid ball with mass M_S, radius R_S, and constant density. Express the energy of the star as a function of the radius and the angular momentum (be careful: the velocity at the equator should be non-relativistic!). Set the irreducible mass of the black hole to M_S, and choose some interesting values, for example the solar mass, etc.

References

1. S. W. Hawking and G. F. R. Ellis, *The Large Scale Structure of Space-Time*, Cambridge University Press, Cambridge, UK, 1973.
2. M. Heusler, *Black Hole Uniqueness Theorems*, Cambridge University Press, Cambridge, UK 1996.
3. E. T. Newman, E. Couch, K. Chinnapared, A. Exton, A. Prakash and R. Torrence, J. Math. Phys. **6** (1865) 918.
4. C. W. Misner, K. S. Thorne and J. A. Wheeler, em Gravitation, Freeman, San Francisco, CA, 1973.
5. B. Carter, Commun. Math. Phys. **10** (1968) 280.
6. J. H. Krolik, *Active Galactic Nuclei*, Princeton University Press, Princeton, NJ, 1999.
7. K. S. Thorne, R. H. Price and D. A. Macdonald, (eds.) *Black Holes: The Membrane Paradigm*, Yale University Press, New Haven, CT, 1986.
8. S. Chandrasekhar, *The Mathematical Theory of Black Holes*, Clarendon Press, Oxford, 1983.
9. R. Penrose, Nuovo Cim. 1, special number (1968) 252.
10. J. M. Bardeen, B. Carter and S. W. Hawking, Commun. Math. Phys. **31** (1973) 161.
11. S. W. Hawking, Phys. Rev. Lett. **26** (1971) 1344.

12. S. W. Hawking, Commun. Math. Phys. **43** (1975) 199; R. M. Wald, *Quantum Field Theory in Curved Spacetimes and Black Hole Thermodynamics*, The University of Chicago Press, Chicago, IL, 1994.

13. J. Frank, A. King and D. Raine, *Accretion Power in Astrophysics*, Cambridge University Press, Cambridge, UK, 2002.

Index